A Laboratory Manual for
GENERAL BOTANY

A Laboratory Manual for
GENERAL BOTANY
SIXTH EDITION

Margaret Balbach

Illinois State University

Lawrence C. Bliss

University of Washington

prepared with the assistance of

Harold E. Balbach, Ph.D.

 SAUNDERS COLLEGE PUBLISHING

Philadelphia   New York   Chicago
San Francisco   Montreal   Toronto
London   Sydney   Tokyo   Mexico City
Rio de Janeiro   Madrid

Address orders to:
383 Madison Avenue
New York, NY 10017

Address editorial correspondence to:
West Washington Square
Philadelphia, PA 19105

This book was set in Souvenir by Caledonia Composition.
The editors were Michael Brown, Bonnie Boehme, Don Reisman, Lynne Gery, and Ruth Low.
The art & design director was Richard L. Moore.
The text design was done by Nancy E. J. Grossman.
The cover design was done by Richard L. Moore.
The new artwork was drawn by Floyd Giles.
The production manager was Tom O'Connor.
This book was printed by Connecticut Printers.

A Laboratory Manual for General Botany

ISBN 0-03-058514-7

1234   111   987654321

CBS COLLEGE PUBLISHING
Saunders College Publishing
Holt, Rinehart and Winston
The Dryden Press

PREFACE

As with the previous edition, this sixth edition of A LABORATORY MANUAL FOR GENERAL BOTANY is dedicated to presenting basic botany as a useful, biologically meaningful, and scientifically pertinent topic. The manual has been completely revised with many new figures, many old figures redrawn, and with an index.

The philosophy remains to present the basic, classical botanical principles and topics in such a way as to relate them not only to the general concepts of our botanical heritage but also to the practical, applied interest in plants.

The exercises fulfill all the needs of the traditional introduction to the principles of laboratory botany. Their organization also makes them directly usable in the individualized instructional laboratory format. Further, the introduction to each exercise gives more background and perspective than is typical for a laboratory manual. This makes the manual especially useful in those circumstances when lecture topics are not always closely aligned to the specific needs of the laboratory coverage.

Changes in this edition were made in response to suggestions and critiques received from a survey of users of the manual and from reviewers of the manuscript. Other changes, of course, have resulted from the authors' efforts to improve the book. Because many universities and schools now have one 3-hour laboratory session per week, we have combined some of the shorter exercises into longer ones to accommodate many users who reported the need to assign several exercises or portions of several to complete one session. While no topics have been dropped, many have been reorganized so there are fewer exercises (31 instead of 36).

A chief feature of A LABORATORY MANUAL FOR GENERAL BOTANY is its flexibility, a feature that has been enhanced in the Sixth Edition. Since some botany students have already had a course in biology, we have tried to arrange exercises to facilitate their selection or deletion. Our surveys indicate that, in addition to instructors who use most of the exercises, there are others who utilize the strengths of one or the other of the two initial parts. In some universities and schools, Part I, Structure and Function of Seed Plants, is the focus of study. In other institutions Part II, Survey of the Plant Kingdom, is the main emphasis of study. Whether you prefer to cover both areas or to emphasize one, the manual is flexible enough to be fully representative of both areas of study.

Part I concerns the Structure and Function of Seed Plants. Structure is always interpreted as

regards its selective advantage to the plant, and economic importance is stressed where it is applicable. Nothing is presented as an empty fact.

Within Part I we have made the following changes or modifications. Because primary growth in roots involves the apical meristem, cell division is appropriately studied at the same time. Therefore, Cell Division (Exercise 5 of previous edition) has been combined with Roots I: Types of Root Systems and Growth of Root Tip (Exercise 6 of previous edition) into Exercise 5 of the current edition. Herbaceous Stems (Exercise 7 of current edition) consolidates Exercises 8 and 9 of the previous edition. Exercises 10 through 13 of the previous edition have been condensed and recombined into three exercises (8 to 10 of current edition). Woody stems are introduced in Exercise 8 with the fairly simple anatomy of the gymnosperms. This is followed by study of the more complex angiosperm wood in Exercise 9. Twig features, wood grain, and tree rings (for dating) are combined in Exercise 10 because of their general structural relationships. These wood features are more readily interpreted if the student already has some knowledge of wood anatomy. It is for this reason that the sequence of woody stem exercises was changed in this edition. The instructor can, however, choose any sequence, because each of these three exercises is reasonably independent of the others.

The study of the flower (Exercise 13 of new edition) has been largely rewritten with the goal of leading the student to a discovery of the flower's functional relationship to insects, birds, and other vectors. Thus, flower form and structure in the second half of the exercise are made more interesting and meaningful.

Part II is a survey of the plant kingdom. Life cycles, often so puzzling to beginning students, continue to be presented as an efficient solution to the universal problems of survival. Each successive plant group is evaluated as to its progressive features over those of the preceding group.

Within Part II the following changes have been made. Blue-green algae have been removed from the alga exercise (Exercise 26 of previous edition) and combined with bacteria in new Exercise 19, reflecting the natural separation of these procaryotic organisms from the eucaryotic. This shifting of the blue-green algae required balancing the work of the remaining exercises on algae. We feel that the new groupings of brown, red, and golden-brown algae in Exercise 24 and the grouping of green algae, euglenoids, and lichens in Exercise 22 is a fairly well-balanced workload for the student. Thus, new Exercises 19, 22, and 24 simply represent regroupings of some genera formerly in Exercises 22 through 28 of the previous edition.

Ferns and other lower vascular plants (Exercises 30 and 31 of previous edition) have been consolidated into one study (Exercise 26 of new edition). In response to users' requests, *Selaginella* is now included, as are new drawings of life cycles of all five representative genera.

The exercise on cone-bearing seed plants (Exercise 27 of new edition) has been extensively restructured with new art to help the student better understand how the cone features relate to the microscopic detail of the gametophyte. Similarly, the exercise on flowering seed plants (Exercise 28 of new edition) has new art that shows the relationship of the flower to the gametophyte in a simpler fashion than did the art of Exercise 33 of the previous edition.

Part III includes the chapters on genetics, ecology, and taxonomy. In Part III, new Figure 29.1 is a labeled and annotated version of Figure 34.1 of the previous edition. We think it will help clarify the comparisons shown.

The Appendix of the Fifth Edition listed the materials needed for activities in each exercise and gave directions for general scheduling. The section entitled The Instructor's Guide serves this purpose in the Sixth Edition. We believe its new name is more descriptive of its function. The Guide has been revised and completely updated with special attention to materials recommended to see that they are currently available from biological, horticultural, and forestry suppliers. Addresses and phone numbers of suppliers are listed at the end of the Guide.

ACKNOWLEDGMENTS

We are especially grateful to the senior author's husband, Dr. Harold E. Balbach, Environmental Biologist, U.S. Army Construction Engineering Research Laboratory, for his invaluable assistance, suggestions, and advice in the writing and reorganizing of this edition (as for the previous edition) and for his skillful technical and editorial assistance throughout the months of preparation of the manuscript. The Appendix of the previous edition was also prepared by him, and he has revised it for inclusion in this edition as the Instructor's Guide.

We gratefully acknowledge the diligent reviewers of the manuscript: Clarence W. Gehris, State University of New York College at Brockport; Gene Pratt, University of Wyoming, Laramie; Linda R. Berg, University of Maryland; and B. J. D. Meeuse, University of Washington. Their critiques were thoughtful, constructive, and helpful in many ways. We also wish to thank the reviewers of the Fifth Edition, whose efforts continue to be reflected in the strengths of the Sixth Edition. In this regard, we acknowledge Jordan J. Choper, Montgomery College; Harold E. Eversmeyer, Murray State University; Hugo A. Ferchau, Western State College of Colorado; James F. Matthews, University of North Carolina; William H. Miller, William Rainey Harper College; B. John Syrocki, State University of New York College at Brockport; James J. Tobolski, Indiana University–Purdue University; and T. J. Watson, Jr., University of Montana.

Much appreciation is extended to Floyd Giles, Professor of Horticulture, University of Illinois, whose artistic skills have improved the quality, clarity, and usefulness of the diagrams. Thanks to Karen L. Bliss for her help in indexing.

We appreciate the efforts of Michael Brown, the biology editor of Saunders College Publishing, in committing resources to significant improvements in the art and design of the Sixth Edition. Thanks to everyone at Saunders who oversaw the implementation of these improvements. A special thanks to Lynne Gery, the project editor, who guided the manual through the final stages of production.

M.B.
L.C.B.

CONTENTS

PART I

Structure and Function of Seed Plants

THE WHOLE PLANT

INTRODUCTION

We live in a world dominated by flowering plants. Flowering plants are the ones most commonly known and utilized by today's consumer, as they were by our prehistoric ancestors. They occur as woody-stemmed trees and shrubs, and as soft-stemmed herbs. Herbs include a wide variety of plants, such as weeds, cereal grains, lawn grasses, garden flowers and vegetables, certain fruits, and many house plants.

There are two major groups of flowering plants, the dicots and the monocots (or, more formally, the dicotyledons and the monocotyledons). Dicots are those plants having two food-storage organs (cotyledons) in the seed, monocots have one.

Most of these plants go through the life stages shown in Figure 1.1.

I. THE MATURE DICOT PLANT

When you look at the general body organization of a mature dicot plant, you see that there is one main stem with smaller stems branching from it. This branching habit usually gives the mature plant a bushy, rounded form. The branching form of most shade trees illustrates this.

A. *THE VEGETATIVE PLANT BODY: STEMS, LEAVES, AND ROOTS*

Vegetative parts of a plant are those concerned with photosynthesis, growth, and storage. In photosynthesis, the green plant absorbs light and converts it to a usable chemical form of energy. Essentially the plant traps sun energy and puts it in storage in sugar

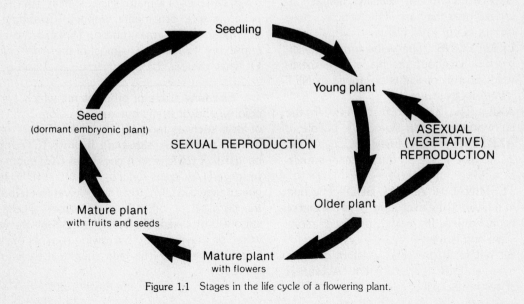

Figure 1.1 Stages in the life cycle of a flowering plant.

molecules. To do this, the plant needs water, CO_2, and light. To get enough of these three essentials, the plant needs extensive absorptive surfaces to absorb water from the soil, CO_2 from the air, and energy from the sun. Leaves are the absorptive surface for light and CO_2, although in many plants, stems also function in this way. Roots provide the absorptive surfaces for water.

The following are suggested examples of dicots to study: two-month-old bean plant (*Phaseolus vulgaris*); two-month-old cucumber plant (*Cucumis sativus*); "Tiny Tim" or any dwarf variety of tomato (*Lycopersicon* sp.); *Coleus*; ornamental pepper (*Capsicum annuum*); Christmas cherry (*Solanum pseudocapsicum*); rubber plant (*Ficus elastica*); or any suitable species of house plant or potted woody dicot; or any leafy shrub, such as lilac (*Syringa vulgaris*), privet (*Ligustrum* sp.), and so on.

ACTIVITY 1 **Examine** the plants available and **identify**

1. STEM The main ascending axis or stalk that bears the leaves.
2. NODE The place on the stem where a leaf occurs.
3. INTERNODE The segment of the stem between any two successive nodes. Stem elongation is due primarily to elongation of the internodes.
4. LEAF Each is composed of two parts, (a) the BLADE—the expanded flat portion, and (b) the PETIOLE—a stalk that attaches the blade to the stem. In some plants the blade is divided into a number of segments, called LEAFLETS.
5. COMPOUND LEAF A leaf whose blade is divided into leaflets.
6. STIPULES Green leaflike blades attached near the base of the petiole. These do not occur in all plants.
7. LEAF VEINS The veins form a meshwork throughout the blade. The meshwork of a dicot leaf is described as NET VENATION.
8. AXIL The angle formed between the stem and the upper side of the petiole.
9. AXILLARY or LATERAL BUD A bud in the axil. It will produce a leafy branch or a flower, or both.
10. TERMINAL or APICAL BUD The bud at the top of the main stem or at the end of each branch. Its growth produces more stem and leaves.
11. BRANCH (if present) A stem growing from another (parent) stem. A branch arises in the axil of a leaf.
12. ROOTS Underground parts composed either of many fibrous strands or of one or more thick parts (as in a tap root).

Label Figure 1.2, using the capitalized terms defined in the preceding description.

B. FLOWERS, FRUIT, AND SEED

The Dicot Flower

Typically, the number of petals and sepals of the flower of a dicot plant is two or four or (usually) five. Or, the number can be some multiple of four or five, such as twelve or ten.

ACTIVITY 2 **Study** the flowers of any number of representative dicot plants. Suggested plants are three-month-old bean plant (*Phaseolus vulgaris*) or cucumber (*Cucumis sativus*); any dwarf tomato variety (*Lycopersicon* sp.); house plants, such as *Oxalis* sp., *Fuchsia* sp., flowering maple (*Abutilon* sp.); or any available garden plants, such as *Petunia* sp., snapdragon (*Antirrhinum* sp.), etc. **Find** (1) the colored PETALS, (2) green leafy-looking SEPALS. (In some plants sepals and petals look alike.) Color attracts birds and insects. Not all flowers are colorful and aromatic.

Count and **record** in Table 1.1 the number of petals and sepals in the flowers of the dicots available for your study. Have the laboratory instructor explain to you or show you how to count them if some of these plants have double flowers.

The Fruit and the Seed

ACTIVITY 3 **Examine** plants bearing fruit, such as a three-month-old bean plant (*Phaseolus vulgaris*); dwarf varieties of tomato, such as "Tiny Tim" (*Lycopersicon* sp.); ornamental pepper (*Capsicum annuum*); or Christmas cherry (*Solanum pseudocapsicum*). Fruits such as most of these are colorful Of what value is the color? _____

_____.

Examine fruits of other plants which are not colorful and/or fleshy but are dull-colored and hard or dry, such as those of elm (*Ulmus* sp.); tree of heaven (*Ailanthus altissima*); hickory (*Carya* sp.); beggar tick (*Bidens* sp.); cocklebur (*Xanthium* sp.); milkweed (*Asclepias syriaca*); peanut (*Arachis hypogaea*); ragweed (*Ambrosia* sp.); velvet leaf (*Abutilon theophrasti*); or other available types. These are variously modified, aiding in dispersal by the wind, water or gravity or by burrowing rodents or other animals when the fruit gets caught in their fur or feathers.

Find the seeds inside these fruits.

Table 1.1. NUMBER OF PETALS AND SEPALS IN FLOWERS OF DICOTS

PLANT		NUMBER OF PETALS	NUMBER OF SEPALS
Common name	Botanical name		

The Dicot Seedling

Seeds germinate into young plants, called *seedlings*.

ACTIVITY 4 **Examine** seedlings of the bean plant (*Phaseolus vulgaris*); cucumber (*Cucumis sativus*); or tomato plant (*Lycopersicon* sp.) that have been grown in a glass or plastic tumbler, jar, or beaker with only water and some sort of physical support supplied to them. Since only water has been supplied, from where have the seedlings obtained the energy and nutrients for this growth? _____
_____

II. THE MATURE MONOCOT PLANT

When you look at the general body organization of a monocot plant, you see that it consists of one main stem that may have many leaves but no side stems. Hence, the body of most monocot plants lacks the full shape of the dicot plant. Many monocots that appear full and bushy gain this form by having several clustered stems growing from an underground part.

A. *THE VEGETATIVE PLANT BODY*

ACTIVITY 5 **Examine** representative monocot plants, such as species of spider plant (*Chlorophytum* sp.); wandering Jew (*Zebrina pendula*); spiderwort (*Tradescantia* sp.); purple heart (*Setcreasea pallida*); day-flower (*Commelina* sp.); *Philodendron*; dumbcane (*Dieffenbachia* sp.) water

hyacinth (*Eichhornia* sp.); or any lawn or pasture grass; or commercial grain such as corn (*Zea mays*) or oats (*Avena sativa*). **Find**

1. The leaf **BLADE** It is relatively narrow, flat, and with parallel veins.
2. **SHEATH** The base of the blade. It partially or completely encloses the stem.
3. **NODES** The swollen-looking joints in the stem. Swollen nodes are characteristic of most monocots.
4. **VEINS** which run parallel to the leaf margin.
5. **ROOTS** Usually fibrous. Many monocots also develop swollen, food-storage parts in their roots, such as in the spider or airplane plant (*Chlorophytum* sp.).

Note: Some monocots, such as *dumbcane* (*Dieffenbachia* sp.) and *Philodendron*, have venation like that of many dicots.
Label Figure 1.3, using the capitalized terms defined in the preceding description.

B. *FLOWERS AND SEEDS*

The Monocot Flower

In many monocots, the petals and sepals look alike and when this is the case, they are called tepals. Typically, the sepals and petals (or tepals) of the flower of a monocot plant each number three or some multiple of three.

ACTIVITY 6 **Examine** the flowers of representative monocots, such as spiderwort (*Tradescantia* sp.), wandering Jew (*Zebrina* sp.), or *Rhoeo* sp. If in

Table 1.2 NUMBER OF FLORAL PARTS IN FLOWERS OF MONOCOTS

| PLANT | | NUMBER OF: | | |
Common name	Botanical name	Petals	Sepals	Tepals

season, the following could also be studied: lily (*Lilium* sp.); *Iris* sp.; *Gladiolus* sp.; tulip (*Tulipa* sp.); or *Crocus* sp.

Count the number of petals and sepals or tepals in the flowers of the monocots available and **record** in Table 1.2.

The Monocot Seedling

ACTIVITY 7 **Examine** seedlings of corn, wheat, or oats that have been grown in glass or plastic tumblers, jars, or beakers with only water and some physical support supplied to them. (1) Does the seedling show the same leaf and stem characteristics as the mature plant? *Yes* *No.* (2) Is there a leaf sheath? *Yes* *No.* (3) Does the leaf have parallel sides? *Yes* *No.*

ACTIVITY 8 **Fill** in Table 1.3.

Table 1.3 COMPARISON OF MONOCOTS AND DICOTS

	MONOCOTS	DICOTS
Stem Appearance of nodes		
Branching (common or uncommon)		
Leaf venation (parallel or net)		
Petiole (distinct or absent)		
Sheath (distinct or absent)		
Root system (tap or fibrous)		
Flower Number of petals		
Seed Number of cotyledons		
Some representative plants		

EXERCISE 1   **Student Name** _____

QUESTIONS

1. If you found a plant in which the basal part of its leaves enclosed the stem like a sheath, which

   would you have, a monocot plant or a dicot plant? _____

2. The leaves and roots of a plant represent very large surface areas designed for what general

   function? _____

3. The conversion of light to a usable form of energy is called? _____

   _____

4. A sugar molecule made by a green plant might be thought of as a "suitcase molecule." Suit-
   cases are portable, tangible objects that can be unlocked and their contents used. Sugar is a

   portable, finite, tangible object. What's "locked" in its molecular makeup? _____

   Is sunlight a portable, tangible "thing" directly usable by protoplasm? _____

5. If you had only the flower of a plant, what observation could you make to determine if you had

   a monocot or a dicot? _____

   _____

6. The structure that protects the seeds and is modified, aiding in dispersal of the seeds is

   the _____.

7. Most dicot plants are branched. What structure on a plant produces a branch? _____

   _____

8. What is the basis for the names monocot and dicot? _____

   _____

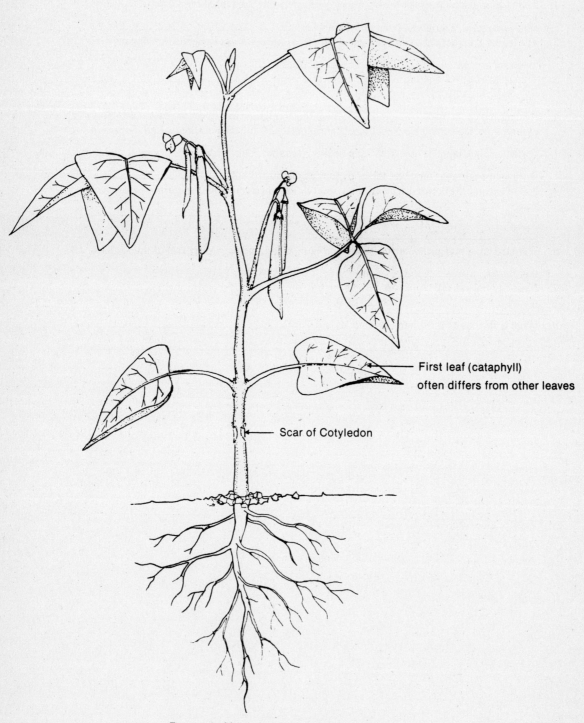

First leaf (cataphyll) often differs from other leaves

Scar of Cotyledon

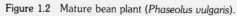

Figure 1.2 Mature bean plant (*Phaseolus vulgaris*).

EXERCISE 1   **Student Name** _____

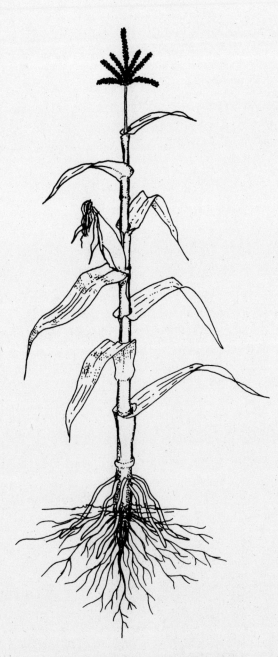

Figure 1.3   Mature corn plant (*Zea mays*).

SEED STRUCTURE AND GERMINATION

INTRODUCTION

Seeds are the great staple food of the world. They feed more people and livestock than any other type of food. Of all the world's harvested acreage, 70 percent is devoted to growing the seeds of cereals. The most important of these are wheat, rice, and corn. Other valuable seeds for food are the soybean, the common bean, rye, oats, sorghum, and millet. All seeds are embryonic plants supplied with the nutrients needed for germination. Seed nutrients ensure the growth of the new plant during its seedling stage when it is most vulnerable to the hazards of weather and other factors. All peoples have exploited seeds for these nutrients.

Nutrients are supplied to the embryo of the ripening seed by the parent plant in a special tissue, called the *endosperm*. It supplies nutrients and growth-promoting hormones. As the seed is ripening* on the parent plant, the embryo of the seed digests some or all of the endosperm. It then restocks the nutrients in its own storage organs, called *cotyledons*.

In some dicot flowering plants all the endosperm is digested during seed ripening, and two plump cotyledons occupy most of the ripe seed. In other dicots only part of the endosperm is digested, so the ripe seed contains some endosperm as well as two cotyledons. In monocots the ripe seed contains an embryo with one cotyledon and variable amounts of endosperm, depending on the species.

---

*Note: Seed ripening is not the same as germination. Ripening is the growth that produces the seed; germination is the growth of the seed that produces the seedling.

I. THE DICOT SEED

A. *THE BEAN SEED—A DICOT SEED WITHOUT AN ENDOSPERM*

ACTIVITY 1 **Examine** seeds of the pinto, kidney, or lima bean (*Phaseolus* sp.) that have been soaked in water overnight or for several hours. **Find**

1. SEED COAT The skinlike protective covering of the seed.
2. HILUM An elliptical scar on the concave side of the seed. It marks the point of attachment of a stalk that attached the seed to the inside of the fruit (pod).
3. RAPHE A small ridge at one end of the hilum. It is a remnant of the stalk.
4. MICROPYLE A small circular scar at the other end of the hilum. It marks the hole where the pollen tube entered just prior to fertilization.

Label the external parts of the bean seed in Figure 2.1, using the capitalized terms defined in the preceding description.
Peel off the seed coat. **Find**

1. COTYLEDONS The two halves of the seed. These are modified storage leaves. Because cotyledons are already part of the living embryo, the food they contain can be quickly digested during germination.
2. EMBRYONIC AXIS You will find this attached to one of the cotyledons. It is easily identified by its two miniature leaves. These are attached near the COTYLEDONARY

NODE to the shoot of the axis, the EPI-COTYL. The epicotyl and leaves together form the PLUMULE. The remainder of the axis is the HYPOCOTYL, the tip of which is the RADICLE, or embryonic root. This tiny embryonic axis utilizes the foods of its two cotyledons during germination and grows into the seedling. The epicotyl forms the shoot and the radicle forms the primary root.

Label the internal parts of the bean seed shown in Figure 2.2, using the capitalized terms defined in the preceding description.

B. GERMINATION OF THE DICOT SEED

Imbibition—Water Absorption and Swelling of the Seed

As long as the seed is dry, it will not germinate. But when moisture is available, and the seed coat is porous, the seed will imbibe water. This nonosmotic uptake of water by dry matter is IMBIBITION, the first phase of germination. If the seed is not dormant and if oxygen is present, it will then germinate.

ACTIVITY 2 **Compare** seeds of beans, peas, and other dicots in the dry state versus those that have been kept moistened for several days in wet rolled paper towels or on moist paper in petri dishes, kept in a warm (21°C) place.

Although small, the dry seed contains large amounts of starch, fats, and oils, the proportions varying with different species. The dehydrated state of these foods permits the storage of these large quantities. Imbibition builds up tremendous water pressure, which is usually sufficient to split the coat and allow the emergence of the growing embryo.

ACTIVITY 3 **Observe** paper cups containing plaster of paris into which dry seeds were mixed before it hardened. Describe the condition of the plaster of paris. _____

Seedling Stages of the Bean— Epigean Type Germination

ACTIVITY 4 **Examine** bean seedlings of different ages that have been grown in sand or soil flats. Look at Figure 2.3 as a guide.

Three-Day-Old Seedling The growth of the radicle has formed the PRIMARY ROOT. Al-though the shoot (epicotyl) has enlarged, it has not yet left its position between the COTYLEDONS. What are two advantages of the root being the first organ to grow? _____

Label Figure 2.3 (A), using the capitalized terms from the preceding paragraph.

Six-to-Eight-Day-Old Seedlings These have just broken through the soil surface. **Identify**

1. HYPOCOTYL It grows in an arched fashion and pulls the cotyledons and plumule up above the soil line. In this way the delicate growing tip is protected from damage during emergence.
2. COTYLEDONS These may now be green and are still close together. They may or may not yet have emerged from the split SEED COAT.
3. PLUMULE The two LEAVES of the plumule may or may not extend from between the cotyledons.
4. PRIMARY ROOT This now has LATERAL or BRANCH ROOTS.

Label Figure 2.3 (B) and (C), using the capitalized terms from the preceding description.

Ten-Day-Old-Seedlings In this stage, the hypocotyl has straightened up and the cotyledons are spread apart and well above ground. This type of germination, which elevates the cotyledons above ground, is called EPIGEAN (epi-"above"; geo-"earth") GERMINATION. Most dicots exhibit this type of germination.

Find young plants with just one pair of leaves. **Identify**

1. COTYLEDONS These fleshy organs may now be wrinkled and somewhat shriveled. Why? _____
_____ .
2. HYPOCOTYL The stem portion below the cotyledons.
3. FIRST TRUE LEAVES These are just above the cotyledons.
4. TERMINAL BUD This is the cluster of very tiny leaves at the tip of the stem, which is between the first true leaves.
5. ROOTS There are many BRANCH or LATERAL (SECONDARY) ROOTS now, which have arisen from the original PRIMARY ROOT.

Label Figure 2.3 (D), using the capitalized terms from the preceding description.

Seedling Stages of the Pea—
Hypogean Type Germination

ACTIVITY 5 **Examine** successive stages of seedlings of the garden pea (*Pisum sativum*) and look at Figure 2.4 as a guide. Are the cotyledons above the ground in the later stages? *Yes No.* In the younger stages is there a loop in the axis that is growing upward? *Yes No.* This arched axis is the EPICOTYL. It functions like the hypocotyl of the bean seedling in that it elongates and tunnels a path through the soil, dragging the plumule (but not the cotyledons) upward. The hypocotyl does not elongate and remains below ground. This type of seedling growth, in which the cotyledons remain below ground, is called HYPOGEAN (*hypo*-"below", *geo*-"earth") GERMINATION.

Label the different seedling stages of the pea in Figure 2.4, using the same terms as those used for the bean seedling in Activity 4.

C. A DICOT SEED WITH AN ENDOSPERM

Seeds of the honey locust tree (*Gleditsia triacanthos*), like those of many other dicots, contain an endosperm as well as two cotyledons. These honey locust seeds have a very hard coat, which must be perforated with a file or knife to allow imbibition to occur.

ACTIVITY 6 **Study** seeds of honey locust that have been perforated and allowed to soak in water for a couple of days. Cut the seed in half across its long axis. Note the two different colored zones in the cut surface. The outer, clear, light yellowish brown zone is the endosperm. The two cotyledons make up the central, yellowish green zone.

Dissect another seed so that you can separate the endosperm from the cotyledons intact. Note the size and shape of the cotyledons. Are they as large and plump as those of the bean seed? *Yes No.* Much of the food for germination of the seedling is stored where? _____ _____ Is there any food also stored in the cotyledons? *Yes No.* Is there an embryonic axis? *Yes No.*

II. THE MONOCOT SEED

A. THE CORN GRAIN

Although the corn grain is really a fruit, there is barely any fruit development, and the bulk of the grain is the seed.

ACTIVITY 7 **Examine** a corn grain that has been soaked for several hours. **Find**

1. SCUTELLUM The white, oval-shaped depression. Basically, it is a modified cotyledon and functions to digest the endosperm just as cotyledons do in dicot plants.
2. EMBRYONIC AXIS. Faintly visible as a linear depression in the center of the scutellum.
3. ENDOSPERM The outer, yellow portion of the grain.

Label the external features of the corn grain in Figure 2.5 A. **Obtain** a soaked corn grain. Hold it flat on the table and make a lengthwise cut through it with a razor blade [see Figure 2.5A for line of cut]. **Find**

1. SCUTELLUM The white portion of the seed.
2. ENDOSPERM The yellow portion of the seed.

Use a hand-lens or dissecting microscope to identify the rest of the seed parts. Look at Figure 2.5B as a guide. **Find**

3. COLEOPTILE A tubular sheath pointed at one end. It is a modified part of the cotyledon and protects the plumule during germination.
4. PLUMULE The growing shoot tip of the embryonic axis.
5. RADICLE The rudimentary root. It is the root tip of the axis.
6. COLEORHIZA A tubular sheath enclosing the radicle. It protects the radicle in the early stages of germination.
7. MESOCOTYL The part of the embryonic axis that lies between the radicle and the plumule.

Label the internal parts of the corn grain in Figure 2.5B, using the capitalized terms defined in the preceding description.

B. SEEDLING STAGES OF THE CORN GRAIN

ACTIVITY 8 **Examine** germinating corn grains and the successive stages of seedling development. Look at Figure 2.6 as a guide. **Find**

1. PRIMARY ROOT This is the first organ to emerge. In its early stages it is enclosed and

protected by the coleorhiza, which you probably won't see.

2. **COLEOPTILE** A colorless tubular sheath that encloses the mesocotyl. It is seen growing up from the seed, or directly opposite the primary root.

3. **MESOCOTYL** The stem portion enclosed by the coleoptile and bearing at its tip the plumule.

4. **PLUMULE** The folded first leaf that emerges from the coleoptile sheath.

When the coleoptile reaches the surface, it splits open, and the leaves of the plumule emerge. Later, new crown roots form at the base (first node) of the stem. In monocot seeds the cotyledon functions mainly in the digestion and translocation of the foods in the endosperm to the growing axis.

Is the cotyledon elevated above ground? *Yes   No.*   What type of seed germination is this called? _____

Label the stages of the corn seedling development in Figure 2.6, using the capitalized terms defined in the preceding description.

EXERCISE 2   **Student Name** _____

Figure 2.1   External features of the bean seed.

Figure 2.2   Internal parts of the bean seed.

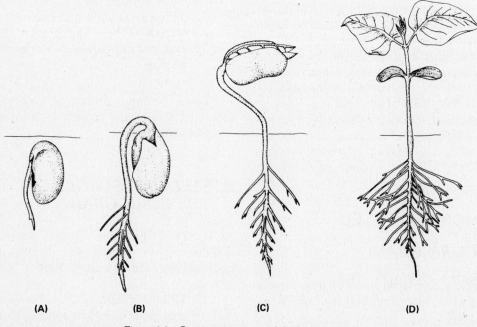

(A)   (B)   (C)   (D)

Figure 2.3   Germination stages of the bean seed.

EXERCISE 2   Student Name _____

QUESTIONS

1. What is the selective advantage of the abundant supply of foods in the seed? _____
_____
_____

2. There are at least two advantages to the storage of foods in the dehydrated state in seeds. What are these?
   (a) _____
   _____
   (b) _____
   _____

3. Primitive land plants like the ferns don't supply their embryos with any food reserves. What problems does such an embryo face compared to that of a seed plant? _____
_____

4. How do grains lend themselves so readily to worldwide trade? _____
_____
_____
   Why are they in such demand? _____
_____
_____

5. Epigean germination is characteristic of the majority of which group of flowering plants?
_____

6. What does the radicle of the seed form? _____
_____

7. Dicots and monocots differ in the way by which the delicate plumule is protected as it is being elevated through the soil. The dicot process involves _____
_____
_____
   The monocot process involves _____
_____
_____

8. In which phase of seed germination was the plaster of paris ruptured? _____
_____

9. Is the endosperm part of the embryo? _____

10. Is a cotyledon part of the embryo? _____

11. Do cotyledons develop during seed ripening or during seed germination? _____
_____

12. Distinguish between seed ripening and seed germination on the following bases.
    a. Ripening takes place usually when the seed is located where? _____
    _____
    b. Germination takes place when the seed is where? _____
    _____
    c. During ripening of the bean seed, is the endosperm being formed or digested?
    _____

13. Do all dicot seeds germinate like the bean seed? _____

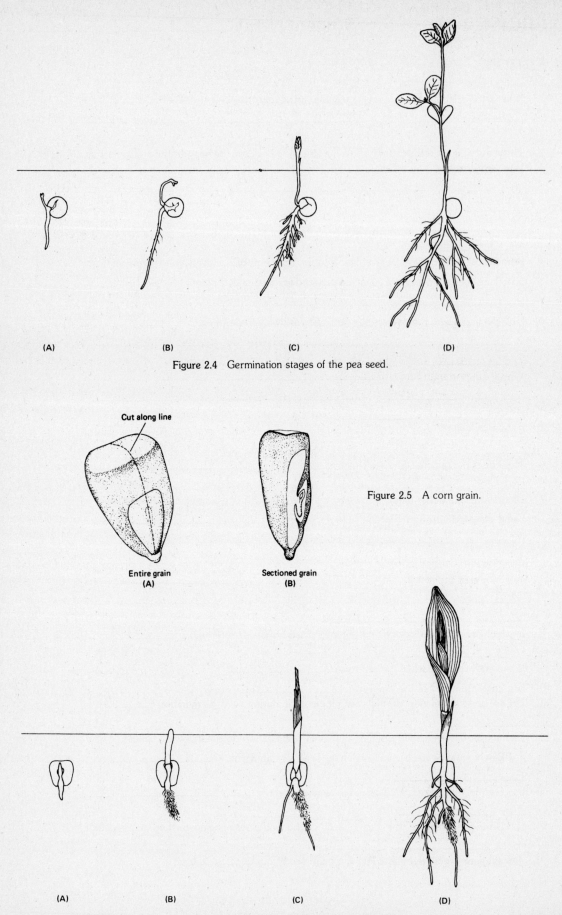

Figure 2.4   Germination stages of the pea seed.

Cut along line

Entire grain
(A)

Sectioned grain
(B)

Figure 2.5   A corn grain.

(A)

(B)

(C)

(D)

Figure 2.6   Germination stages of the corn seed.

FACTORS THAT INFLUENCE SEED GERMINATION

EXERCISE 3

INTRODUCTION

Three major factors influence seed germination. They are (1) seed viability, (2) seed dormancy, and (3) environmental conditions. If germination is to occur, certain conditions in all three factors must be met. (1) The seed must be viable, that is, the embryo must be alive and capable of growth; (2) any dormancy conditions within the seed (which may chemically or physically prevent germination) must have disappeared or been eliminated; and (3) the seed must be exposed to favorable conditions of moisture, temperature, and oxygen. (Some seeds also require light or darkness.)

I. SEED VIABILITY

A viable seed is one that contains a living embryo capable of germinating. The U.S. Federal Seed Act requires that all lots of seed entered in trade must be tested for their viability beforehand. The most commonly used test is the Germination Percentage Test. Another one is the Tetrazolium Test. Both tests can be performed in the classroom laboratory.

A. GERMINATION PERCENTAGE TEST

The germination percentage is the percentage of seed tested that produces normal seedlings under optimal conditions. When you buy a package or a lot of seed, the information on the label will usually include the minimum germination percentage.

In licensed laboratories the germination percentage test is performed on 400 seeds. For classroom accuracy, you can use only 100 seeds of any one type divided into lots of 25. The total germination of the four lots is the germination percentage.

ACTIVITY 1 **Select** seeds of different plants, such as lettuce (*Lactuca sativa*); radish (*Raphanus sativus*); spinach (*Spinacia oleracea*); corn (*Zea mays*); bean (*Phaseolus vulgaris*); pot marigold (*Calendula officinalis*); celosia (*Celosia argentea* v. *cristata*); dahlia (*Dahlia pinnata*); nasturtium (*Tropaeolum majus*); or morning glory (*Ipomoea purpurea*).

Work in groups of four students. Any four students will test 100 seeds of one type of plant. All materials and the work area must be as clean as possible to reduce contamination with fungi. Each group of students will be given a beaker containing a bleach type disinfectant made of nine parts water and one part commercial bleach (5 percent sodium hypochlorite). This gives a solution of about 0.5 percent sodium hypochlorite.

Soak the seeds in the bleach for 5–10 minutes. While the seeds are soaking, each student should soak two paper towels in some of this bleach solution. Reserve one for preparation of the seeds and use the second one to wipe clean the table work area.

Each student prepares 25 seeds as follows. (1) Spread out the moistened paper towel. (2) Space five seeds along one side of the towel and fold the edge over the row of seeds. (3) Space another row of five seeds and roll the first fold over them. Continue adding rows. Do not make tight rows. Five rows are needed. (4) Tie the rolls at each end with a string or twist tape. You have now made a "rag-doll tester." (5) Put the rag-doll testers in plastic bags and store in a warm place (about 21°C). Make certain that the testers do not dry out during the germination period.

Table 3.1 GERMINATION PERCENTAGE TEST RESULTS

PLANT NAME	NUMBER OF SEEDS TESTED	NUMBER OF SEEDS GERMINATED	GERMINATION PERCENTAGE

After ten days unfold the rag-doll testers and **count** the number of seeds that have germinated. **Compute** the germination percentage for each species tested. Record in Table 3.1. Some will unavoidably be lost due to fungal contamination.

B. TETRAZOLIUM TEST

This is a quick test to determine if the embryo in a seed is alive and thus potentially capable of germination. The living tissue of viable seeds respires and changes the colorless tetrazolium into a red dye that stains the living parts. Because the test is not foolproof, it is not used for germination labeling on seed tags in commercial trade. However, it is valuable because it is quicker than a germination test, and also does not use up as many seeds.

ACTIVITY 2 **Select** seeds of corn (*Zea mays*) that have been soaked in water overnight in the dark. (1) With a razor blade cut each seed in half and place the cut surface on a filter paper in a petri dish. (2) Moisten the paper with 0.1 percent tetrazolium chloride. (3) Set the dish aside for 15–30 minutes in the dark. (4) Prepare and test soybeans or pinto beans in the same way. (5) Examine the seeds for any red coloration. Any of the following results indicates

a viable seed: embryo entirely red; embryo mostly red; endosperm also red. Dead seeds are indicated by a pink embryo or an embryo remaining white.

II. PRACTICAL CONSEQUENCE OF DIFFERENCES IN SEEDS AND SEED GERMINATION

Commonly, large discount stores in the north temperate zones of the United States display inexpensive grass seed in 3- to 5-pound bags sold at the same price as 1-pound bags of other grass seed mixes. These less expensive mixes are usually composed mostly of annual rye-grass, whereas the more expensive mixes contain more of the bluegrass varieties and fescues. In the Gulf states of the United States, Bahiagrass is less expensive than Bermuda grass as a lawn grass.

ACTIVITY 3 Ignoring the relative merit of one grass over the other, and given the information in Table 3.2, **calculate** for each grass (1) the approximate number of seeds in a pound (or kilogram) that will germinate, and (2) the number of pounds of seed needed per 1,000 square feet (or the number of kilo-

Table 3.2 COMPARISON OF PRICE AND COST OF "EXPENSIVE" VERSUS "INEXPENSIVE" GRASSES

TURFGRASS	APPROXIMATE NUMBER OF SEEDS PER POUND (PER KILOGRAM)	GERMI-NATION (%)	APPROXIMATE NUMBER OF SEEDS IN A POUND (KILOGRAM) THAT WILL GERMINATE	NUMBER OF SEEDS NEEDED PER 1000 FT^2 (PER 100 M^2)	NUMBER OF POUNDS (KILO-GRAMS) OF SEED NEEDED PER 1000 FT^2 (PER 100 M^2)
"Expensive" Grasses Kentucky bluegrass (*Poa pratensis*)	2,177,000/lb (4,799,000/kg)	75		2,736,000/1000 ft^2 (2,944,000/100 m^2)	
Bermuda grass (*Cynodon* sp.)	1,787,000/lb (3,921,000/kg)	80		2,304,000/1000 ft^2 (2,479,000/100 m^2)	
"Inexpensive" Grasses Annual ryegrass (*Lolium multiforum*)	227,000/lb (500,000/kg)	95		1,736,000/1000 ft^2 (1,868,000/100 m^2)	
Bahiagrass (*Paspalum notatum*)	166,000/lb (366,000/kg)	75		1,152,000/1000 ft^2 (1,240,000/100 m^2)	

grams per 100 square meters)—that is, the seeding rate.

Results and conclusions of your calculations.

1. If the price of Kentucky bluegrass is $5.00 per pound and the price of annual ryegrass is $0.90 per pound and you have a lawn of 5,000 square feet to seed, what would be the cost to you to seed your lawn with the bluegrass? _____ with the ryegrass? _____

2. If you saw a sign in the store advertising the Bahiagrass at 4 pounds for $3.60 and Bermuda grass at $5.25 per pound, which would be the better buy? _____

3. Do "sales" of seed really always mean less outlay of your money? *Yes No.*

4. What is one important fact about different grass seeds that is worth knowing? _____

5. Do differences in germination percentages make up for differences in seeding rates? *Yes No.*

6. Seed tags and packages of seeds are required by law to show the germination percentage. What other information would be helpful, especially for turfgrass seed? _____

III. SEED DORMANCY

Seed dormancy is the failure of a seed to germinate even though optimal environmental conditions exist. It is due to chemical or physical barriers within the seed. In most wild (and even cultivated) plants this is of value. For example, it is very common for seeds to be produced at the end of the summer or early fall. Conditions at this time are often perfect for germination, but if germination took place, there would not be time for much seedling growth before the first killing frost. It is, therefore, a survival advantage to the species if its seeds remain dormant at this time.

Seed dormancy is caused by many factors, but three of the most common ones are (1) dormant embryo, (2) impermeable seed coat, and (3) chemical inhibitors within the seed. These three factors may exist separately or in combination in a seed. The easiest of these to demonstrate is dormancy due to an impermeable seed coat.

Dormancy Due to an Impermeable Seed Coat

Impermeable seed coats prevent water and oxygen absorption. Without water, the seed's reserve foods are not hydrolyzed (made soluble) and the digestive enzymes are not activated, so the embryo cannot grow.

ACTIVITY 4 **Set up** rag-doll testers as you did for the germination percentage test. Use seeds of the honey locust tree (*Gleditsia triacanthos*) collected in the autumn or use another locally available and suitable species.

Proceed as follows: (1) Scarify the seeds by scraping each with a file until an opening is visible. (2) Do this for several seeds and wrap them in a rag-doll tester. (3) Leave an equal number of seeds intact and wrap them in a separate tester. (4) Label the testers and store them in plastic bags in a warm (21°C) place for five to seven days. Make certain the testers do not dry out during the germination period.

After five to seven days, unfold the testers and count the number of seeds that have germinated. **Compute** the germination percentage. **Record** in Table 3.3.

IV. ENVIRONMENTAL CONDITIONS INFLUENCING SEED GERMINATION

If a seed is viable and not dormant, it should germinate when exposed to favorable environmental conditions. There are four such conditions: (1) adequate and available moisture, (2) adequate and available oxygen, (3) suitable temperature, and (4) (for some seeds) light or darkness. We will consider only oxygen and temperature requirements.

Table 3.3 GERMINATION TESTS OF SEEDS WITH INTACT IMPERMEABLE SEED COATS VERSUS IDENTICAL SEEDS WITH SCARIFIED SEED COATS

	NUMBER OF SEEDS TESTED	NUMBER OF SEEDS GERMINATED	GERMINATION PERCENTAGE
Intact Seeds			
Scarified Seeds			

A. OXYGEN REQUIREMENT FOR GERMINATION

ACTIVITY 5 **Prepare** a few *dry* corn seeds in two rag-doll testers. Get two jars (baby- or junior-food jars are just right) and put one rag-doll tester in each. Add enough water to each jar so that the testers are thoroughly wet and there is about 6 mm of water in the bottom of the jars. Immediately close one jar with a metal lid or some impervious stopper. Leave the second jar open to the air, or cover with filter paper (to prevent mold growth). Set both jars in a warm place for five to ten days. Make certain that the unsealed jar doesn't dry out during this time. After five or ten days, examine each jar. Has germination occurred in the sealed jar?   *Yes   No.*   In the unsealed jar?   *Yes   No.*   This is a very simple demonstration of the need for aeration for germination to occur. What is the gas in the air that promotes germination? _____

B. TEMPERATURE REQUIREMENT FOR GERMINATION

The experienced gardener living in the north temperate zone knows that you don't have to wait until the really warm days to set out the seeds of all the different vegetables you plan to grow. Some can be planted earlier than others. This is because different plants have different minimum temperatures at which their seeds will germinate. For instance,

peas (*Pisum sativum*) and turnips (*Brassica rapa*) can be planted outdoors six weeks before the average last freeze date of spring; lettuce (*Lactuca sativa*) can be planted five weeks before the average last freeze date of spring; and spinach (*Spinacia oleracea*) and beets (*Beta vulgaris*), three weeks. However, for cucumbers (*Cucumis sativus*), green peppers (*Capsicum frutescens* v. *grossum*), pumpkins (*Cucurbita pepo*), and watermelon (*Citrullus vulgaris*) you must wait to plant two weeks after the average last freeze date, and for sweet corn (*Zea mays*), about three weeks after the last average freeze date. The relative resistance of the seedlings to frost is also a factor in choosing planting dates for some of these crops.

ACTIVITY 6 **Obtain** seeds of peas, turnips, or cabbage and prepare them for germination in rag-doll testers. Put seeds of only one seed type in each tester, label, and date it. Do the same for seeds of cucumber, green pepper, eggplant (*Solanum melongena*), and sweet corn. Store all these in plastic bags in a refrigerator maintained about 10°C (somewhat warmer than a typical household refrigerator). Store another set of each of these types in some warm place or a germination chamber if one is available.

Between one and two weeks later, check the rag-doll testers to observe whether or not germination has occurred. The days required for germination under optimum conditions are listed in Table 3.4, and these should be consulted and used as a time guide for making your observations. **Record** your findings in Table 3.4.

Table 3.4   THE EFFECT OF TEMPERATURE ON SEED GERMINATION AND THE GERMINATION RESPONSE OF DIFFERENT PLANTS TO LOW TEMPERATURE

PLANT	STANDARD GERMINATION TIME (DAYS AT OPTIMUM TEMPERATURE)	OBSERVED GERMINATION TIME (NUMBER OF DAYS AT LOW (____ °C) TEMPERATURE)	GERMINATION PERCENTAGE	OBSERVED GERMINATION TIME (NUMBER OF DAYS AT HIGHER (____ °C) TEMPERATURE)	GERMINATION PERCENTAGE
Cabbage	6–9				
Pea	7–10				
Turnip	5–10				
Cucumber	7–10				
Green pepper	10–14				
Eggplant	10–21				
Sweet corn	5–12				

EXERCISE 3   **Student Name** _____

QUESTIONS

1. Give two advantages of delay in seed germination.

   (a) _____

   _____

   (b) _____

   _____

2. What are the three major factors that influence seed germination? _____

   _____, (b) _____, (c) _____

   _____

3. What is a viable seed? _____

   _____

4. What is the standard, legally recognized test for the determination of viable seeds?

   _____

5. Would you expect the seeding rate to be high or low for seeds with low germination

   percentages? _____

6. What else, besides germination percentage, determines the seeding rate for plants such as

   turfgrasses or crops such as rice and wheat? _____

   _____

7. Suppose a tetrazolium test on a test batch of oak (*Quercus* sp.) seeds is positive. You then
   plant the seeds of this lot, but you get zero germination. What is the most probable ex-

   planation? _____

8. Give two specific uses of water imbibed by the seed. (a) _____

   _____; (b) _____

9. Why is reseeding necessary in a field flooded by rains? Do the seeds burst from imbibing

   too much water? _____ What is the limiting or missing factor in this situation:

   _____

10. In the laboratory, you filed a hole in the seed coat of the honey locust seed. In nature, how do

    you suppose the seed coat is made permeable by the time spring arrives? _____

    _____

11. How does the minimum temperature for germination of different vegetable species help

    explain the recommended times for planting seeds in the spring? _____

    _____

    _____

12. What role does soil temperature and soil moisture play in the time required for seed ger-

    mination? _____

    _____

USE OF THE MICROSCOPE FOR STUDY OF PLANT CELLS

EXERCISE 4

INTRODUCTION

Plants, like most living things, are composed of cells. The cell is the basic building unit. To study most cells you need a microscope. Today you will look at a few cells and learn how to use the microscope.

Remember first, though, that your textbook probably contains illustrations of the parts of a cell as photographed through an electron microscope. Many cell components, such as mitochondria, ribosomes, and endoplasmic reticulum, are so small that, unlike larger objects, they do not bend enough light waves for us to be able to see them; hence, you cannot see any of this kind of minute detail through the standard laboratory microscope.

Consider, for example, your experiences in the laboratory. For some objects, you can see enough detail if you simply use a hand-lens to magnify the parts so that they appear to be 2 to 15 times larger. For smaller objects, such as the cells of a leaf, which would be visible only if magnified 20 to 1,000 times, you must use a light microscope. This microscope focuses light beams on the object and then enlarges its image by means of glass prisms. But for objects that are so small that they must be magnified 2,000 to 50,000 times to be seen, you must use something other than light waves: electron beams are used. Such minute features, which are detectable only by electron beams, are collectively referred to as the ultrastructure of matter.

I. THE MICROSCOPE

A. MICROSCOPE PARTS

Figure 4.1 is a diagram of one type of microscope in general use in elementary botany courses.

Refer to this diagram for aid when following directions for using the microscope. Not all microscopes are exactly like the one shown. Among the differences you may encounter are: a mirror instead of a substage light; different position of the fine adjustment knob; a rotating diaphragm disk with different size holes instead of a substage condenser lens system; upright or differently slanted ocular lens and tube; two objectives rather than three on the nosepiece.

B. SETTING UP THE MICROSCOPE

ACTIVITY 1

1. Use two hands to carry the microscope, and hold it upright while you carry it.
2. Set the microscope down with the arm toward you and the stage away from you (the arm may be away from you if your microscope has a reverse-angled tube).
3. Turn the nosepiece so that the lowest power objective (the shortest one) is in line with the body tube. It should snap into position.
4. Turn on the switch for the substage light, or, if your microscope has a mirror, position it so that it reflects light up through the condenser and body tube.
5. Look through the ocular with nothing on the stage to check that bright light is passing fully through the lenses to your eye. If you don't see a full circle of bright light, your instructor will show you how to adjust the light source and condenser (and mirror, if your microscope has one).
6. If the lenses or mirror are cloudy or dusty, wipe them gently with lens paper. Never use tissue paper or cloth.

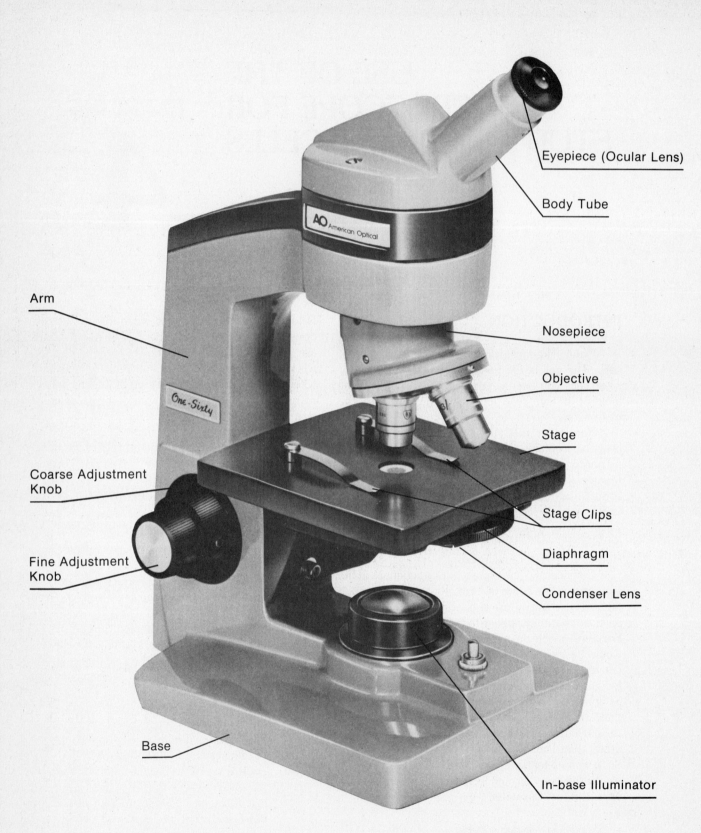

Eyepiece (Ocular Lens)

Body Tube

Nosepiece

Objective

Stage

Stage Clips

Diaphragm

Condenser Lens

Arm

Coarse Adjustment Knob

Fine Adjustment Knob

Base

In-base Illuminator

Figure 4.1 A standard compound light microscope. (Courtesy of American Optical Corporation.)

C. FOCUSING THE MICROSCOPE

ACTIVITY 2   **Obtain** a slide that has three or four silk, wool, or cotton fibers of different colors wetted and crossed over each other and held down by a cover glass.

1. Place the slide on the stage of the microscope so that the specimen material is centered over the hole in the stage.
2. Clip the slide in place with the stage clips.
3. Turn the 10× objective into place.
4. Turn the coarse adjustment knob to lower the objective until it will not go lower. Look into the ocular and slowly turn the coarse adjustment knob to increase the distance between the stage and the low-power objective. Soon the specimen should come into view.
5. If you find that your objective is more than 10 or 15 mm above the slide and you still haven't seen anything in the field of view, then start over. Decrease the distance between the objective and the stage as in step 4.
6. Repeat step 4, and this time *very very* slowly turn the knob.
7. Focus the specimen as sharply as possible with the coarse adjustment knob.
8. If you need to see more minute differences between various parts of the fibers, turn the fine adjustment knob very, very slightly. Turn it no more than 1 mm at a time! Turn it backward (toward you) or forward (away from you) until the item is clearly in focus.

Under low power, are *all* the fibers in focus at one time, that is, are they all in *one focal plane?* Yes   No.   Are all sections of any one fiber in focus at any one time (that is, in any one focal plane)? Yes   No.

Focus carefully with the fine adjustment knob. Does this help to show you which is the top fiber where two cross?   Yes   No.   **Admit** more light by opening the substage condenser unit (or by rotating the diaphragm disk unit to a larger hole). Are all the fibers clearly seen?   Yes   No. Reduce the light by closing the substage condenser (or by rotating the diaphragm to a smaller hole). Are the fibers sharply or less sharply visible? _____

Rotate the nosepiece to put the high-power objective (43× magnification) in line with the body tube. Can you see an entire fiber?   Yes   No.   Do you see more detail in each fiber? Yes   No.   Are all the fibers in focus at one time (that is, in one focal plane):   Yes   No.

The *greater the number of different objects* in focus at one time (that is, in one focal plane), the greater the depth of focus. The *fewer the number of, or portions of, objects* in focus at one time, the *less the depth of focus.* At which magnification is there greater depth of focus, low or high power? _____

D. PREPARING A WET MOUNT

ACTIVITY 3   **Make** a wet mount. Use elodea (*Anacharis canadensis*) or other appropriate material.

1. Use forceps (tweezers) and pick off one or two of the youngest leaves near the tip of the elodea plant.
2. Place the leaf on the center of a clean glass microscope slide.
3. With a pipette (dropper) put one drop of water over the leaf.
4. Balance a cover glass at one edge of the drop of water at 45 degrees. Wait until the water spreads across the edge of the cover glass. Then, with a dissecting needle slowly lower the cover glass on the water. If air bubbles are trapped, tap the cover glass gently.
5. Use this slide for your study of the cell—coming up next.

II. THE PLANT CELL

Study the following list of terms as a guide while you work.

A. SOME TERMS AND DEFINITIONS FOR THE PLANT CELL

PROTOPLAST   The organized, living contents of the cell.

CELL WALL   A nonliving, porous, semi-rigid "casing" for the living protoplast. It thickness varies with the different functions of different cells.

CYTOPLASM   The viscous, granular ground-mass part of the protoplast. Usually looks gray or colorless. In mature cells occurs as a thin layer lining the wall and extending as fine **STRANDS OF CYTOPLASM** in a meshwork throughout the cell. In young cells it fills the cell.

VACUOLE   A more or less clear area in the cell not occupied by cyto-

CELL SAP
Watery solution filling the vacuoles.

ORGANELLE
Any one of the separate bodies in the cytoplasm with a specialized structure and function.

NUCLEUS
The organelle that contains the chromosomes. Usually visible as a round body, darker colored than the cytoplasm.

CHLORO-
PLAST
An organelle that contains chlorophyll and functions in photosynthesis. Appears as a bright green, oval body and in varying numbers depending on the cell type.

PLASMA
MEMBRANE
AND
VACUOLAR
MEMBRANE
A vital, living membrane system enclosing the cytoplasm and regulating passage of material into and out of the cytoplasm. Called the *plasma membrane* where it covers the cytoplasm adjacent to the cell wall; called the *vacuolar membrane* (or tonoplast) where it covers the cytoplasm adjacent to vacuoles.

CYTO-
PLASMIC
STREAM-
ING
A flowing motion of the cytoplasmic granules. Chloroplasts and other organelles are pushed around in it.

B. A LIVING GREEN CELL— ELODEA LEAF CELL

ACTIVITY 4   **Use** the wet mount of elodea leaf that you have just prepared (in Section 1.D). This leaf isn't very thick or complicated. **Note** the midrib of the leaf. It has long, slender cells. On the margin of the leaf, **find** the thornlike spur cells. **Focus** on or near the margin of the leaf. Here there are few layers of cells, and you can see the cellular details better.

Study the leaf cells and answer these questions.

1. What is the most abundant organelle in the cells? _____

2. Are these organelles moving or stationary? (You'll have to look at more than one cell to answer this.) _____

3. What force is responsible for this? _____
_____

4. Do the four sides (walls) of a cell always remain visible as you focus up and down, or

do they disappear in different planes of focus? _____

5. **Focus** up and down carefully with the fine adjustment. Are the cells of the different layers exactly aligned one over the other? *Yes   No.*

6. If a wall suddenly goes out of view (as you turn the fine adjustment), this means that your focal plane is either above or below the wall. Do you agree?   *Yes   No.*

7. **Turn** the fine adjustment very slightly in *one direction* as you look at one small area. Do you see the walls come in and out of focus?   *Yes   No.*

8. What does this tell you about the number of layers of cells? Is there one layer or more than one? _____

Focus on a spur cell on the leaf margin. **Reduce** your light by rotating the diaphragm disk to a smaller hole or by moving the lever on the substage condenser. The nucleus should be fairly visible in the spur cell. Can you see nuclei in the green cells of the leaf?   *Yes   No.*

Label the drawing of the leaf cells of elodea (Figure 4.2), using the terms defined in Section II.A.

C. A LIVING NONGREEN CELL—EPIDERMIS CELL OF AN ONION BULB

ACTIVITY 5

1. **Prepare** a wet mount of onion-leaf epidermis tissue. **Obtain** from the instructor a piece of an onion bulb scale. The scale has a thin epidermal layer that is easily removed. **Use** forceps to peel off a portion of the epidermis from the *concave* surface of the scale. Put the tissue on a clean glass slide. Add a drop of water, and spread out the tissue so that there are no folds in it. Cover with a cover glass and examine.

2. **Staining the cell.** For better definition of the nuclei (and only after you have examined the cell in its living condition), place one or two drops of iodine at the edge of the cover glass. Draw the iodine under the cover glass by holding a small piece of paper toweling at the opposite edge of the cover glass. Your instructor will demonstrate this procedure if necessary. The iodine will kill the cells but stain the nuclei.

Examine the onion cells and answer these questions.

1. Are these cells arranged the same way as in the elodea leaf?   *Yes   No.*

2. Are there any chloroplasts: *Yes   No.*
3. Give a reason why you would or would not expect to find chloroplasts in these cells.

_____

_____

_____

4. Are nuclei visible?   *Yes   No.*
5. Is the cytoplasm visible?   *Yes   No.*
6. Which has the darker color, the nucleus or the cytoplasm? _____

_____

7. Is the cytoplasm abundant or is it in fine strands? _____

_____

8. Can you see vacuoles?   *Yes   No.* What is the clue that tells you if these cells are mature or young? _____

_____

9. How many layers compose this tissue, one or several? _____

10. Are the cytoplasmic granules in motion? *Yes   No.*   Do they jiggle back and forth or show a directional flow? _____ _____ If the granules simply jiggle, you are seeing what is called "Brownian movement." This is a motion of the granules resulting from their collisions with water molecules. Brownian movement occurs in any aqueous medium, living and nonliving.

D. MORE EXAMPLES OF CELL TYPES

Up to now you have seen simple, unmodified cells. These are in no way representative of all plant cells. But they do show you the basic major components of the plant cell—wall, nucleus, cytoplasm, and vacuoles. Animal cells never have a wall or vacuoles. Plants have many kinds of cells, and you will study these in later exercises.

ACTIVITY 6   **Examine** slides set up on demonstration microscopes. These will give you an idea of the variety of cell types. Examples you might see are

1. Cells specialized for support in mature woody stems—for example, SCLEREN-CHYMA FIBER CELLS from macerated basswood (*Tilia* sp.).
2. Cells specialized for support in young soft stems—for example, COLLENCHYMA FIBER CELLS from a freshly made thin section of a *Coleus* stem.
3. Cells specialized to change shape—for example, GUARD CELLS from the epidermis of the leaf of wandering Jew (*Zebrina pendula*) or *Rhoeo discolor*.
4. Cells specialized to conduct water—for example, TRACHEIDS and VESSEL ELEMENTS from macerated wood of pine (*Pinus* sp.) and oak (*Quercus* sp.) or basswood (*Tilia* sp.).

EXERCISE 4   **Student Name** _____

QUESTIONS

1. What is the name of the magnifying lens that your *eye* looks into? _____

2. Which adjustment knob should you use when you are first trying to locate simply the material on the slide and bring it into the field of view? _____. Which objective lens? _____

3. When first locating material on the slide, it is bad practice to look into the ocular as you decrease the distance between stage and objective. (a) True; (b) false. Circle (a) or (b).

4. What are the two components of plant cells that are not found in animal cells: (a) _____ _____; (b) _____

5. Ordinarily you *never* need to give the *fine adjustment knob* more than a millimeter turn. (a) True; (b) false. Circle (a) or (b).

6. Define an organelle: _____ _____ _____

7. What is the name of the part of the microscope that is used to rotate the objectives? _____

8. Do young cells have vacuoles? _____

9. How is the cytoplasm distributed in a mature cell? _____ _____

10. Do all plant cells have chloroplasts? _____

11. Is the cell wall alive? _____

12. Is the plasma membrane alive? _____

13. Is the vacuolar membrane alive? _____

14. In a living cell, could the nucleus occur in the vacuole? _____

15. If cytoplasmic granules show a jiggling, jerking motion, it is referred to as _____ _____

16. The high-power objective magnifies the specimen 43×, and this magnified image passes up the body tube to the ocular lens. If your ocular lens has a magnification power of 5×, what is the final magnification of the image that the eye sees? _____; what would be the final magnification with an ocular lens of 10× magnification? _____

17. What material fills the vacuole? _____

18. Are the vacuolar and plasma membranes (a) separate, distinct structures, or (b) one continuous membrane bounding all portions of the cytoplasm? Circle (a) or (b).

19. Which can occur in a living cell—(a) only cytoplasmic streaming; (b) only Brownian movement; (c) both (a) and (b)? Circle one.

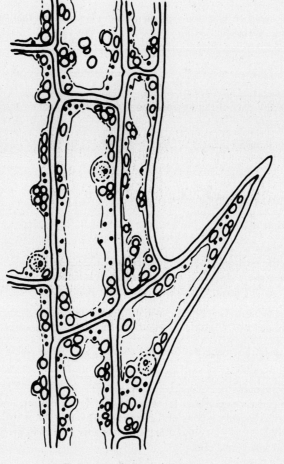

Figure 4.2 Leaf cells of elodea.

THE ROOT AND
THE APICAL MERISTEM   EXERCISE

5

INTRODUCTION

For most of the plant kingdom, there is no life without soil, and no soil without life. Roots play an important link between the two. Secretions of roots, as well as their binding action and aerating effect on soil, help preserve a good soil and the variety of organisms in it. Components of the soil sustain the organisms.

I. THE MAJOR TYPES OF ROOT SYSTEMS

1. Tap root system (Figure 5.1)
   a. One predominant root commonly growing straight downward, with smaller lateral or branch roots.
   b. Originates from the seed radicle.
2. Fibrous root system (Figure 5.2)
   a. Numerous equal-sized roots growing in many directions.
   b. Originates from the seed radicle.
   c. In general, but not always, more shallow-growing than tap root systems.
   d. Most trees have fibrous root systems within 1 meter of the soil surface.
3. Adventitious root system (Figure 5.3)
   a. Fibrouslike roots growing from stems and leaves.
   b. Originates from the stem or leaf and not the seed radicle.
   c. In most monocots (like the mature corn plant and all turfgrasses) it forms the major root system; in many others it supplements the primary root system.

d. The ability of stems to form adventitious roots is the key to the highly successful method of propagating ornamental plants by stem cuttings.

The root system of most plants is not exclusively one or the other of these three types but some modification of them. This is because root growth is influenced by many variables—temperature and available moisture; soil type (sands, clays, loams, and silt); competition with other plants; and cultural practices such as plowing and root pruning.

ACTIVITY 1 **Examine** the plants available for study and determine the root system of each. **Record** in the spaces provided on the next page the names of the plants exhibiting the root systems described.

Suggested plants for study: dandelion (*Taraxacum officinale*), any lawn grass or grassy weed, a mature corn plant (*Zea mays*); mature barley (*Hordeum vulgare*), turnip (*Brassica rapa*); beet (*Beta vulgaris*), carrot (*Daucus carota* v. *sativa*), bean (*Phaseolus vulgaris*), *Coleus*; English ivy (*Hedera helix*), nasturtium (*Tropaeolum majus*), African marigold (*Tagetes erecta*), *Rhoeo* sp.; wandering Jew (*Zebrina pendula*), *Kalanchoe* sp.

Suggested cuttings of plants that have been growing for several weeks in water, sand, or rooted in commercially prepared peat cubes or blocks: stems of geranium (*Pelargonium* sp.), *Coleus blumei*, *Kalanchoe* sp., bloodleaf (*Iresine* sp.), *Impatiens* sp., *Peperomia* sp., petiole of African violet (*Saintpaulia ionantha*), leaf of snake plant (*Sansevieria* sp.).

Example: Plants with Tap Root System

_____   _____

_____   _____

_____   _____

_____   _____

Example: Plants with Fibrous Root System

_____   _____

_____   _____

_____   _____

_____   _____

Example: Adventitious Root Formation and Plants with
             Adventitious Root Systems

_____   _____

_____   _____

_____   _____

_____   _____

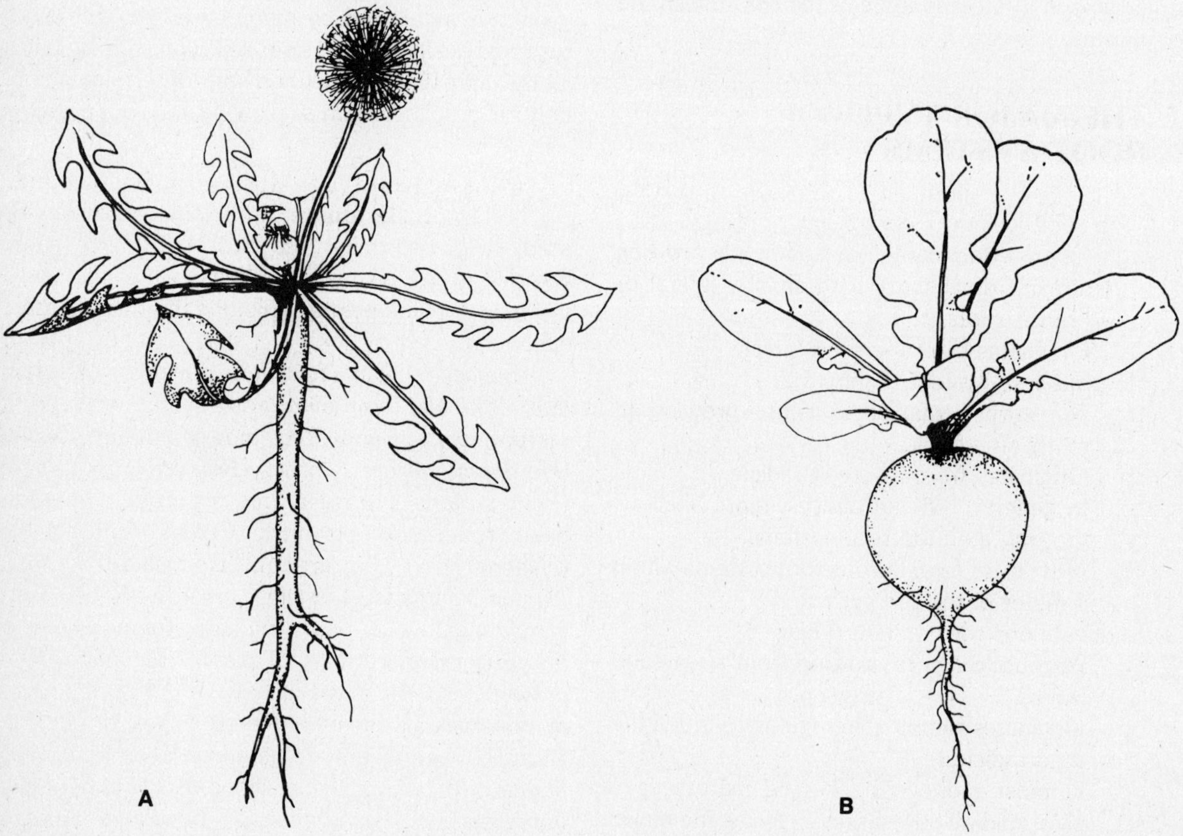

Figure 5.1 Examples of tap root system. *A,* Dandelion
(*Taraxacum officinale*). *B,* Turnip (*Brassica rapa*).

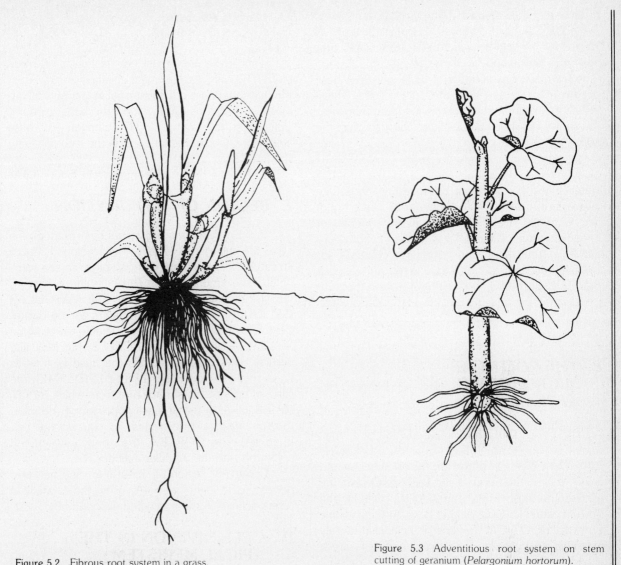

Figure 5.2   Fibrous root system in a grass.

Figure 5.3 Adventitious root system on stem cutting of geranium (*Pelargonium hortorum*).

II. ROOT HAIRS

Although old parts of roots are absorptive to some degree, most water absorption is by the root hairs located near the tip of the root. These young absorptive tips are known as feeder roots. Feeder roots of most trees lie within 20 to 40 cm of the soil surface. At this depth, oxygen and moisture can readily reach the roots.

ACTIVITY 2   **Examine** the roots of young seedlings of radish (*Raphanus sativus*) that have been germinated on moist filter paper in closed petri dishes. Note the white "fuzz"-like growth on the roots. This "fuzz" is really a mass of ROOT HAIRS. Each root hair is a thin, tubelike extension of an epidermal cell. These hairlike cells extend out into the soil and adhere to soil particles from which they absorb moisture. All together, they greatly increase the area of absorption of the root.

Examine for root hairs the roots of other plants or stem cuttings of plants that have been rooted and grown in peat cubes. Suggested plants might be coleus (*Coleus blumei*); geranium (*Pelargonium* sp.); common garden bean (*Phaseolus vulgaris*); cucumber (*Cucumis sativus*); or soybean (*Glycine max*).

III. GROWTH REGIONS OF THE ROOT TIP

A. *THE ROOT CAP— MACROSCOPIC APPEARANCE*

In your studies of seed germination you saw how the tip of the shoot is protected as it emerges through the soil either by a coleoptile (as in corn) or by the tunneling action of the arched growth of the hypocotyl (as in bean). The root tip, too, needs protection as it progresses through the soil. A thimblelike mantle of cells, the **ROOT CAP**, provides this protection.

The root cap consists of large cells that bear the brunt of friction with the soil particles. These cells rupture and exude a slimy fluid that lubricates the root tip so that it passes more easily through the soil.

ACTIVITY 3 **Examine** the tips of the roots of water hyacinth (*Eichhornia crassipes*), water lettuce (*Pistia* sp.), or other suitable material. Use a hand-lens or dissecting microscope. If root samples of American green alder (*Alnus crispa* v. *mollis*) are available, they show interesting red-colored root caps. Can you give an explanation for the presence of root caps on water hyacinth? Aren't they unnecessary? _____

_____

B. THE ROOT CAP— MICROSCOPIC APPEARANCE

ACTIVITY 4 **Obtain** a prepared slide of a longitudinal section through the root tip of an onion (*Allium cepa*). **Use** low power on the microscope and locate the pointed end of the section. **Orient** the slide so that this pointed end is at "six o'clock" (the bottom center) of your field of view. At the extreme tip is the ROOT CAP, a zone of somewhat loosely arranged corky cells. These root cap cells are a shield for the meristem. They protect its delicate dividing cells from the damage encountered as it grows between abrasive soil particles.

Label the region of the root cap in Figure 5.4A.

C. REGION OF THE APICAL MERISTEM

The APICAL MERISTEM is behind the root cap. Its cells are more densely packed and arranged in orderly rows parallel to the long axis of the root. **Note** that many of these cells are almost square in outline. Some have very large nuclei, almost completely filling the cell (as in Figure 5.4B). Some have nuclei in some stage of division. These cells were undergoing division when the root tissue was sectioned and stained to make this slide. Hence the cells are now "caught dead" in some phase of cell division.

Label the region of the apical meristem in Figure 5.4A.

D. REGION OF ELONGATION

This is the region just behind the apical meristem. The cells have enlarged chiefly in length. Most of the elongation is due to internal pressures built up by an increase of water within the large vacuoles that occupy most of the cell. The elongation of these cells increases root length and is the force pushing the tip of the root forward.

Label the region of elongation in Figure 5.4A.

E. REGION OF MATURATION

In this region (which may not be present on the slide) the enlarged cells have now developed certain structural features, and you can begin to see some differences among them. The cells are organized into three tissue systems: (1) The VASCULAR CYLINDER, which usually appears as the darker central portion. Here you may be able to see some XYLEM (water-conducting) CELLS that look like vertical rows of loose spirals (because their walls have spiral thickenings). (2) The EPIDERMIS, the layer of cells at the surface and from which ROOT HAIRS arise. In the process of preparing the slide, the root hairs are destroyed, so you will not find them. (3) The CORTEX, the zone between the epidermis and the vascular cylinder.

Label the region of maturation and its component parts in Figure 5.4A, using the capitalized terms defined in the preceding description.

IV. CELL DIVISION IN THE APICAL MERISTEM

Cells are unique. Not only are they the basic building units of an organism, but they also carry within their nuclei the "blueprints"—the directions for building the organism. These directions are stored in the nucleus in a code composed of DNA (deoxyribonucleic acid). DNA is the principal component of the chromosomes.

When a cell divides and produces two daughter cells, it is essential that this genetic information be accurately duplicated and that none be lost. Herein lies the importance of the events that occur during the three phases of cell division: interphase, mitosis, and cytokinesis. The chromosome duplicating process occurs in the lengthy interphase portion of the cell cycle. The orderly and equal distribution of the chromosomes to two daughter nuclei occurs during the brief mitotic phase of the cell cycle. Cell wall formation between the duplicate nuclei occurs during cytokinesis, creating two separate cells. The three phases of cell division are summarized in Table 5.1

Table 5.1 PHASES OF CELL DIVISION

INTERPHASE followed by⟶	MITOSIS followed by⟶	CYTOKINESIS
Three stages recognized: First growth (G₁): protein and RNA synthesis Synthesis (S): chromosomes become doubled by the replication of DNA Second growth (G₂): continuation of some protein synthesis **The genetic information now exists in duplicate form.**	An orderly and equal distribution of the duplicated genetic information into two nuclei.	Division of the cytoplasm into two more or less equal portions by the formation of a cell wall between the duplicate nuclei.

Active cell division occurs (usually) seasonally in certain parts of the plant. These parts are the tips of roots and shoots, lateral buds, and a tissue called the *vascular cambium,* whose cell divisions result in an increase in diameter of the stem and root. Cell division also occurs in any growing organ such as leaves, flowers, fruit, and seeds. A *cork cambium* is responsible for providing the corky layer(s) present in the barks of trees. Tissue in which cell division occurs on a temporary or permanent basis during the life of the plant is known as a *meristem.* Meristems are therefore located in stem and root tips, in the cambium, in buds, and similar tissues. Today's work consists of observing the meristem of a root tip and learning to recognize the stages of cell division.

ACTIVITY 5 Locating the phases of cell division in the Apical Meristem.

A. INTERPHASE

Look for cells with very large nuclei that almost fill the cell. The features of interphase are:

1. Dark-stained **CHROMATIN GRANULES** distributed throughout the nucleus. These granules are all that can be seen of the duplicating chromosomes.
2. **NUCLEAR MEMBRANE** Encloses the nuclear material.
3. **NUCLEOLI** One or more small, dark-stained bodies, larger than the chromatin granules. They are centers of RNA synthesis.

Label the appropriate stage in Figure 5.4B, using the terms capitalized in the preceding description. **Title** the figure as to the stage it represents.

B. MITOSIS (OR KARYOKINESIS)

Find the four stages of mitosis, (a) prophase, (b) metaphase, (c) anaphase, and (d) telophase.

(a) Prophase

Look for cells with the following features:

1. Chromatin material appears as a tangled loose mass of dark-stained, long threads. These are the **CHROMOSOMES.** Each consists of two duplicate strands, called **CHROMATIDS** (also **SISTER CHROMATIDS**), held together at one region, called the **CENTROMERE.**
2. **NUCLEOLI** still present but decreasing in size.
3. **NUCLEAR MEMBRANE** becoming less distinct as it now begins to break down.
4. **SPINDLE FIBERS (MICROTUBULES)** faintly visible as fine lines diverging from opposite poles of the nucleus and radiating toward the midline (equatorial region).

Label the appropriate stage in Figure 5.4B, using the terms capitalized in the preceding description. **Title** the figure as to the stage it represnts.

(b) Metaphase

Look for cells in which the dark-stained chromosomes are clumped in the mid (equatorial) region of the cell and fine lines extend from the chromosomes to the poles and continuously from pole to pole. In the configurations that you see are

1. **SISTER CHROMATIDS** These have condensed into tightly coiled, thick, fingerlike objects.
2. **EQUATORIAL PLANE** This is not an object in the cell. It is the midregion of the cell where **CENTROMERES** of the joined sister chromatids are all aligned in one plane, the **EQUATORIAL PLANE.** Here each centromere is attached to four or more **SPINDLE FIBERS,** which run perpendicular to the plane. Because conditions become very crowded in this equatorial plane, the arms of the **CHROMATIDS**

themselves extend off this plane in all directions.

Label the appropriate stage in Figure 5.4B, using the terms capitalized in the preceding description. **Title** the figure as to the stage it represents.

(c) Anaphase

Look for cells in which there are two clusters of chromatids lying close to, but not at, the poles of the cell. Fine lines extend across the cell between the chromatid clusters. The features you see here are:

1. **CHROMATIDS** clumped together. Many appear V-shaped, with the point of the V toward the pole of the cell. By use of fine focusing, you will see that most of the chromatids have this shape. This is because they are physically being pulled to the poles by the spindle microtubules.
2. **SPINDLE FIBERS (MICROTUBULES)** The shortening of these fibers pulls the sister chromatids apart and carries each to an opposite pole of the cell.
3. **CENTROMERE** This is the point of attachment for the spindle fibers. The centromere thus is the "leading" section of the migrating chromatid.
4. **ARMS** of the **CHROMATID** These are the chromatid parts following along behind the centromere.

The equal distribution of genetic material to daughter nuclei occurs in anaphase. The duplicate sister chromatids have been split apart at their centromere, and each is drawn to an opposite pole by contracting fibers. Once the chromatids have become so separated, they are essentially individual chromosomes.

Label the appropriate stage in Figure 5.4B, using the terms capitalized in the preceding description. **Title** the figure as to the stage it represents.

(d) Telophase

Look for cells in which there are two dense, dark clumps and between them a faint black line. The features you see here are:

1. **NEWLY FORMED NUCLEI** Each is composed of the same number and the same kind of chromosomes as the original parent cell. The **CHROMOSOMES** are beginning to return to their long, uncoiled, linear form.
2. **A CELL PLATE** is forming between the nuclei. It serves as the cementing middle layer between the cell walls that will be laid down later.

3. **NUCLEOLI** reappear in each daughter nucleus.
4. **NUCLEAR MEMBRANE** reappears, surrounding each nucleus.
5. The **REMNANT SPINDLE FIBERS** will soon disappear.

Label the appropriate stage in Figure 5.4B, using the capitalized terms in the preceding description. **Title** the figure as to the stage it represents.

C. CYTOKINESIS

Cytokinesis is the phase that follows nuclear division. During cytokinesis, a cell wall forms between the nuclei, and the cytoplasm of the parent cell is divided.

Look for pairs of cells that are brick-shaped with large, round nuclei filling about one third to one half the cell. These cells and other features that you see are:

1. **DAUGHTER CELLS.**
2. **NEW CELL WALL,** which separates the daughter nuclei.
3. **DUPLICATE NUCLEAR MATERIAL,** which occupies each nucleus.

Label the appropriate stage in Figure 5.4B, using the terms capitalized in the preceding description. **Title** the figure as to the stage it represents.

V. DEMONSTRATION OF THE REGIONS OF PRIMARY GROWTH IN THE ROOT

ACTIVITY 6 **Examine** ink markings on the root of a germinating seed. Either you or the instructor could have prepared these seeds a day or two earlier. The procedure is as follows.

Germinate seeds of bean (*Phaseolus vulgaris*) or pea (*Pisum sativum*) in moist vermiculite or sand until the primary root is about 3 to 4 cm long. Then gently blot the root dry. Lay the seed flat and place a ruler alongside the root. Starting with the root tip as your zero point, mark off intervals on the root 2 mm apart with a thread wetted with India ink. Gently touch the inked thread to the root at the measured intervals. Look at Table 5.2. The region between any two marks can be recorded as regions 1, 2, and so on. Record the date on which the marks were made. Draw your marks on the root of the seed shown in Figure 5.5A, and label the regions between the marks as 1, 2, and so on. Pin the germinating seed to the underside of a cork and insert it into the top of a jar with some water in the bottom. An air hole in the cork would ensure adequate oxygen. The root

Table 5.2 RECORD OF ROOT TIP GROWTH

MARKINGS MADE BETWEEN	REGION	DATE MARKINGS MADE	AMOUNT OF INCREASE IN LENGTH OF REGION (MM)	DATE OF SECOND MEASUREMENT
Tip–2 mm	1			
2–4 mm	2			
4–6 mm	3			
6–8 mm	4			
8–10 mm	5			

should be above the water. After a day or two, inspect the marks on the root. Measure the distances between the marks and record in Table 5.2. Record the date also. In Figure 5.5B draw the positions of the marks as they now appear. Label the intervening regions, 1, 2, and so on. Draw dotted lines between Figure 5.5A and B, connecting the corresponding marks.

Has the distance increased between all the marks or just between some? _____ In which numbered region or regions do you see any change? _____. Which numbered region shows the most increase? _____. Is it just behind the root tip or much farther up the root? _____. What is the name of this region? _____.

VI. MYCORRHIZA

ACTIVITY 7 **Examine** slides of roots of an orchid or other plant species that show mycorrhizae. A **MYCORRHIZA** is the association of a root and fungus growing together in a physiological union that is frequently mutualistic. The fungus derives nutrients from the root cells, and the root depends to some extent on the fungus for absorption of soil moisture and minerals. Mycorrhizae probably occur in the majority of woody plants and in many herbaceous plants. In these plants, root hairs may be absent, and their role in absorption is in large part carried on by the fungus, which in some cases has absorptive filaments extending far out into the soil. There are two types of mycorrhiza, ectotrophic and endotrophic. In an ectotrophic mycorrhiza the fungus grows over the root as a sheath, and many fungal filaments grow between the cells of the cortex of the root. In an endotrophic mycorrhiza, no fungal sheath is formed over the root, and many of the fungal filaments extend into living cells of the root cortex as well as between them.

Study the slides showing the mycorrhizae and determine which type is present. Does the root have an outer mass of intertwining tubular filaments completely enveloping it? *Yes No.* If so, do these filaments also extend between the cells of the root? *Yes No.* If you have answered *Yes* to both questions, then the mycorrhiza type present is _____. If you have answered *No* to both questions, then look again closely at the root cells. Each cell has a very large nucleus. In the remainder of the cell are there many threadlike, intermingled filaments? *Yes No.* What are these filaments? _____ What type of mycorrhiza is this? _____

(A)   **(B)**

Figure 5.5 Growth of the root tip.

EXERCISE 5   **Student Name** _____

QUESTIONS

1. Which type of root system does not originate from the radicle of the seed? _____

   _____

2. Do most trees have tap root systems or fibrous root systems? _____

   _____

3. How deep are the feeder roots of most trees? _____

4. Suppose you have just purchased a new house on a wooded lot and the builder recommends that you add a foot of topsoil over the ground so you will have a better lawn. This may be good for a lawn, but why would it be bad for large trees? _____

   _____

5. New plants can be started from stems cut from mature plants. What makes this possible?

   _____

   _____

6. Which is responsible for the increase in root length, (a) increased cell numbers in the meristem region, or (b) lengthening of the cells in the region of elongation? Circle (a) or (b).

7. Do all regions of the root behind the meristem increase in length equally? _____

8. Is the absorptive capacity of a root the same along its entire length? _____

9. What is the function of root hairs? _____

   _____

10. Which does a root hair consist of, (a) one elongate cell or (b) several cells? Circle (a) or (b).

11. Transplanting (that is, digging up a tree or shrub and replanting it somewhere else) is usually a physiological shock to the plant. Typically it wilts, unless special care and preparation are given to it before and after it is transplanted. Which particular roots on the plants are damaged in the transplant? _____

12. Define mycorrhiza. _____

    _____

13. Define meristem tissue. _____

    _____

    _____

14. Do chromosomes duplicate before or during mitosis? _____

15. What is the main component of chromosomes? _____

16. One of the characteristics of a meristematic tissue is that it is always in a state of active cell division. (a) True; (b) false. Circle (a) or (b).

17. During what phase of cell division is the nuclear material doubled? _____

    _____

18. Name two places in the plant (other than root and shoot tips) where meristematic tissue is located (a) _____; (b) _____

19. Name the part of the chromatid that is attached to the spindle fibers _____

20. What should be the chromosome number in daughter nuclei formed by mitotic division of cells having the following chromosome numbers: 16 _____; 22 _____; 68 _____ 8 _____

21. Are the stages of mitosis synchronized in a meristem? _____

22. In what material is the coded information of a cell stored? _____

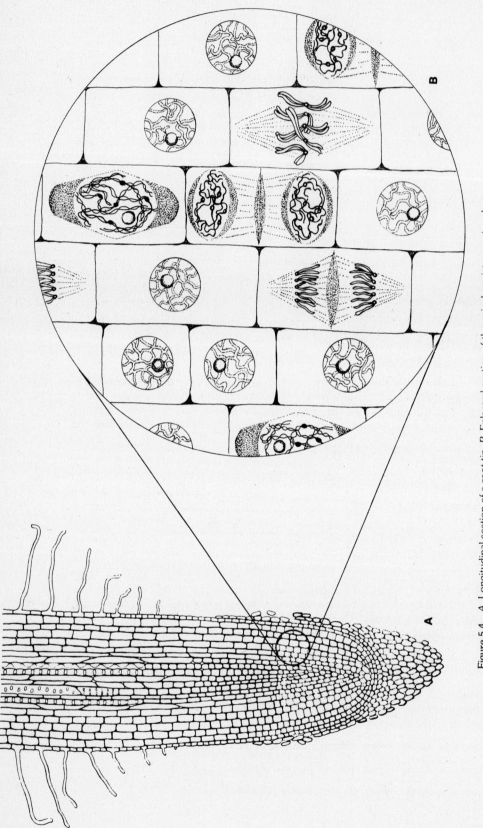

Figure 5.4 *A*, Longitudinal section of a root tip. *B*, Enlarged section of the apical meristem region showing the phases of cell division.

INTERNAL STRUCTURE
AND FUNCTION
OF ROOTS

EXERCISE
6

INTRODUCTION

Checking plant roots is routine practice with growers of greenhouse crops and container-grown nursery stock. The grower samples the crop by knocking a few plants out of their pots in order to look at the root tips on the soil ball.

Why are the root tips important, and why are they good indicators of the plant's general condition? Because their color shows if the plant is healthy, and their number indicates the degree of growth, and thus, also, absorption. If the plant is healthy and actively growing, many of its roots will show up as tiny white protuberances on the surface of the soil ball.

No matter what the age or size of a plant, uptake of water and nutrients from the soil occurs mainly in the region of root hairs at the tips of roots (see Exercise 5). These young, absorptive regions at the tips of all the roots on a plant comprise what is collectively known as the "feeder roots" of the plant. The older parts of the root absorb very little and serve mainly for anchorage and as passageways (or for storage) of materials moving through the root system.

Figure 6.1 illustrates how any one cell laid down by the meristem gradually becomes farther and farther removed from the tip. When any given cell (as the one designated at A) is still close behind the meristem (as in B), it is active in absorption. But, after many weeks, months, or years, any cell that once was near the tip is far removed from the tip and is located in an old part of the root, as in D of Figure 6.1. In trees and shrubs these old parts are woody. Almost no absorption takes place in these woody parts. But every old root, woody or not, always has its young growing tip where cells are active in absorption. Thus, even an old tree

with a massive woody root system has young feeder roots.

In this exercise we will see how the nature and arrangement of tissues in the feeder roots govern absorption and translocation.

Feeder roots of trees resemble (both externally and internally) the feeder roots of soft-stemmed herbs. Since it is not too easy to dig up and study tree roots, we will examine the root of a little herb, the buttercup (*Ranunculus* sp.), and also the root of the greenbrier vine (*Smilax* sp.).

In the feeder root we find a double cylinder composed of: (1) an inner vascular cylinder "sealed off" from the soil, and (2) an outer cylinder permeated with air, water, and mineral ions from the soil. The outer cylinder controls entry of these materials into the vascular cylinder. Inside the vascular cylinder, movement of materials is upward to the stem and leaves and downward from them into the root.

I. TISSUES OF THE ROOT OF A HERBACEOUS DICOT

ACTIVITY 1 **Obtain** a prepared slide of a cross section of a young root of a dicot plant such as buttercup (*Ranunculus* sp.) and study the following parts.

A. *THE OUTER CYLINDER— COMPOSED OF EPIDERMIS, CORTEX, AND ENDODERMIS*

Epidermis

Focus on the outermost layer of cells, the EPIDERMIS. Some epidermal cells grow far out into

41

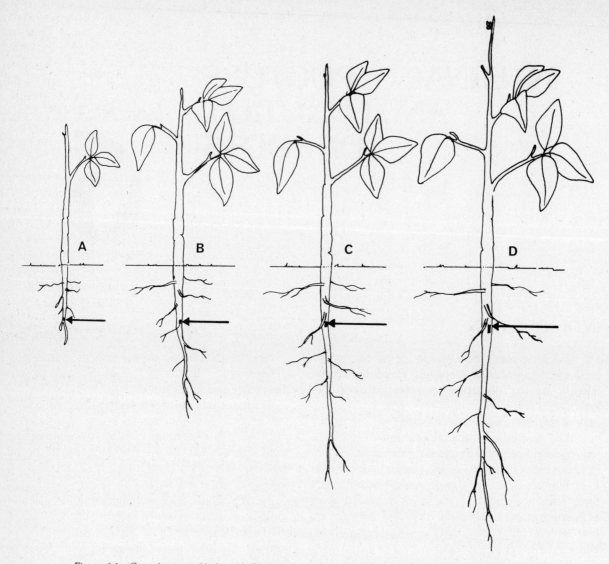

Figure 6.1 Growth stages (*A* through *D*) showing how a cell (indicated by the arrow) that is formed in the tip of the root (at stage *A*) gradually becomes farther and farther removed from the tip (stages *B–D*).

the soil. This increases the root's area of contact with the soil. What do you call such epidermal cells?

_____

In more mature root regions, a thin **CUTICLE** may cover the epidermis. If one is present, you will see a clear, pink-stained coating on the epidermis. **Note** the walls of the epidermal cells. These are highly permeable to the soil solution. Water of the soil solution moves into the cell interior, as well as through and along the walls. But the mineral ions are thought to move mainly through the walls of the epidermal cells. **Label** the epidermis in Figure 6.2A and 6.2B.

Cortex

The **CORTEX** is the very large zone of rounded **PARENCHYMA** cells situated between the epidermis and the vascular cylinder. The vascular cylinder has a prominent star-shaped central axis of

(red-stained) cells. **Look** at the cortex. Note that it is a big zone with large, loosely arranged **PARENCHYMA** cells and much free space between them (intercellular spaces). This construction serves three general functions: (1) it provides space for a high volume of soil solution to move quickly into the root; (2) the plasma membranes of these large cells all together form a large surface area for absorption; and (3) their respiration supplies the energy needed to move mineral ions into the cytoplasm of the cells.

Ions move through the cytoplasm of cortex cells across to the vascular cylinder. Water moves mostly through and along the walls of the cortex cells and not through the cytoplasm. Thus, water absorption and transfer are independent of that of mineral ions. What advantages does this have to the plant?

_____

_____

_____

_____

Does the cortex have an outer zone of smaller cells with thicker walls? *Yes No.* If so, then your root section was cut through an older part of the root where absorption has more or less ceased.

Note if there are any (blue-stained) particles in the cortex cells. These particles are **STARCH GRANULES,** storage products translocated into the root from the shoot system.

Label the cortex in Figure 6.3A and 6.3B, using the terms capitalized in the preceding description.

Endodermis

Look at the inner boundary of the cortex and **focus** on the distinct ring of smaller, rectangular cells surrounding the central vascular cylinder. This ring of cells is the **ENDODERMIS,** the innermost layer of the cortex. If your root section was cut from a *young* portion of the root, most of the endodermal cell walls will be thin. If the walls are thick, then you have an *older* root section. It is described in the second paragraph below.

On a young root section, **look** at the walls between any two adjacent endodermal cells. These are the radial walls. They are thick because they have a layer of suberin—an impermeable material—concentrated in bands, called **CASPARIAN STRIPS.** These strips are thought to form a barrier to the passage of anything moving through the cell walls. They occur in all the walls except the tangential ones (those that face the cortex and vascular cylinder). Materials are thus routed through the porous tangential walls and into the cytoplasm of the endodermal cells (Figure 6.2). The living cytoplasm thus controls what passes through it. In this way, anything approaching entry into the vascular cylinder is first "screened" by the endodermis.

If your slide contains an older section of the root where less absorption is taking place, as shown in Figure 6.2A, most of the endodermal cells will have thick walls. A few thin-walled **PASSAGE CELLS** may allow some movement of materials into the vascular cylinder. Note also that the passage cells lie opposite each of the ridges of the star-shaped central axis of (red-stained) xylem cells. This channels the water into the xylem and helps to prevent its passage into the phloem. **Label** the endodermis in Figure 6.2A, indicating those cell walls with the casparian strip, and, if the root is an older section, the passage cells. **Label** also the endodermis in Figure 6.2B.

B. THE CENTRAL VASCULAR CYLINDER

The **VASCULAR CYLINDER** is the central core of the root. It functions in lengthwise conduc-tion and (in many plants) in growth too. It is made up of three tissues, and sometimes a fourth. The three tissues always present are the **PERICYCLE,** the **PRIMARY XYLEM,** and the **PRIMARY PHLOEM.** In a mature buttercup root, and in those plants destined to have some secondary root growth, the fourth tissue is the **VASCULAR CAMBIUM.**

Pericycle

It will not be easy for you to recognize the **PERICYCLE** zone because there is nothing distinctive about the cells to describe to you. So, look for thin-walled cells that lie just within the circle of the endodermis. These are the pericycle cells, forming a zone usually only one cell wide. They are parenchyma cells. Though not distinctive in structural features, they are nevertheless important because, unlike the cells of most surrounding tissues, they retain the ability to resume active cell division. In other words, they are meristematic, and they give rise to branch roots.

In some species the pericycle also gives rise to shoots called root sprouts or root suckers. Many perennial plants, including certain trees and shrubs, send up these sprouts from their roots. It is a natural way for the plant to reproduce itself vegetatively and thus to expand the area of ground it occupies. The tendency for a plant to produce this type of growth is known as "suckering," and certain plants are more disposed to it than others. It helps them to succeed in natural surroundings, but when grown in lawns or shrub borders, this type of plant can be a nuisance. The sprouts appear in the lawn or other places where they aren't wanted and must be cut down. Certain dogwood (*Cornus* sp.) shrubs, the white poplar (*Populus alba*), and the tree-of-heaven (*Ailanthus*) are examples of suckering plants. The pericycle also gives rise to a cork cambium, which, in turn, gives rise to a cork tissue in the roots of woody plants. **Label** the pericycle in Figure 6.2A.

Primary Xylem

In three-dimensions, the **PRIMARY XYLEM** (as seen in Figure 6.2) is a fluted cylinder extending the length of the root. In cross section it is a mass of (red-stained) cells in the center of the root. It has three, four, or five ridges reaching to the pericycle. **Note** the large, thick-walled cells, called **VESSEL ELEMENTS.** Vessel elements are open-ended, barrel-shaped, dead cells devoid of all cytoplasm. They are joined end to end to form water tubes, called **VESSELS. Note** that vessel elements (or cells) are different sizes. The smaller cells are called **PROTOXYLEM.** They developed lignified walls early in their growth, and this prevented their further enlargement. The larger cells are called META-XYLEM. They grew for a longer time before their walls became lignified. Metaxylem cells are concentrated in the center of the primary xylem. **Label** the PRIMARY XYLEM in Figure 6.2A and 6.2B using the terms capitalized in the preceding description.

Primary Phloem

Look for small semicircular clusters of cells nested between the ridges of primary xylem. These are the PRIMARY PHLOEM. Phloem consists of two cell types, SIEVE TUBE ELEMENTS and COMPANION CELLS. These cell types are difficult to distinguish from each other in this section. Companion cells are the smaller cells that are more dense looking; they may have a nucleus that fills most of the cell. Sieve tube elements are larger in diameter and lack nuclei. They are vertically elongate, conducting cells stacked one above the other, forming SIEVE TUBES. Strands of cytoplasm extend between them through their porous end walls, called SIEVE PLATES. Sucrose and other materials synthesized by the plant move through the tubes in the connecting cytoplasmic strands. Foods may move up or down, depending on the changing needs of the plant. **Label** at least two of the clusters of primary phloem in Figure 6.2A, using the terms capitalized in the preceding description.

Vascular Cambium

Cambial cells may not be easy to identify. **Look** in the region between a cluster of phloem cells and the concave part of the xylem. **Find** a band of flat, rectangular cells. This band is the VASCULAR CAMBIUM. In the buttercup root these cambial cells may produce only very few secondary xylem and phloem cells. But in roots that become woody, the vascular cambium produces much secondary tissue,

and the root comes to resemble a shoot in cross section. **Label** the vascular cambium in Figure 6.2A.

II. THE ORIGIN AND GROWTH OF BRANCH OR LATERAL ROOTS

ACTIVITY 2 **Examine** a slide of a young willow (*Salix* sp.) root showing the origin of a branch root from a primary root. In which tissue of the root do branch roots originate? _____ **Look** at the pattern of the cells in the branch root. In what other study of the root did you see this particular pattern? _____ The newly formed branch root cells elongate and push the root through the cortex, out into the soil.

III. TISSUES OF THE ROOT OF AN HERBACEOUS MONOCOT

Monocot anatomy is more complex than that of dicots. There are more variations. This makes it difficult to choose any one plant as typical of all.

ACTIVITY 3 **Examine** at least one monocot root so you can see some of the differences from a dicot. **Examine** a prepared slide showing a cross section of the root of a monocot, such as wheat (*Triticum* sp.); *Iris* sp.; corn (*Zea mays*); or greenbrier (*Smilax* sp.).

Keeping this slide on your microscope, **compare** this monocot root with your dicot root (Figure 6.2) and **complete** Table 6.1.

Is this root a double cylinder as in dicots? *Yes No.* **Note** the circle of large META-XYLEM cells, and tapering radially outward from them the smaller PROTOXYLEM cells. All together they form a multirayed ring of primary xylem. The small clusters of cells nested between the protoxylem points of the xylem ridges are the PRIMARY PHLOEM.

Table 6.1   COMPARISON OF THE INTERNAL ANATOMY OF HERBACEOUS MONOCOT AND DICOT ROOTS

FEATURE	DICOT	MONOCOT			
	Buttercup	Corn	Greenbrier	Wheat	Iris sp.
Number of cylinders					
Number of primary xylem ridges (few or numerous)					
Pith (present or absent)					
Endodermis—indicate in which group it is thicker, monocots or dicots					
Vascular cylinder size—larger or smaller than cortex					

The numerous ridges of protoxylem are characteristic of monocot roots. A large PITH occupies the center of the VASCULAR CYLINDER. Was there a pith in the dicot root?   *Yes   No.*   **Look** at the ENDODERMIS. Its cell walls have developed more suberin and more cellulose than is found in dicot root sections of similar age. This is typical in monocots where roots last a long time but never develop secondary growth. Just inside the endodermis is the PERICYCLE, which in monocots is several layers thick. In mature roots it will be composed of thick-walled cells. In most dicots this is not the case. How many layers of pericycle did you see in the dicot root? _____ The CORTEX in *Zea* or *Smilax* is smaller than the VASCULAR CYLINDER, but in other monocots it may be larger. These size differences are related to the number of protoxylem ridges. The more numerous the xylem ridges, the larger the pith and vascular cylinder and the smaller the cortex. Conversely, the fewer the xylem ridges, the smaller the pith and vascular cylinder and the larger the cortex. To check this, **look** at slides of roots of other monocots, such as wheat (*Triticum* sp.) or *Iris* sp. Are the number of protoxylem ridges as numerous as in corn or greenbrier?   *Yes   No.*   Which is the more massive, the cortex or the vascular cylinder? _____

Label the cross-section drawing (Figure 6.3) of the monocot root of *Smilax*, using the terms capitalized in the preceding description.

IV. TISSUES OF THE WOODY ROOT OF A DICOT

ACTIVITY 4   **Study** a prepared slide of a cross section of a woody root of a dicot plant, such as basswood (*Tilia* sp.) or tulip tree (*Liriodendron tulipifera*). This sample section is a young woody root only a few years old. When this root was less than one year old, it had the same structure as the buttercup root. Here, though, SECONDARY XYLEM has been laid down by the vascular cambium in concentric annual rings similar to those in stems. The regularity of the rings in these very young woody roots is, however, not very clear cut and may be difficult for you to recognize. SECONDARY PHLOEM lies outside the secondary xylem and is not as distinct as the clusters of primary phloem you saw in younger roots. Is the endodermis present?   *Yes   No.*   Is the epidermis still intact?   *Yes.   No.*   Is the cortex present?   *Yes   No.*   Your answers will depend on the particular plant you are examining. Cortex is eventually replaced by cork or wood, but how soon this happens varies with different species. In a section of root this old, is absorption taking place to the same degree as in young root sections?   *Yes   No.*   In some root sections you may find CORK already present. It has been produced by a cork cambium that arises in the outer area of the roots of all woody plants when they reach two or three years of age.

EXERCISE 6   Student Name _____

QUESTIONS

1. Does the woody portion of a tree's root system play the major role in active absorption?

   _____

2. How deep are the feeder roots of most trees and shrubs? _____

3. You are having a swimming pool built in your backyard and the contractor piles the excavated soil on the ground under some of your trees. Why is this bad for the health of

   most trees? _____

4. What part of the root is concerned with lengthwise conduction? _____

   _____ , with lateral conduction? _____

5. What layer of cells regulates the entry of water and other materials into the vascular

   cylinder? _____

6. The absorption and movement of water in the root is independent of that of mineral ions. (a) True; (b) false. Circle (a) or (b).

7. What two features of the cortex make it a most suitable tissue for the *influx* of the soil

   solution? (a) _____ ; (b) _____

   _____

8. What is the passage route of water across the cortex? _____

   _____

9. What is the passage route of mineral ions across the cortex? _____

   _____

10. Are mineral ions passively absorbed or do the cortex cells have to work to "pull" them in?

    _____

11. In what two ways does the large size of the cortex cells promote the absorption of mineral ions?
    (a) _____

    _____ ; (b) _____

12. Besides absorption and lateral conduction, what other function does the cortex serve?

    _____

13. What prevents the water that has freely flowed through the cortex from simply moving into

    the vascular cylinder? _____

    _____

14. Buttercups aren't such important plants. Why did we study buttercup roots? _____

    _____

15. Sieve tube members are alive, but vessel members are dead. (a) True; (b) false. Circle (a) or (b).

16. Which tissue in the root gives rise to secondary (lateral) roots? _____

    _____

17. How does water get past the endodermis when most of the endodermal cell walls are

    "waterproof"? _____

    _____

18. In general, the vascular cylinder of monocot roots has many fewer protoxylem ridges than that of dicots. (a) True; (b) false. Circle (a) or (b).

19. Other than by seeds, how do some plants increase their numbers in an area? _____

    _____

    _____

20. What role does the cytoplasm of the endodermal cells play? _____

    _____

    _____

21. The vascular cambium is a meristematic tissue. What other tissue in the root is meri-

    stematic? _____

22. Dicot roots have a pith. (a) True; (b) false. Circle (a) or (b).
23. All monocot roots look very similar. (a) True; (b) false. Circle (a) or (b).

24. Is a root sprout a shoot or a root? _____

25. When a dicot root grows old and woody, its internal structure is essentially similar to
    that of a stem. (a) True; (b) false. Circle (a) or (b).

EXERCISE 6   **Student Name** _____

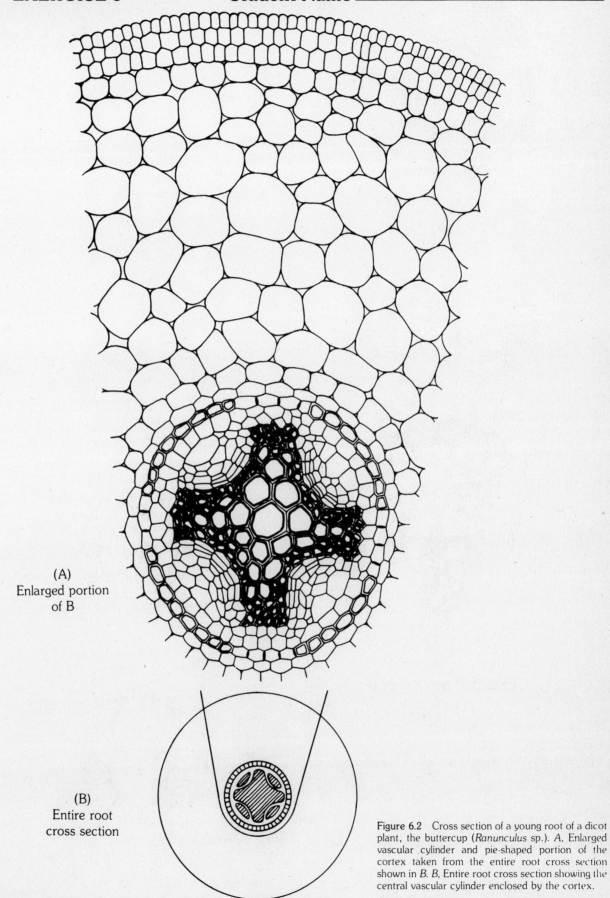

(A)
Enlarged portion
of B

(B)
Entire root
cross section

Figure 6.2   Cross section of a young root of a dicot plant, the buttercup (*Ranunculus* sp.). *A,* Enlarged vascular cylinder and pie-shaped portion of the cortex taken from the entire root cross section shown in *B. B,* Entire root cross section showing the central vascular cylinder enclosed by the cortex.

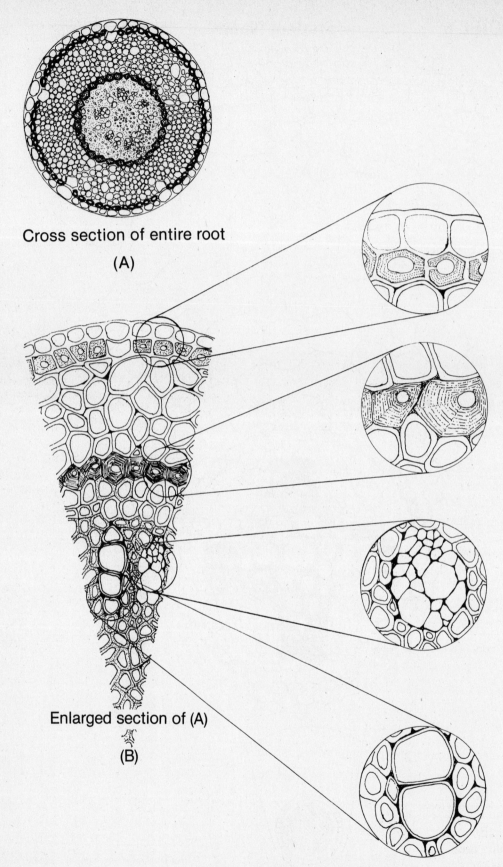

Cross section of entire root

(A)

Enlarged section of (A)

(B)

Figure 6.3 Cross section of a young root of a monocot plant, the greenbrier vine (*Smilax* sp.)

HERBACEOUS STEMS

INTRODUCTION

The term herb is popularly interpreted to mean a plant grown for some medicinal purpose or for its sweet scent or flavor. But botanically an herb is any seed plant that does not develop woody, persistent tissue. Herbs may be annuals (live one year) and come up from seed, or perennials (live a few or many years) and come up from old underground parts.

Herbs use most of their photosynthate (the food synthesized) to make leaves, stems, roots, fruits, and seeds, and (if perennials) other food storage organs for use the next growing season. Woody plants use much of their photosynthate to build a permanent, woody framework supporting a canopy of leaves. They also build a large root system. In a sense, the woody plant construction is designed mainly for perpetuating the individual plant. Year after year, you pass the same trees, summer and winter, on your way to school or work. By contrast, the herbaceous plant construction is designed mainly for perpetuating the population. Year after year, you may pass by an abandoned field; in winter dead plant stems stand in the field, but in summer the same kind (same population) of plants may grow, such as grasses, goldenrod, or Queen Anne's lace. We consider the herb design a more specialized way of life. Humans have benefited from the herbaceous plant design because it is chiefly from the herbs that we derive our food.

The growth of herbs is mostly of the kind we call primary. Primary growth produces new stems, leaves, and buds. Characteristically, the stems are soft. *All* plants, both herbs and woody plants, have primary growth. But in woody plants the soft tissues of the stem laid down during primary growth are later covered by other tissues that make the stem hard. These tissues are the result of secondary growth (called secondary because it follows primary growth in time).

Many dicot herbs, such as the sunflower (*Helianthus* sp.) and the castor bean (*Ricinus* sp.), do have some amount of secondary growth and are hard-stemmed. Monocots are mostly herbs. A few, most notably the palms and the Joshua tree (*Yucca brevifolia*) are treelike but do not have a woody structure such as that found in dicot trees.

I. HERBACEOUS DICOT STEMS —PRIMARY GROWTH

A. GENERAL EXTERNAL APPEARANCE

ACTIVITY 1 **Review** briefly the terminology relating to the dicot plant body given in Activity 1 of Exercise 1.

Obtain a representative dicot herbaceous plant such as *Coleus blumei* or Swedish ivy (*Plectranthus* sp.). **Examine** the extreme tip of the main stem or of the side branches. You should be able to find a tiny tight cluster of very small leaves that are almost indistinguishable. This is the region of the TERMINAL or APICAL MERISTEM, the zone where all new growth originates.

Gently **bend** or **squeeze** the stem. Its softness reflects the soft PRIMARY TISSUES formed by the apical meristem.

Find the small side branches along the stem. These grew from the lateral buds, which were located in the axils of the leaves. Each lateral bud contained an apical meristem (just as in the main stem), and the branch is the result of the primary growth produced by this apical meristem.

Look at the length of the internodes. They are longer between the older (lower) leaves than between the younger (upper) leaves. Stem elongation is due mostly to the elongation of the cells in the internode, particularly those at the base of the internode. Thus it is the lengthening of internodes that brings about stem elongation.

B. MICROSCOPIC APPEARANCE OF PRIMARY GROWTH ZONE OF THE DICOT STEM

ACTIVITY 2 **Study** a prepared slide with a longitudinal section of the stem tip of *Coleus blumei*. On this slide you will see the microscopic appearance of the cells and tissues that were described in Section I.A.

The small, undeveloped leaves and stem section seen on your slide and shown in Figure 7.1 was made from a longitudinal cut through the tiny, tight cluster of leaves at the tip of the plant shown in the upper right of Figure 7.1.

This section of the plant is so small that the largest leaves on the slide are the smallest leaves that you can just barely see in the tight cluster at the tip of the plant.

Find the APICAL MERISTEM, which is the central dome or crest of cells near the top of the section. It is flanked by two partly developed outgrowths, the LEAF PRIMORDIA. A LEAF PRIMORDIUM (singular) is a leaf in its initial growth stage. It originates from dividing subsurface cells on each side of the dome-like apical meristem. These cells divide faster than those in the center of the dome, and consequently the young leaves grow and extend upward beyond the stem apex (meristem). Depending on where the technician's knife cut through the stem tip, you may or may not see the first (youngest) pair of leaves. Commonly the knife cut misses them, so you see only the next older pair of leaves. These will be very long, and extend far above the meristem.

Note the definite, orderly arranged layers of cells in the apical meristem. Certain herbicides kill plants by altering this orderly arrangement.

One apical meristem layer produces the EPIDERMIS. Another meristem layer, the PROCAMBIUM, is easy to identify because it appears as two narrow, darker traces of longitudinally elongate cells. These traces extend up into the embryonic leaves but are best seen in sections lower down the stem. The PROCAMBIUM gives rise to the PRIMARY XYLEM and PRIMARY PHLOEM. If enough of the stem has been mounted on the slide, you can see xylem cells in the more mature sections. However, this usually isn't the case.

The central region of the stem is known as the PITH. The region between the vascular trace and the epidermis is known as the CORTEX. Cortex

and pith are derived from a tissue called ground meristem.

Look for small crest or dome-shaped bulges of tissue in the angle (axil) between a leaf and the stem. Each of these is a LATERAL or AXILLARY BUD PRIMORDIUM. When these bud primordia mature and grow, what do they produce? _____

Note the stump of tissue below and on the outside of the base of each leaf. Each of these is the remnant of the base of the leaf (at the next lower node), which was cut off by the technician's knife when the slide was made. In a section this young, the nodes are close together.

Label the longitudinal section of the stem tip in Figure 7.1C, using the terms capitalized in the preceding description.

II. PATTERN OF STEM GROWTH OF THE MONOCOT HERB

ACTIVITY 3 **Examine** monocot plants, such as corn (*Zea mays*); any turfgrass or grassy weed (if growing in season); or any of the common house plants such as dumbcane (*Dieffenbachia* sp.), spiderwort (*Tradescantia* sp.), wandering Jew (*Zebrina* sp.), *Rhoeo* sp., or *Philodendron* sp. Is there a terminal bud on any of these plants? *Yes No.* Are there any axillary buds? *Yes No.* Are there branches on any of these plants? *Yes No.*

Pull apart the leaves of a young corn plant leaving at least three leaves, or use common house plants such as purple heart (*Setcreasea pallida*), wandering Jew (*Zebrina* sp.), or *Rhoeo* sp. Note the way the base or SHEATH of each LEAF extends down over the INTERNODES. After removing all the leaves, you will be left holding a stem with knobby ridges along it. What are these ridges? _____

Label the monocot stem in Figure 7.2, using the terms capitalized.

Each leaf seems to be wrapped around the next younger, inner leaf. The elongation of each leaf is due to the division and expansion of cells at the base of the leaf sheath. The stem lengthens because the internode lengthens owing to expansion of new cells formed by the INTERCALARY MERISTEM at the base of the internode.

The stem apical meristem of a grass, such as corn, can be visualized as a mound of cells sitting at the bottom of a well. The sides of the well are formed by the tubelike leaves growing upward from the flanks of the mounded apical meristem. Each new leaf that is formed by the meristem comes up in the center of the "tube" of the previously formed leaf.

Now, if you have followed this description, and it is still a bit confusing, you can understand why we don't ask you to study a microscopic section of the growing tip of a monocot stem!

ACTIVITY 4 **Complete** Table 7.1 showing the comparison of the external features of the stem of monocot and dicot herbs.

III. ABNORMAL GROWTH OF THE APICAL MERISTEM CAUSED BY AN HERBICIDE

ACTIVITY 5 **Observe** young bean plants or other dicot plants that have been sprayed with the herbicide 2,4-D. Which part(s) of the plant look abnormal? _____ Do the leaves look normal? *Yes No.* The herbicide is a super auxin chemical. It overstimulates the apical meristem, and one of the results is malformed leaves. The bending and twisting of the stem and petiole (caused by uneven growth induced by the 2,4-D) is another characteristic symptom of 2,4-D injury and is known as EPINASTY.

IV. HERBACEOUS DICOT STEMS—INTERNAL FORM AND FUNCTION

In the previous activities of this exercise, you examined microscopic sections of stems that were cut lengthwise to show how the tissues and organs are related to primary growth.

Now you will look at microscopic sections of stems that have been cut crosswise (transverse cuts). These transverse cuts may at first seem incomprehensible to you. But don't get lost in the details of the anatomy. The main point in studying these transverse sections is to demonstrate that the arrangement of the tissues is an efficient design that integrates four different functions of the stem: (1) support, (2) conduction, (3) storage, and (4) photosynthesis.

ACTIVITY 6 **Examine** a slide showing the cross section of a young stem of a representative dicot herb such as sunflower (*Helianthus* sp.) or geranium (*Pelargonium* sp.). **Look at** Figure 7.3 as a guide.

A. SUPPORTIVE FUNCTION OF THE STEM

In an herb, support is achieved by: (1) fibrous tissue (collenchyma cells and sclerenchyma cells), (2) the arrangement of fibrous tissue, and (3) turgor pressure. Fibrous tissue occurs near the outer region of the stem. This arrangement creates a cylinder—the strongest type of columnar support acknowledged by engineers.

Fibrous Tissue and Its Functional Arrangement

The Collenchyma Layer **Find** the COLLENCHYMA LAYER, a band of thick-walled cells just within the EPIDERMIS, the surface layer of cells. COLLENCHYMA CELLS are thick-walled, elongate, and particularly suited to support the stem

Table 7.1 COMPARISON OF THE STEMS OF DICOT AND MONOCOT HERBS

	FEATURE	DICOT	MONOCOT
EXTERNAL	Branches (present or absent)		
	Terminal buds (present or absent)		
	Axillary buds (present or absent)		
	Primary growth (present or absent)		
	Secondary growth (present in many or absent in most)		
	Location of the apical meristem		
INTERNAL	Vascular bundle arrangement		
	Vascular cambium (present or absent)		
	Specific supporting structures or tissues		
	Filler tissue		
	Storage tissue		

when it is young and growing. They give strong, stable support, yet are plastic (stretchable), adjusting to stem elongation. In mature stems, the collenchyma cell walls are hard. Sometimes collenchyma occurs in strands visible as ridges along stems or petioles. The "strings" you can pull from a celery stalk are strands of collenchyma. **Label** the collenchyma in Figure 7.3A.

Sclerenchyma–Bundle Cap Fibers   Focus on the FIBROVASCULAR BUNDLES. These are individual strands of tissues running lengthwise in the stem and appearing in cross section as cell clusters, called VASCULAR BUNDLES. The bundles are arranged in an orderly ring close to the outer zone of the stem.

On its outer side, each bundle is capped by a mass of very thick-walled SCLERENCHYMA CELLS, known as the BUNDLE CAP FIBERS. The rest of the bundle consists of the vascular tissue (xylem and phloem). The arrangement of the bundles near the periphery of the stem forms a cylinder. The tough sclerenchyma cap reinforces each strand (bundle) as it extends the length of the stem. This pattern is comparable in principle with the construction found in reinforced concrete pillars, where steel rods are embedded in the concrete to increase its strength.

These SCLERENCHYMA CELLS (also known as FIBERS) are very long with thick, rigid, but highly elastic walls. They can take pulls and pushes, but will rebound to their original length. The fibers intermesh and form a cablelike system. This gives the stem strength yet also permits it to yield flexibly to the stress of wind and the weight of leaves and bobbing fruit. The sclerenchyma in some plants is a valuable material. We use it to make rope, twine, and fabrics. The sclerenchyma of the marijuana plant is the hemp fiber of commerce used for cordage. The sclerenchyma of the *Corchorus* plant is the jute fiber used for making rough fabrics such as burlap. The sclerenchyma of the *Linum* plant is the fine fiber of linen. **Label** the bundle cap with the sclerenchyma cells in Figure 7.3A.

Turgor Pressure

Turgor pressure is the pressure developed by the water present in plant cells. It keeps plant parts (leaves, flowers, and herbaceous stems) rigid or turgid.

Note the large thin-walled PARENCHYMA CELLS that fill in the center of the stem and also the area around each bundle. These large cells contain water that builds up a turgor pressure that keeps the stem firm. **Label** a parenchyma cell in Figure 7.3A.

B. STORAGE FUNCTION OF THE STEM

Look again at the thin-walled parenchyma cells that fill in all the areas not occupied by sclerenchyma or fibrovascular bundles. Besides their role in turgor pressure, parenchyma cells have two other functions—(1) they store materials, and (2) together they serve as lightweight filler material that keeps the fibrovascular bundles in place. Large water-filled vacuoles make the cells look almost empty of contents. Starch grains (stained blue or purple) in many cells indicate their food-storage role.

Different "place names" have been given to the parenchyma tissue where it fills in different parts of the stem. In the very center of the stem, within the circle of fibrovascular bundles, parenchyma is given the name PITH. Between the fibrovascular bundles, it is named PITH RAYS. Between the epidermis and the outer edge of the ring of fibrovascular bundles, it is named CORTEX. The cortex typically has supporting tissues besides parenchyma cells. **Label** the different parenchymatous regions of the stem in Figure 7.3A, using the terms capitalized in the preceding description.

C. CONDUCTIVE FUNCTION OF THE SYSTEM

Locate again the ring of fibrovascular bundles. The conducting tissues in the bundle are the XYLEM and PHLOEM. Between them lies the VASCULAR CAMBIUM.

The Phloem

Locate the zone of thin-walled cells that lies just inside the bundle cap. This is the PRIMARY PHLOEM. **Note** that there are two sizes of cells present; the larger ones are the SIEVE TUBE ELEMENTS, and the smaller ones are either COMPANION CELLS or PHLOEM PARENCHYMA CELLS. Sieve tube elements are long cells stacked one above the other, forming a pipelike SIEVE TUBE. The sieve tube cells have perforated end walls, called SIEVE PLATES, through which cytoplasmic strands extend from one cell to the next. Food and other materials are carried in these strands. Sieve plates are not commonly visible. COMPANION CELLS are specialized cells whose nuclei are thought to have some control over the functioning of the cytoplasm in the sieve tube cell, which has no nucleus. PHLOEM PARENCHYMA cells serve as filler tissue and for storage. **Label** all the phloem cell types in Figure 7.3A, using the terms capitalized in the preceding description.

The Xylem

The **PRIMARY XYLEM** makes up the innermost tissue of each fibrovascular bundle. Most xylem cells are angular in outline and have thick walls. These are the water-conducting cells, **TRACHEIDS** and **VESSEL ELEMENTS**. Vessel elements are larger in diameter than tracheids. Vessel elements are barrel-shaped cells stacked one above the other, forming a pipelike **VESSEL**. Water moves up through the vessels and also through the more slender tracheids, but at a slower rate. Located here and there among the tracheids and vessel elements are smaller, thinner walled, rounded **XYLEM PARENCHYMA** cells. **Label** the xylem cells in Figure 7.3A, using the terms capitalized in the preceding description.

The Vascular Cambium

Find the VASCULAR CAMBIUM, a thin layer of four-sided, flattened looking cells located between the phloem and the xylem. **Note** that this band of flattened looking cells occurs not only within the fibrovascular bundle but also in the spaces (pith rays) between the bundles. The cambium within the bundle is called **FASCICULAR CAMBIUM**; the cambium lying between the bundles is called the **INTERFASCICULAR CAMBIUM**. As the sunflower grows, the interfascicular cambium forms secondary xylem and phloem in the pith ray regions, thus "filling in" the spaces between the bundles. Eventually, instead of having separate bundles of xylem and phloem, the plant has a single complete ring of xylem and phloem tissue in the stem. **Label** the two types of cambia in Figure 7.3A, using the terms capitalized in the preceding description.

D. PHOTOSYNTHETIC FUNCTION OF THE STEM

The stems of herbs are usually green due to the presence of chloroplasts. **Examine** the outermost cells of the stem, the **EPIDERMIS**, and also those cells forming the **COLLENCHYMA BAND**, and other cells in the vicinity. When they are young, collenchyma cells have chloroplasts, but depending on the age of the plant and on how the tissue was stained and prepared for the slide section, you may or may not see the chloroplasts. The epidermis contains openings, **STOMATES**, flanked by two **GUARD CELLS**. Chloroplasts may be visible in the guard cells. Carbon dioxide and oxygen pass through the stomates. You may or may not be able to see stomates and guard cells.

Stem photosynthesis is mostly supplementary in mature plants, but it contributes significantly to the sustenance of seedlings and many desert plants. In Figure 7.3A **indicate** which of the cells are photosynthetic cells.

ACTIVITY 7 **Examine** twine, rope, bags, or cloth made from jute, hemp, or linen fibers—all products of stems of herbaceous dicots.

III. THE STEM OF THE HERBACEOUS MONOCOT

Monocots have evolved from herbaceous dicots, and herbaceous dicots evolved from woody dicots. As you have seen, herbaceous dicots show partial or almost complete decline of the vascular cambium, and monocots show a total absence of this tissue. Because monocots lack a vascular cambium, most are herbs.

ACTIVITY 8 **Examine** a slide showing a cross section of the young stem of corn (*Zea mays*) or of lily (*Lilium* sp.). Using low power, **locate** the tissues and cells as described in the following outline of stem functions: *conduction, support, storage,* and *photosynthesis.* **Look at** Figure 7.4 as a guide.

A. CONDUCTIVE FUNCTION OF THE STEM

Are the vascular bundles arranged in a ring as in a dicot? *Yes   No.* Are they in some regular pattern? *Yes   No.* Is there a pith? *Yes   No.* Is there a cortex? *Yes   No.* Are the vascular bundles all the same size? *Yes   No.* Are the bundles in the center larger than those near the periphery? *Yes   No.* Are the bundles more densely clustered at the periphery? *Yes   No.*

Each of these vascular bundles is a leaf trace or several combined leaf traces. If the stem was cut across a node, a leaf trace may be seen diverging outward.

In monocots the network of vascular strands in the stem is very complex and has not yet been as adequately deciphered as in dicots. In the stems of some monocots, such as wheat (*Triticum* sp.) and rye (*Secale* sp.), the vascular strands are distributed in a ring as in dicots, but more frequently their complicated network shows up in cross section in a scattered pattern, as in corn.

Change to high power and **study** the details of a single vascular bundle. **Locate** the following tissues and cell types.

The Xylem

Primary xylem tissue consists of the thick-walled cells, the VESSEL ELEMENTS, many of which in series form the tubelike VESSEL. Two or three of these vessel elements commonly occur in an arrangement that could imaginatively be compared to the eyes and nose of a face (if you imagine that the entire vascular bundle resembles a face). Between the larger vessel elements are smaller vessel elements and also thinner walled XYLEM PAREN-CHYMA. Typically, there is a large round hole (forming the mouth of our imaginary face). This hole (called a lacuna) was caused by the collapse of cells stretched to the breaking point during the time the stem was elongating. **Label** the cell types in the xylem of the vascular bundle in Figure 7.4A, using the terms capitalized in the preceding description.

The Phloem

Primary phloem tissue is located to the outside of the primary xylem in what would be the forehead of the "face." Two types of cells are clearly visible here, the larger SIEVE TUBE ELEMENTS and the smaller COMPANION CELLS. **Label** the cell types in the primary phloem of the vascular bundle in Figure 7.4A, using the terms capitalized in the preceding description.

The Sclerenchyma Sheath

Thick-walled sclerenchyma cells (fibers) surround the xylem and phloem, forming a protective sheath that also serves to help support the stem.

B. SUPPORTIVE FUNCTION OF THE STEM

You have already noted that each vascular bundle is enclosed by a SCLERENCHYMA SHEATH, which is a strong protective sleeve composed of long sclerenchyma fibers (cells). The scattered arrangement of the vascular bundles, while basically due to the complex vascular network, effectively "installs" the fibrous support tissue (the sclerenchyma) in a manner somewhat comparable to that of embedded steel rods in steel-reinforced concrete columns. These "rods" of support fibers are stabilized in place by the large, rounded PARENCHYMA CELLS, which together serve as filler. Individually, each parenchyma cell stores food and other materials.

One interesting fact is that the plant fibers of commerce come from the *stems* of dicot herbs but from the *leaves* of monocot herbs. The sheathed vascular strands (bundles), like the ones seen here in corn, are removed from the *leaves* of the monocot *Agave* grown commercially in Mexico. The extracted strands are known as sisal fiber. Sisal is used to make twine, sacks, rugs, and fiberboard. Such bundles are also removed from the petioles of the monocot *Musa textilis*, which is grown commercially in the Philippines. These fibers are known as Manila hemp, which is used in making very strong cordage and twine. Years ago Manila hemp fibers were used to make many kinds of strong brown paper, and thus gave their name to the Manila envelope.

Examine the extreme outer edge of the stem. Here are many small SCLERENCHYMA CELLS. In older stems these cells are even more numerous. They, together with the densely crowded, fibrous sheathed, small vascular bundles, form the hard RIND of the corn stalk. **Label** the sclerenchyma sheath of a vascular bundle and the peripheral sclerenchyma cells in Figure 7.4A.

C. STORAGE FUNCTION OF THE STEM

Examine the large, thin-walled PAREN-CHYMA CELLS that fill in all stem areas not occupied by the vascular bundles and sclerenchyma. These parenchyma cells are collectively referred to as the GROUND PARENCHYMA. The large clear area in each cell is the VACUOLE that stores water and other materials. Turgor pressure in these cells helps keep the stem tissues firm. Are there starch granules present? *Yes No.* In the sugar cane plant (*Saccharum officinale*) (which is a large grass, like corn) these parenchyma cells are filled with sucrose, which is extracted by crushing the stem between fluted rollers. The young corn stem is similarly filled with sugary water, but as the plant matures it moves most of its sugars and other foods into the kernels of the ear. **Label** the ground parenchyma of Figure 7.4A.

D. PHOTOSYNTHETIC FUNCTION OF THE STEM

As with herbaceous dicots, monocot herbs have green, photosynthetic stems. The outermost parenchyma cells contain chloroplasts, and the epidermis has stomates with guard cells. However, only an occasional cross section on a slide will show stomates or guard cells, so your chances of finding any are slim. Similarly, chloroplasts do not show up readily in these prepared sections. **Indicate** which parts of the stem in Figure 7.4A are photosynthetic cells.

ACTIVITY 9 **Examine** products made from sisal or Manila hemp.

ACTIVITY 10 **Complete** Table 7.1 comparing the internal features of stems of dicot and monocot herbs.

EXERCISE 7   **Student Name** _____

QUESTIONS

1. Are practically all woody flowering plants dicots or monocots? _____

2. Do monocots have axillary buds? _____

3. Is the formation of leaf primordia and stem tissues orderly or random? _____
   _____

4. How would you recognize 2,4-D injury to a plant? _____
   _____
   _____

5. What is the mode of action of 2,4-D, and what tissue does it affect? _____
   _____
   _____

6. Elongation of monocot stems is due mainly to elongation of cells located where? _____
   _____

7. Does primary growth cause increase in girth or length? _____

8. Give two examples of monocots that have a tree growth form. (a) _____
   _____; (b) _____

9. Why does your lawn continue to send up blades of grass even though you mow it all
   summer? _____
   _____

10. If you ran the mower over your dad's young tomato plants would they continue to grow?
    _____ Explain: _____
    _____

11. Do monocot plants branch? _____

12. Do monocot plants have a vascular cambium? _____

13. Does secondary growth cause increase in girth or length? _____

14. In general, why are herbaceous plants more valuable to man as a food source than woody
    plants? _____
    _____

15. What is the botanical definition of an herb? _____
    _____
    _____

16. What are the three ways in which support is achieved in an herbaceous stem? (a) _____
    _____; (b) _____;
    (c) _____

17. Name three commercial fibers made from the sclerenchyma fibers of the stems of dicot
    herbs. (a) _____; (b) _____;
    (c) _____

18. Name two commercial fibers that are obtained from the sheathed vascular strands of the
    leaves or petioles of monocots. (a) _____
    _____; (b) _____
    _____

19. Parenchyma tissue in herbaceous stems serves which two purposes? (a) _____
    _____; (b) _____

20. What is meant by the primary body of a plant? _____
    _____
    _____

21. Do woody plants have a primary body? _____

22. Monocots have evolved from (a) woody monocots; (b) herbaceous dicots; or (c) woody dicots? Circle (a), (b), or (c).

23. In construction engineering of monocot herbs, is support in the stem achieved by the same general means as in herbaceous dicots? _____ Whether you answered *Yes* or *No*, specify what means are used. _____

    _____

    _____

24. Biologically, which is the more progressive way of life, the woody plant design or the herb design? _____ Which is the better design for perpetuating the population?

    _____

25. Sometimes during the summer, your tomato plants and garden flowers wilt, but they usually regain turgor if you water them, or if it rains. In the autumn, an early-season overnight frost will wilt your tomato plants. Even though the next few days are warm and you give the plants adequate water, they won't regain turgor. Why? _____

    _____

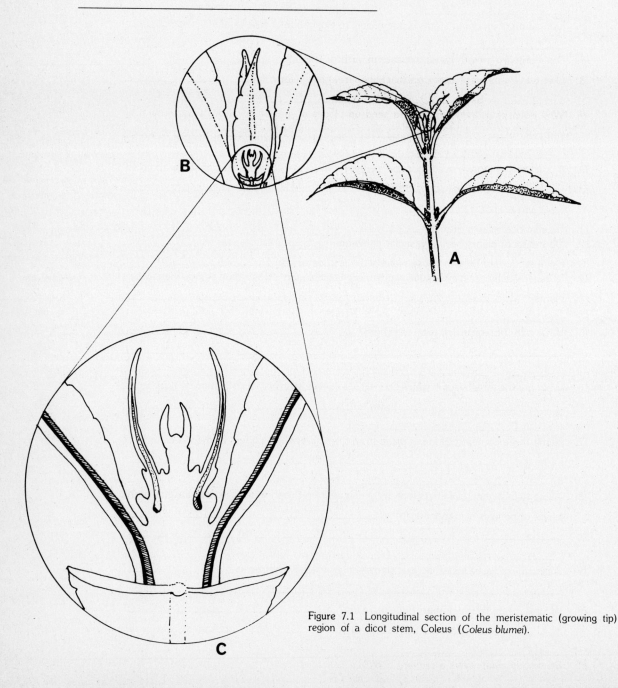

Figure 7.1 Longitudinal section of the meristematic (growing tip) region of a dicot stem, Coleus (*Coleus blumei*).

EXERCISE 7   **Student Name** _____

Figure 7.2   Stem features of a monocot plant (*Tradescantia* sp.)

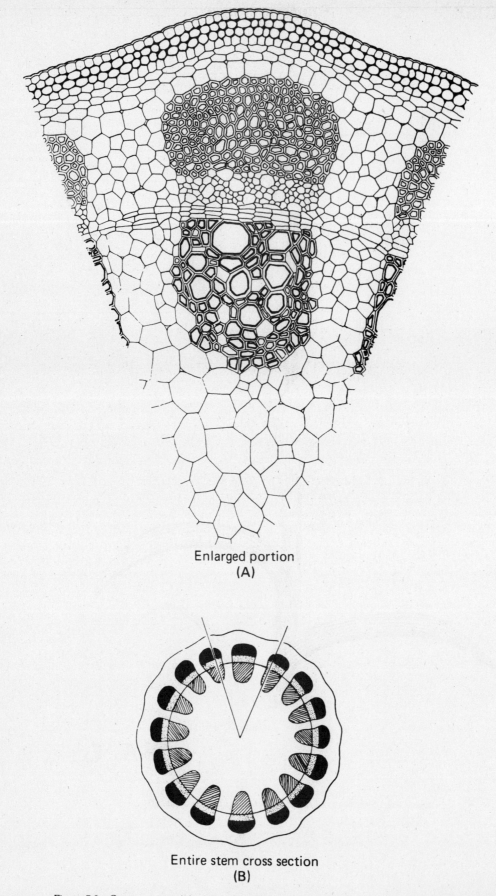

Enlarged portion
(A)

Entire stem cross section
(B)

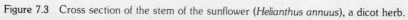

Figure 7.3   Cross section of the stem of the sunflower (*Helianthus annuus*), a dicot herb.

EXERCISE 7   **Student Name** ———————————————

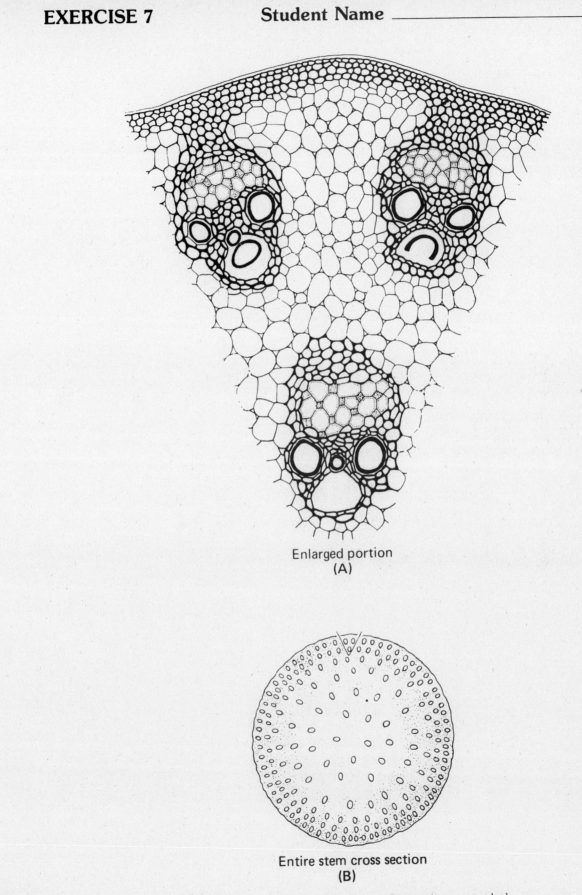

Enlarged portion
(A)

Entire stem cross section
(B)

Figure 7.4   Cross section of the stem of corn (*Zea mays*), a monocot herb.

WOODY STEMS I: Introduction to Woody Stem Anatomy—The Gymnosperm Stem

INTRODUCTION

A woody plant is a permanent structure that yearly increases its leafy canopy through primary growth. Supporting this canopy requires a reinforcement of the woody framework. Secondary growth (growth that increases width) provides this reinforcement.

As the plant grows more massive through secondary growth, most of its cells are buried by the added layers and cannot be supplied with oxygen.

It would seem then that there might be a physiological limit to the final size of a woody plant, but in fact the plant has no such limit. This dilemma has been "solved" by the way plants evolved. Only the few layers of cells nearer the outside are alive, while those nearer the center are dead. But most of these dead cells remain structurally intact and furnish the additional support needed each year.

The *woody stem* has three basic parts, each composed as shown.

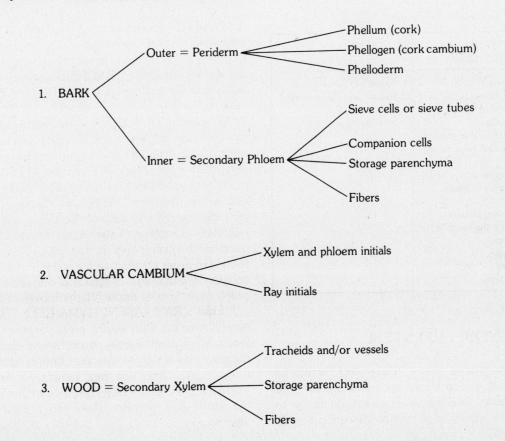

1. BARK
   - Outer = Periderm
     - Phellum (cork)
     - Phellogen (cork cambium)
     - Phelloderm
   - Inner = Secondary Phloem
     - Sieve cells or sieve tubes
     - Companion cells
     - Storage parenchyma
     - Fibers

2. VASCULAR CAMBIUM
   - Xylem and phloem initials
   - Ray initials

3. WOOD = Secondary Xylem
   - Tracheids and/or vessels
   - Storage parenchyma
   - Fibers

The main functions of these parts are *conduction, storage, growth,* and *protection,* each as indicated:

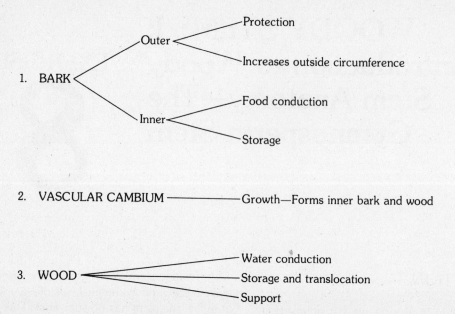

1. BARK
   - Outer
     - Protection
     - Increases outside circumference
   - Inner
     - Food conduction
     - Storage

2. VASCULAR CAMBIUM ——————— Growth—Forms inner bark and wood

3. WOOD
   - Water conduction
   - Storage and translocation
   - Support

In the stem, most cells are elongate rather than cube-shaped and are oriented in one of two ways, transverse or longitudinal. Longitudinally oriented cells are those oriented parallel to the long axis of the stem. They make up the **AXIAL SYSTEMS,** which form the bulk of the stem. Cells oriented crosswise to the long axis of the stem make up the **RAY SYSTEM.**

LONGITUDINAL AXIAL SYSTEM CELLS	TRANSVERSE RAY SYSTEM CELLS
1. Tracheids 2. Vessel elements (form tubes called vessels) 3. Sieve cells 4. Sieve tube elements (form tubes called sieve tubes) 5. Fibers 6. Axial xylem parenchyma 7. Axial phloem parenchyma 8. Xylem and phloem initials of the vascular cambium 9. Phellogen 10. Phellum 11. Phelloderm	1. Xylem ray parenchyma 2. Phloem ray parenchyma 3. Ray initials of the vascular cambium

I. GYMNOSPERM STEM

Gymnosperms [pines (*Pinus* sp.), hemlocks (*Tsuga* sp.), spruce (*Picea* sp.), and the like] are ancient plants. Internally their stems are similar to those of the woody angiosperms (flowering plants),

but they are generally much simpler and more homogeneous. They have no vessels or sieve tubes and relatively little parenchyma, and many lack fibers. For this reason you will find it easier to begin your study of woody stems with a gymnosperm.

A. XYLEM CELL TYPES IN A GYMNOSPERM

ACTIVITY 1   **Examine** a prepared slide of macerated wood of some gymnosperm such as pine (*Pinus* sp.). This will enable you to see the general shape of the water-conducting TRACHEIDS and their side walls; you will also see the RAY PARENCHYMA cells.

Locate a TRACHEID. A tracheid is an elongate, slender, needlelike cell with long, tapering end walls. It is a dead cell, lacks a protoplast, and has reinforced walls. **Note** the circular PITS on the (radial) side walls. Pits are permeable cavities in the wall. In the plant, tracheid cells overlap each other, with the tapered end walls of one cell touching the radial wall of another. Water passes from tracheid to tracheid through pores in the pits. This zigzag routing of water means that the water moves slowly compared with the rates typical of angiosperms, in which water moves through large tubelike vessels.

Find a RAY PARENCHYMA CELL. This is a parenchyma cell with a living protoplast that stores foods. In the plant, the ray parenchyma cells border the tracheids at right angles and connect with them through pits. They are much shorter cells than tracheids, their length being only one to three times the width of a tracheid. Their end walls are not tapered.

B. CROSS SECTION OF THE WOODY STEM OF A GYMNOSPERM

ACTIVITY 2 **Examine** a prepared slide showing a cross section of a three-year-old (or older) stem of pine (*Pinus* sp.). Don't get lost in the details of the anatomy. **Look** at Figure 8.1 as a guide and **label** the parts as you proceed. The *main* point of this study is for you to see that the arrangement of the axial and radial systems is an efficient design. It integrates five different functions—*support, conduction, storage, protection,* and *growth.*

The Wood

In ordinary use the term wood means any hard part of a tree or shrub. But botanically, WOOD is SECONDARY XYLEM only.

The Water-Conducting Function of the Wood Water is conducted in the axial tracheid system. **Focus** on the central portion of the stem where you can easily recognize concentric rings of very orderly arranged cells. These cells are the TRACHEIDS, which are more rectangular than round in cross section. The rings are known as ANNUAL RINGS or GROWTH RINGS, and each represents one year's formation of secondary xylem. Each year the vascular cambium adds a new ring around that of the previous year. **Note** the gradation in size of the cells in any one ring. Tracheids at the inner margin of each ring are much larger than those at its outer edge. The larger tracheids were formed early in the growing season and are known as EARLYWOOD or SPRINGWOOD. The smaller tracheids, formed later in the growing season, are known as LATEWOOD or SUMMERWOOD. The sharp contrast between the last formed latewood cells of one growing season and the first formed earlywood cells of the following season delineates the boundary of a growth ring. How old is your section of wood? _____ How many growth rings are there? _____. Are all these rings secondary wood? _____. Explain: _____
_____
_____

_____. What is the functional advantage to the plant of the larger sized cells in the earlywood?
_____
_____

Most of the water moves up the plant in the outermost ring, that is, the current season's tracheids. The more recent of the previous years' annual rings contain water, but the plant does not rely on them for its current needs of upward flowing water. Do any of the tracheids have a protoplast?

Yes _____ No _____. **Label** a growth ring in Figure 8.1. **Label** Figure 8.1C, using the terms capitalized in the preceding description.

The Support Function of the Wood **Switch** to high power and examine the walls of the tracheids. **Note** that the walls are relatively thick. These are the SECONDARY WALLS, which give the cells a rigid reinforcement. They are also encrusted with a material known as lignin, which makes them hard and dense. Which has the thicker walls compared to the overall cell size, (a) latewood, or (b) earlywood? (Circle one.) Which do you suppose has the greater strength, (a) latewood, or (b) earlywood? (Circle one.) Tracheids thus perform two functions: (a) _____
and (b) _____. **Label** the secondary wall of a tracheid in Figure 8.1B or C.

The Storage Function of the Wood **Note** the single-file rows of thin-walled cells that radiate intermittently across the growth rings somewhat like the spokes of a wheel. These radial rows are called RAYS. Each ray is a sheet of parenchyma cells that extends vertically in the stem. Ray cells store sugars, fats, and other foods. In the spring these stored materials are transferred to the actively dividing vascular cambium or to the apical meristems of the buds.

Note also that many of these rays extend into the secondary phloem, which lies outside the secondary xylem. Rays in the secondary phloem have the same structure and function as those in the secondary xylem. As the tree enlarges, new rays are constantly formed by the vascular cambium so that the density of rays is maintained approximately the same. **Label** the xylem ray in Figure 8.1C.

Other Features of the Wood—Pith and Resin Canals **Focus** on the exact center of the stem within the innermost growth ring. This central core is the PITH. Originally it contains parenchyma cells, and these may or may not still be present in the stem section you are examining. The life span of the pith cells varies with the species. With age, pith cells commonly accumulate tannins and crystals (waste products). In many plants, the pith becomes hollow or chambered. **Label** the pith cells in Figure 8.1D.

Switch to low power and **observe** the large cavities scattered throughout the stem. These are the RESIN CANALS or RESIN DUCTS (not found in all gymnosperms). They extend lengthwise in the stem. **Switch** to high power and **note** that the cavities are lined with parenchyma cells. The parenchyma cells secrete a resinous substance that in pines is known as oleoresin. Its usefulness to the plant is unknown. It may possibly repel or attract insects. Turpentine and rosin are manufactured from oleoresin and are still used in some high-quality paints, though they have been largely replaced by synthetics. **Label** the resin duct in Figure 8.1B.

The Vascular Cambium— Growth Zone of the Stem

The woody part of a stem grows only in girth. The VASCULAR CAMBIUM increases the girth of the stem by forming layer after layer of secondary xylem and secondary phloem around the original primary cylinder. In the temperate zones of the world cambial growth is cyclic, with periods of activity alternating with periods of relative rest. Cambial activity is under the control of hormones that reactivate it each spring. Some of these hormones are known to come from the developing buds and leaves.

Focus on the latewood of the outermost growth ring in the xylem. Immediately outside this latewood you will find a band of flattened looking cells. This is the VASCULAR CAMBIAL ZONE, a band of unexpanded meristematic cells. It is about four to eight cells wide. It consists of one VASCULAR CAMBIUM layer of self-perpetuating cells, called CAMBIAL INITIALS, flanked by layers of dividing daughter cells. Daughter cells on the outside of the cambial layer are called PHLOEM INITIALS and develop into secondary phloem cells; those on the inside of the cambial layer are called XYLEM INITIALS and develop into secondary xylem cells. **Label** the vascular cambial zone in Figure 8.1A.

The Bark

BARK is defined as all those tissues that lie outside of the vascular cambium. It consists of three tissues—secondary phloem, cortex (which may or may not be present), and the periderm (present only in older stems). The phloem is considered to be the INNER BARK and the periderm the OUTER BARK.

The Food-Conducting Function of the Inner Bark (the Secondary Phloem) **Focus** on the region just outside the vascular cambial zone. Here there are angular, thin-walled cells in somewhat orderly rows. Scattered among them are larger, round cells filled with dark-colored material. Cells farther out are flattened and arranged in a zigzag pattern. If you have located tissue fitting these descriptions you have found the phloem. **Locate** the following phloem cell types.

SIEVE CELLS are vertically elongate cells that in this cross section are the angular, thin-walled cells in orderly rows. Sieve cells conduct photosynthates to actively growing buds and young leaves or to roots or storage rays. Sieve cells are named for their numerous SIEVE AREAS, the porous zones in their walls. Strands of protoplasm extending through

these pores connect functioning adjacent sieve cells and ray parenchyma. Food transport in sieve cells is thought to occur in the form of a flowing solution of foods within the protoplasm. The protoplasm itself doesn't flow. Exactly how this is done is still unresolved.

Observe the outer regions of the secondary phloem. These outer, older layers may still be alive but are not all actively functioning. Many of the sieve cells are crushed, and the rays are distorted into a wavy, accordian-pleated shape. Normally only the innermost (youngest) layer of secondary phloem is functional. The outer, older layers are incorporated into the outer bark, which may eventually be sloughed off. **Label** the functional secondary phloem in Figure 8.1A.

The Storage Function of the Inner Bark (Axial and Ray Parenchyma) **Locate** the round, thin-walled cells filled with dark-colored material. These are the AXIAL PHLOEM PAREN-CHYMA CELLS. When alive, these cells store foods transferred to them from adjacent or nearby sieve cells or other parenchyma. When foods are mobilized for growth, these foods are recirculated back through sieve cells or through ray parenchyma to the vascular cambium and developing buds and leaves. **Label** an axial phloem parenchyma cell in Figure 8.1A.

Focus on the inner (youngest) region of secondary phloem and find the large, radially elongated cells with prominent nuclei. The length of one of these cells spans about five of the adjacent sieve cells seen in cross section. These are the RAY PARENCHYMA CELLS. They are lined up in a transverse row, called a RAY. Can you identify any ray in the outer crushed, nonfunctioning phloem? *Yes* *No*. Crushed rays give the old phloem a zigzag pattern. Phloem rays are similar in form and function to xylem rays.

The Cortex The cortex is primary tissue, produced by the apical meristem. In a young, nonwoody stem, one of the functions of the cortex is support, but as the plant ages, this function is accomplished by the wood produced by secondary growth. Certain of its cells may go through a dedifferentiation (lose their distinctive shape) and become meristematic again. These meristematic cells become organized into the CORK CAMBIUM (PHELLOGEN). As the tree gets even older, however, it is commonly old phloem cells that become cork cambium.

The Growth and Protective Function of the Outer Bark—the Periderm The CORK CAM-BIUM produces CORK CELLS (PHELLEM) to the outside of the stem and thick-walled parenchyma cells, PHELLODERM, to the inside. These three tissues, cork cambium, cork, and phelloderm, comprise the **OUTER BARK**—the **PERIDERM**. You may not be able to identify the cork cambium as easily as the vascular system. Mature cork cells are dead and approximately square in outline with thick walls. They are usually arranged compactly with no intercellular spaces. What advantage is this? _____

_____. Cork cell walls are impregnated with suberin, a fatty material that makes them impervious to water. Cork also has thermal insulating qualities. If your pine stem section came from a stem only three years old, the amount of periderm will be negligible.

EXERCISE 8   **Student Name** _____

QUESTIONS

1. Latewood is usually stronger than earlywood. What material is abundant in latewood that makes this wood so strong? _____

   _____

2. Leaves of gymnosperms are designed to limit or impede water loss. This leaf design more or less compensates for what internal feature of the wood? _____

   _____

3. Gymnosperms growing in exposed sites need some kind of protection during winter because a severe wind can kill many of their leaves that turn a brown color, described as winter burn. Explain why the leaves are killed by the wind. _____

   _____

4. Explain how the woody plant body can allow for unlimited growth with no physiological limitations. _____

   _____

   _____

5. Do any animals have comparable unlimited growth? _____

6. Is the annual, new leafy framework of a woody plant the result of primary or secondary growth? _____

7. Which tissue is vital for the survival of a woody plant? _____

8. Define bark. _____

   _____

9. Define a growth ring. _____

   _____

   _____

10. Why isn't there an accumulation of secondary phloem growth rings as there is for secondary xylem? _____

    _____

11. In gymnosperm wood, the main support comes from what cell type? _____

12. What is (are) the function(s) of the ray system of woody plants? _____

    _____

13. The diameter of the woody cylinder is continually increasing. How is it that there is always enough bark to cover the outside? _____

14. If you carved your initials in the bark of a tree when you were ten years old, would these initials be higher up on the trunk when you reach 16 years? _____ Explain your answer? _____

    _____

15. Which system supplies nutrients to the actively growing vascular cambium: the sieve tubes or the ray system? _____

16. Is bud opening and renewed vascular cambial activity correlated? _____ How?

    _____

17. Name two tissues of primary growth that are expendable in the mature tree.

    (a) _____ ; (b) _____

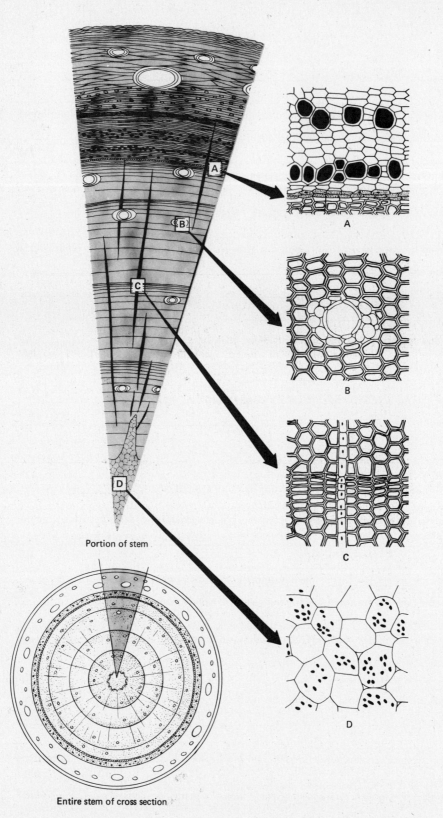

Portion of stem

Entire stem of cross section

Figure 8.1 Cross section of the woody stem of a gymnosperm (*Pinus* sp.).

WOODY STEMS II:
Anatomy of the Advanced Woody Stem—The Angiosperm Stem

INTRODUCTION

The general organization of a woody angiosperm stem is essentially the same as that of a gymnosperm. The main difference between them is that angiosperms have many more specialized cells, particularly those concerned with water and food conduction and with support. In gymnosperm wood, tracheids provide much of the support, in addition to conducting water. In angiosperm wood more fibers are present for support, and, besides tracheids, there are vessel elements (cells) that are specialized for rapid water flow. Also, the sieve tube cells of angiosperms can move foods faster than the sieve cells of gymnosperms.

With these improved cell types, angiosperms (both woody and herbaceous) have become adapted to a wider variety of habitats than gymnosperms. They tend to grow faster than gymnosperms, and they can also afford to have broader leaves whose photosynthetic output exceeds that of the smaller, needlelike leaf of the gymnosperm. Even though this broader leaf is more prone to evaporative water loss, it can be quickly replenished by the rapid water delivery vessel system in the stem. In studying the stem of a woody angiosperm, therefore, you should look for the cell specializations that have helped to make the angiosperms such a successful group of plants.

I. XYLEM CELL TYPES

ACTIVITY 1 **Examine** a prepared slide of macerated angiosperm wood so you can see and compare individual cells in their entirety rather than as segments in a cut section.

Look for four cell types. All are secondary cells produced by the vascular cambium.

1. WOOD FIBERS (sclerenchyma cells) are very long, slender cells with tapering ends. They are support cells. Their walls are thick with many pits, so small, however, that little water passes through them. Most of these very long cells are broken into smaller segments on this slide.
2. VESSEL ELEMENTS are shorter, wider, barrellike cells with thin side walls that have many pits. The end walls have one large circular pit or a few smaller, more rectangular-shaped pits. In intact woods, vessel elements are connected end to end, forming a long and wide tube, the VESSEL. A vessel is capable of rapid conduction of water.
3. TRACHEIDS are long, narrow cells with tapered ends. They resemble wood fibers somewhat but are not as long and, therefore, not as easily broken during the process of macerating the wood. If you find a cell with both tapered ends intact, it is probably a tracheid and not a wood fiber. Tracheids conduct water but not as rapidly as a vessel.
4. XYLEM RAY PARENCHYMA CELLS are short cells with straight-sided end walls. They have a granular appearance due to the presence of cytoplasm. **Note** the pits in their side walls. What is the function of the ray parenchyma? _____

II. WOOD ANATOMY

ACTIVITY 2 **Examine** a slide of oak (*Quercus* sp.) wood showing three sections of cut: cross, radial, and tangential and **look** at Figures 9.1 to 9.3 as a guide.

Focus on the CROSS SECTION, which is recognizable because it consists of portions of growth rings. **Note** the extremely large, circular, or (in some cases) angular cells. These are the VESSEL ELEMENTS, cells "streamlined" for water transport, with wide diameters for high volume flow and without cytoplasm or end walls for fast, unimpeded flow of water. Vessel elements are stacked up one over the other in longitudinal series and joined end to end to form a water tube, the VESSEL. Transport velocities in vessels are rapid, several meters per hour—much faster than in a tracheid system, in which water moves in a zigzag route through small pores of narrow tracheids that border each other but are not joined to form a distinct tubular unit.

Note the appearance of any one growth ring. As in pine, the inner part of each ring is the EARLY-WOOD, the outer part is the LATEWOOD. In angiosperms, earlywood and latewood are not as easily distinguished as in most gymnosperms. In some species, vessels are mostly in the earlywood, and fibers make up most of the latewood. This arrangement is called RING POROUS WOOD. In other species, vessels occur throughout the entire width of the ring. This is called DIFFUSE POROUS WOOD.

Focus on the smaller cells between the larger vessels. These are the WOOD FIBERS or TRA-CHEIDS. Fibers have thick walls and are very small in diameter. Cells of the same small diameter but with thinner walls are tracheids. Fibers are largely responsible for the hardness and strength of angiosperm wood. These cells provide the hardness needed for baseball bats, hockey sticks, skis, and handles for tools. There are angiosperms, however, with very few fibers in their wood, which is, therefore, a soft wood. Balsa wood of the tropical *Ochroma* tree is a good example.

Switch to higher power (if necessary) and find the narrow rows of parenchyma cells that radiate across the growth rings. These are the XYLEM RAYS. A few cells of the XYLEM RAY PAREN-CHYMA system may also be visible in this section. All ray cells are parenchymatous and serve for storage.

Label the vessels, wood fibers, xylem rays, and tracheids in the cross section shown in Figure 9.1.

Move the slide so that the RADIAL SECTION is in view. This section is represented in Figure 9.2. The stem has been cut lengthwise and along a radius. Here the VESSELS appear as wide, vertically elongate cavities. In the intact stem they are much longer, but in this section you see only portions of them. The very narrow, columnar cells with tapered ends are the WOOD FIBERS. **Note** how densely packed the fibers are and how they compose the

bulk of the wood. Some angiosperms like basswood (*Tilia* sp.) and willow (*Salix* sp.) have a large number of vessels per unit volume of wood with fewer fibers than you see here in oak. Wood from such trees is much softer than oak. Although the U.S. Department of Agriculture classifies angiosperm wood as "hard woods" and gymnosperm wood as "soft-woods," many angiosperm woods are actually as soft as, or softer than, gymnosperm wood.

Look for bands of cells running crosswise to the fibers and vessels. These are the XYLEM RAYS. **Label** the vessels, wood fibers, and xylem rays in the radial section shown in Figure 9.2.

Move the slide so that the TANGENTIAL SECTION is in view. In this section the stem has been cut lengthwise in a plane perpendicular to its radius. VESSELS and WOOD FIBERS look the same here as they do in the radial section, but the XYLEM RAYS appear as lens-shaped clusters of cells. Some are wide, massive clusters, some are narrow, slender clusters only one or a few cells wide. Since rays run horizontally in the stem along the various radii, in this vertical tangential cut you are seeing them cut across the long axis in cross-sectional view.

Label the vessels, wood fibers, and xylem rays in the tangential section shown in Figure 9.3.

III. CROSS SECTION OF A WOODY ANGIOSPERM STEM

A. THE PHLOEM

ACTIVITY 3 **Examine** a slide showing the cross section of some woody angiosperm stem such as basswood (*Tilia* sp.) and **look** at Figure 9.4 as a guide. *Tilia* is not representative of all woody angiosperms. **Find** the PHLOEM, which is the darker stained region near the outer part of the stem. It is made up of triangular-shaped sections. Those triangular sections that "point" inward are the PHLOEM RAYS. Other RAYS only a few cells wide form narrow rows of parenchyma cells as they cross the phloem region in general. **Note** that all the rays in the phloem are continuous with rays in the xylem (which is the homogeneous-looking central core with the growth rings).

Wedged between the inward-pointing phloem rays are other triangular sections that point outward. In these sections are bands of heavily stained cells with very thick walls and pin-pointlike lumens (the central part of the cell bounded by the walls). These heavily stained cells are the PHLOEM FIBERS. Alternating with these bands of phloem fibers are narrower bands of larger cells with thinner walls.

These large, angular cells with very little cytoplasm and with thin walls are the SIEVE TUBE ELEMENTS. They are wide, columnar cells connected end to end in a longitudinal series to form SIEVE TUBES. The end walls of sieve tube elements are not completely open as in vessel elements but have large porous end areas, called SIEVE PLATES. Sieve plates are not commonly seen on these slides. Any one of the several sieve tube elements that make up a sieve tube can deliver foods at a rate of some five times its own volume per second. Hence the flow through the entire tube is very fast.

Find the COMPANION CELLS. These are smaller cells usually found at the corners of sieve tube elements, hence the name "companion cells." They have thin walls, considerable cytoplasm, and a nucleus that occupies most of the cell. The nuclei of companion cells are thought to exert some control over the normal functioning of the sieve tube elements that have lost their nuclei in the process of cell maturation. The absence of a nucleus in such a physiologically active cell as the sieve tube elements is inexplicable.

Label the phloem tissue and all its specialized cells in detail diagram B of Figure 9.4, using the terms capitalized in the preceding description.

B. THE XYLEM

The center of the cross section is occupied by a pith and three or more concentric layers of xylem. EARLYWOOD and LATEWOOD are easily recognized. A VASCULAR CAMBIAL ZONE separates the XYLEM and PHLOEM. Basswood is a soft-wooded angiosperm because its xylem has many more vessels than fibers. The structural differences between the various xylem cells are not readily apparent in *Tilia*. It is, therefore, not as good an example as oak for an introductory study of angiosperm wood (secondary xylem tissue). Hence we do not ask that you study the basswood xylem in detail. Simply note that, as in all woody dicots and gymnosperms of the temperate zone, the secondary xylem is in concentric growth rings, and RAY PARENCHYMA radiate across them. **Label** detail diagrams C, D, E, and F of Figure 9.4, using the terms capitalized in the preceding description.

C. THE CORTEX AND PERIDERM OF THE WOODY ANGIOSPERM STEM

Examine the outermost portion of the three-year-old *Tilia* stem.

Cortex

The CORTEX is the zone of loosely arranged, relatively large parenchyma cells found just external to the outermost band of secondary phloem fibers. Depending on the age of the stem, the cortex may have been replaced by the periderm. Is the cortex a primary or a secondary tissue? (Circle one.)

Periderm

The PERIDERM is the dense band of cells just under the stem surface. All the periderm cells are rectangular, with their long axes paralleling the circumference of the stem. From the phloem outward, the PERIDERM consists of (a) four to six layers of PHELLODERM cells that are thick-walled and filled with cytoplasm; the phelloderm is a secondary cortex formed by the cork cambium; (b) one to two layers of thin-walled, nucleated cells filled with cytoplasm and making up the CORK CAMBIAL ZONE; and (c) three to four scalelike layers of dead CORK CELLS. **Label** detail diagram A of Figure 9.4, using the terms capitalized in the preceding description.

ACTIVITY 4 **Complete** Table 9.1.

EXERCISE 9   **Student Name** _____

QUESTIONS

1. In angiosperm wood, the main support comes from what cell type? _____

2. In gymnosperm wood, the main support comes from what cell type? _____

3. What are the water tubes of angiosperms known as? _____

4. What is the name of the cell type that makes up this water tube? _____

   _____

5. What is the name given to the food-conducting tubes of angiosperms? _____

   _____

6. What is the name of the cell type that makes up this food-conducting tube? _____

   _____

7. It is conceded that the water- and food-conducting system of angiosperms is more efficient than that of gymnosperms. What advantages does this give the angiosperm over a gymnosperm? _____

   _____

8. What are the two structural features of the vessel element that make it a superior water-conducting cell, compared to the tracheid? (a) _____; (b) _____

   _____

9. Angiosperms have relatively thin flat leaves. Gymnosperm leaves have thick cuticles and are prism-shaped to reduce water losses. How do you account for this difference? Explain in terms of their wood anatomy. _____

   _____

   _____

Table 9.1   COMPARISON OF THE WOODY STEMS OF ANGIOSPERMS AND GYMNOSPERMS

FEATURE	GYMNOSPERM WOOD	ANGIOSPERM WOOD
Comparative makeup of the wood (simple or complex)		
Chief water-conducting cell		
Comparative diameters of the water-conducting cells (wide or narrow)		
Number of different functions performed by the water-conducting cell		
Chief food-conducting cell		
Comparative diameters of these food-conducting cells (wide or narrow)		
Cell type for support		
Comparative growth rates		

Figures 9.1–9.4 Sections of wood as photographed through the compound microscope.

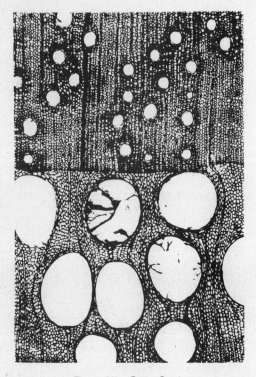

Figure 9.1 Cross Section.

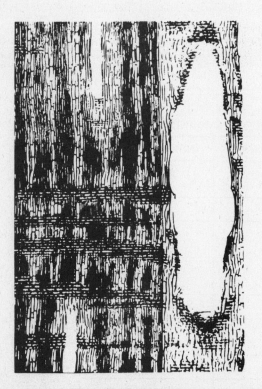

Figure 9.2 Radial Section.

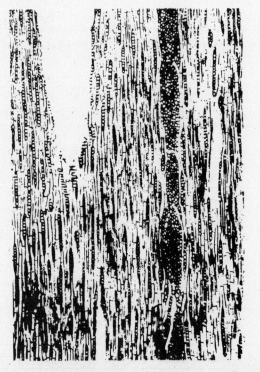

Figure 9.3 Tangential Section.

EXERCISE 9

Student Name _____

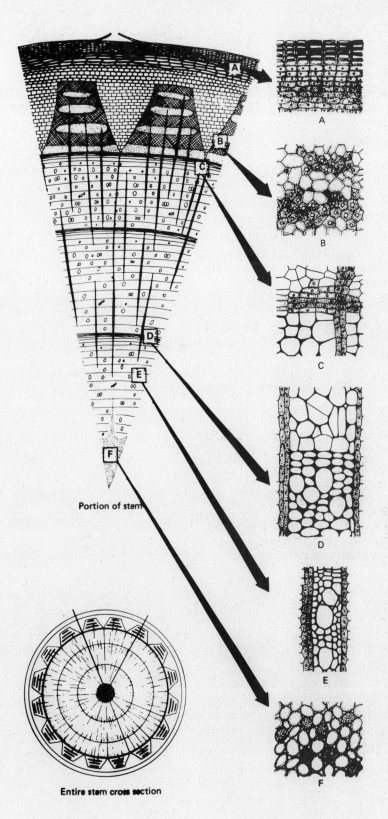

Portion of stem

Entire stem cross section

Figure 9.4. Cross Section of Stem of Woody Angiosperm (*Tilia* sp.).

WOODY STEMS III: External and Internal Features of Growth

EXERCISE 10

INTRODUCTION

The woody plant form has existed on earth for millions of years. It predominated at a time when world climate was universally mild. As the climate fluctuated, and drought and cold spells became common, the tree's way of life was not as well suited for survival as that of the herb. Many did not survive.

Although we consider the life style of an herb to be biologically a more successful way of life, this is not to say that the tree or shrub is a lost cause. In climates where there are no adverse seasons and there is little or no freezing weather, woody plants are overwhelmingly dominant. Also, there are many woody plants that are exceedingly well adapted to survival in severe climates.

In today's climate, woody plants living in zones with drastic seasonal differences survive the cold or dry season by going dormant. However, their woody framework still stands vulnerable to wind, ice, snow, or drought. The herb escapes these forces by not making any permanent woody framework. Instead, it is short-lived and quickly produces the seeds of the next generation.

Herbs are fast-growing plants. Trees, in contrast, grow more slowly. Their growth rate and net photosynthetic rate are slower than that of herbs. They are slow to reproduce, typically requiring years of growth before they first flower and produce the seeds of another generation. Most of their photosynthate and energy are spent on building the woody framework of stems and roots and making a new set of leaves each year.

Thus the record of growing conditions during these many years is preserved in a number of places in the woody plant. The record of climatic or of local environmental conditions of the most recent four or five seasons is visible in certain external features of the twig. The record of growing conditions of seasons of 10 or 50 years and sometimes centuries ago can be seen by examining the internal structure —the growth rings—of the older limbs and trunk of a tree.

I. THE YOUNG WOODY STEM—THE TWIG

Twigs are the youngest parts of the tree or shrub.

ACTIVITY 1 **Examine** leafless twigs taken from dormant trees. Suitable species for study are horse chestnut or buckeye (*Aesculus* sp.), walnut (*Juglans* sp.), tree of heaven (*Ailanthus altissima*), ash (*Fraxinus* sp.), and tulip tree (*Liriodendron tulipifera*). **Identify** the following features. **Look** at Figure 10.1 as a guide.

A. MAIN TWIG FEATURES

Terminal Bud

A TERMINAL BUD occurs at the tip of a twig and is responsible for its growth in length. If your twig has branches, each branch also will have a terminal bud. **Note** the BUD SCALES. These are modified leaves. They enclose and protect the embryonic leaves and apical meristem within the bud. Depending on the bud type or the plant species, the bud may also contain embryonic flowers.

At the onset of the growing season, bud scales flare out and fall off as the young stem and leaves grow. The internodes between successive leaves elongate, thereby increasing the twig length. Some plants, like *Aesculus* sp., have a solitary terminal bud.

Others, like oaks (*Quercus* sp.), ashes (*Fraxinus* sp.), and maples (*Acer* sp.), have two or more lateral buds clustered closely about the terminal bud, giving the appearance of several terminal buds. Still others, like sycamore (*Platanus* sp.), have no terminal buds; nearby lateral buds bring about twig elongation. **Label** the terminal bud(s) on Figure 10.1.

Terminal Bud Scale Scars

A TERMINAL BUD SCALE SCAR is a band of closely grouped scars encircling the twig like so many rings. Each such band of scars marks the position of the terminal bud of a previous year. The scars are those left after the bud scales fell off. **Label** the terminal bud scale scars on Figure 10.1 and indicate with a bracket ONE SEASON OF GROWTH.

Measure (in millimeters) the length of stem produced in each of the last two to four seasons of growth and record the figures.

REGION	LENGTH OF REGION
1. Youngest (or most recent)	_____ mm
2. Next older growth region	_____ mm
3. Next older growth region	_____ mm
4. Next older growth region	_____ mm

Leaf Scars

A LEAF SCAR is a scar left on the twig where the petiole or base of the leaf was once attached. The scar is a corky layer, sealing off the living tissues beneath. Leaf scars are of many different shapes and sizes, depending on the species, and are a key feature used to identify woody plants. Will new leaves arise from these leaf scars: *Yes No.* **Label** a leaf scar in Figure 10.1.

Vascular Bundle Scars

Continue to examine one of the leaf scars on your specimen. **Note** the pinhead-size, corky bumps within the leaf scar. These are the scars of the broken ends of the vascular strands (bundles) that passed between leaf and stem. The numbers and arrangement of these VASCULAR BUNDLE SCARS vary with different species. They, like leaf scars, are useful characters for identifying woody plants during their dormant season when leaves are absent. **Label** the vascular bundle scars in Figure 10.1.

Nodes and Internodes

The leaf scar occurs at a position on the stem called the NODE. The node is defined as that region on a twig where one or more leaves are borne. The portion of the twig between any two successive nodes is the INTERNODE.

Compare the lengths of the internodes of one growing season with those of another growing season. Are they equal in length each year? *Yes No.* **Label** a node, and bracket an internode in Figure 10.1.

Lateral Buds

Note the bud on the upper margin of each leaf scar. Since this bud occurs on the side of the twig, it is called a LATERAL BUD. When the leaf is present, this bud is in the axil (upper angle) between the petiole and the stem, so it is also known as an AXILLARY BUD. Lateral (axillary) buds are composed of the same parts as a terminal bud.

Lateral buds are formed at the same time as the leaves on a twig. But they do not grow when the leaves grow because they are inhibited by a hormone (auxin) from the terminal bud. As the twig lengthens, the distance between lateral buds and the terminal bud increases. This results in less inhibition by the terminal bud, and lateral buds then may grow and produce branches. **Label** a lateral bud in Figure 10.1.

Sometimes lateral buds fail to grow for an indefinite period and become buried in the bark as the tree ages. These are called LATENT BUDS. Under certain conditions they may start growth and form small leafy shoots on old large limbs.

Lenticel

Examine the surface of your twig (or larger limbs, if available). Note that all over the surface there are rounded, pinheadlike or linear, slitlike markings. These are the LENTICELS. Lenticels are raised areas of loose cork. In trees with deeply furrowed barks they are not visible. They are avenues for entry of air into the woody parts just as stomates admit air into leaves. A heavy growth of moss, lichens, or vines on the trunk of a tree can interfere with its lenticels and in extreme cases the tree may die. What other agents might injure trees by entry through the lenticels? _____ _____. **Label** the lenticels in Figure 10.1.

B. SOME COMMON ADDITIONAL TWIG FEATURES

Thorns, Spines, and Prickles

ACTIVITY 2 **Examine** twigs of honey locust (*Gleditsia triacanthos*), hawthorn (*Crataegus* sp.),

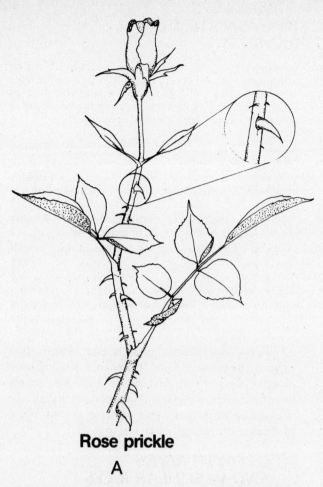

Rose prickle

A

Figure 10.2   Examples of prickles (A), thorns (B) and spines (C).

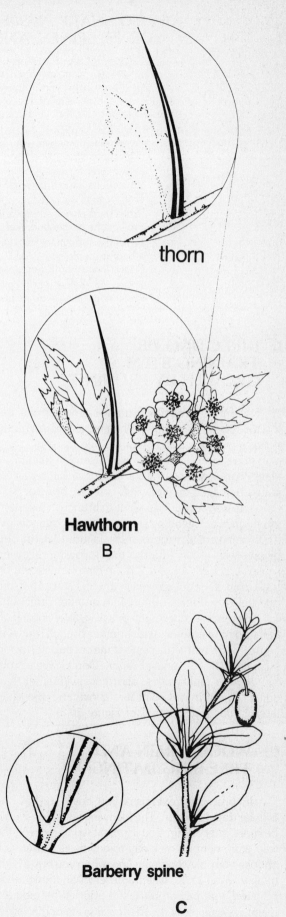

thorn

Hawthorn

B

Barberry spine

C

black locust (*Robinia pseudoacacia*), and crabapple (*Malus* sp.). These and many other woody plants have stems with sharp projections that are either thorns, spines, or prickles. These are simply modified growth forms of something else. Thorns (Figure 10.2B) are modified twigs. Spines (Figure 10.2C) are either modified leaves or modified stipules, depending on the plant type. Prickles are just epidermal or cortical outgrowths, as in roses (Figure 10.2A).

Because thorns are usually objectionable features for shade trees or garden plants, any varieties of a thorny plant that lack these are selected for use as cultivars (cultivated varieties).

Examine the twigs of plants available, and, by using the key given below, determine whether the sharp projections are thorns, spines, or prickles.

A key is a set of clues arranged systematically to lead you to the correct identification of a particular item—in this case, the particular kinds of protuberances on dicot trees and shrubs.

There are keys for the identification of many things in nature (insects, plants, birds, and rocks) as well as for trouble-shooting operating problems in aircraft, automobiles, and computer systems.

KEY TO SHARP PROJECTIONS ON STEMS
(THORNS, SPINES, AND PRICKLES)

Directions for Using Key: This key consists of pairs or couplets of choices, 1a–1b, 2a–2b, and so on. Compare your plant specimen with the descriptions given in couplet 1a–1b. Choose whichever statement (1a or 1b) is more applicable to your specimen and go on to the next statement as indicated by the number at the extreme right. Continue through the key, each time choosing between a pair of choices until you terminate at the name (thorn, prickle, or spine).

1a.	Projections branched (e.g., honey locust)	**Thorns**
1b.	Projections straight or curved, but without side branches	2
2a.	Projections scattered along twig more or less randomly (e.g., rose)	**Prickles**
2b.	Projections always (at least on young twigs) associated with a feature such as a leaf, leaf scar, or bud	3
3a.	Projections always above a leaf or leaf scar (in the position of an axillary bud)	**Thorns**
3b.	Projections near a bud, leaf, or leaf scar, but not always above, or indistinctly so	4
4a.	Projections rough, or with leaf scars or bud scars, or leaves or buds on their surface, or with a bud at the tip	**Thorns**
4b.	Projections smooth and unscarred	5
5a.	Projections below a bud or branch axil, sort of in place of a leaf	**Spine**
5b.	Projections (single or in pairs) below (or beside) a leaf or leaf scar in place of stipules	**Spine**

II. DIRECTING OR TRAINING STEM GROWTH BY PRUNING

ACTIVITY 3 **Practice** pruning if time permits. It's worthwhile to know how. When pruning is correctly done, a tree or shrub can be trained to grow in any given direction. Pruning is also useful in keeping a plant from overgrowing its area and for repair of plants damaged by ice, wind, disease, or insects.

The way to induce a plant to branch in a certain direction is achieved by shortening the young twigs. If a pruning cut is made as at *B* in Figure 10.3, the bud below will grow (during the following growing season) to form a branch in the direction shown by the arrow. If the cut is made as at *C* in Figure 10.3, the bud below will form a branch in another direction.

To remove a large limb or a twig, always cut it off at its point of juncture with another branch (cut *A* in Figure 10.3). Make the plane of the cut parallel to and flush with the remaining branch. Don't leave a stub.

If you simply want to shorten a twig rather than cut it off, cut it at a point just above any one of its lateral buds (cuts B or C of Figure 10.3).

III. WOOD GRAIN AND TREE-RING DATING

In preceding exercises you have studied the cellular detail of wood. The following activities should be more interesting. You'll look at wood more in the way you might do on a day-to-day basis. When you choose your plywood paneling, when you buy a high-quality wood table, when you select anything made of wood, you pay a pretty stiff price. What you are mainly paying for is the particular grain of the wood, the species, and, of course, the labor costs.

Knowledge of wood grain is useful information to you. The physical visible properties of wood have always been admired. And since the 1950s scientists have been using the growth rings of trees to discover the weather of the past. This is known as the science of dendrochronology, or tree-ring dating.

A. GROWTH RINGS AND VASCULAR RAYS

ACTIVITY 4 **Examine** blocks of wood or twigs or limbs cut across the long axis of the stem (transverse or cross cuts). Each light ring you see is earlywood, whose cells are large; each dark ring is latewood, whose cells are small. One pair of dark- and light-

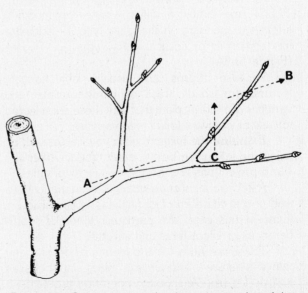

Figure 10.3 Correct cuts to make when removing a branch (cut A) or when shortening a twig and directing (training) its growth (cuts B and C).

WOODY STEMS III **83**

colored rings represents one **ANNUAL RING** (or **GROWTH RING**). In any one growth ring, which is outermost, (a) earlywood; (b) latewood? Circle (a) or (b). Was the amount of growth the same each year? *Yes   No.*   What is the indicator of the amount of growth? _____. What is the tissue that makes up all these rings? _____. Is there any phloem in these rings? _____. What are the lines of cells running across the growth rings like the spokes of a wheel? _____

Label a growth ring in Figure 10.4A on its transverse plane of cut.

B. HEARTWOOD AND SAPWOOD

At the very outer edge of your wood section is a dark brown, rather crusty zone, the bark, which you'll study later. We simply point it out here to help direct you where to look. **Concentrate** on looking at the whole general area of growth rings inside the bark. Can you see any color differences in this zone?   *Yes   No.*   If you have a good sample, the more central of the growth rings will be a darker color. This dark-colored central region is the **HEARTWOOD**. Heartwood no longer contains water or living cells. Various compounds and coloring materials have infiltrated it and protect it from internal decay. As long as no air gets to it, it is practically indestructible. Although dead, it is structurally sound and the backbone of the tree. When used for lumber, it is more durable than **SAPWOOD**.

SAPWOOD is the lighter zone of growth rings. It lies between the heartwood and the bark. The relative widths of sapwood and heartwood vary in different plants. Sapwood is not necessarily a narrower zone than heartwood. In sapwood, some of the storage cells are still alive and some of the vessels and tracheids still contain water. But most of the water moving up the tree is conducted only in the outermost growth ring of the sapwood.

In the standing living tree, sapwood is more resistant to decay than heartwood. When trees are pruned, the heartwood is exposed to air and is the first to succumb to fungus or insect invasion. This is the reason why you should paint cut surfaces of large limbs whenever you prune a tree. Any asphalt-base paint can be used. Don't use creosote; it kills tissues. **Label** the heartwood and sapwood in Figure 10.4A.

C. THE BARK

BARK is the brown outermost zone of your specimen. **Look** carefully. Can you see a crack separating the bark from the sapwood?   *Yes*

No.   This crack marks the position of the **VASCULAR CAMBIUM**. All the soft, delicate vascular cambial cells have died and disappeared since this wood was cut.

BARK is defined as all the tissues that lie outside the vascular cambium. It is made up of an **INNER BARK** and an **OUTER BARK**. The **OUTER BARK** is the outermost protective corky material.

INNER BARK is all the phloem that the plant has made all its life. Without a microscope you can't tell the difference between the inner and outer bark. Let's say that about half the thickness of the bark represents phloem. **Look** how thin the phloem is. It is negligible compared to the amount of xylem. Some of the difference in thickness between xylem and phloem exists because the xylem is hard and uncompressed, whereas the phloem is soft and gets squashed by the expanding xylem.

There's definite engineering "sense" to this. What is the function of all the old xylem, (a) support; (b) water conduction? Circle (a) or (b). Xylem cell walls have lots of cellulose and lignin. Phloem cell walls do not. Is there any point in saving phloem tissue for future structural use?   *Yes   No.* **Label** the bark in Figure 10.4A.

D. THE FIGURE OF WOOD —THE "GRAIN"

The figure (grain) of many kinds of wood is economically valuable because of its beauty. The particular grain is determined by the plane of cut of the tree trunk. In different planes of cut, the xylem rays, the fibers, and the growth rings present very different and identifiable patterns.

Examine the appearance and relationship of the rays and growth rings on each face so you will be able to identify the cut in other structures, where only a single face is visible.

ACTIVITY 5   **Examine** different cuts of wood or blocks of wood showing three planes of cut as in Figure 10.4B, C and D.

When wood is sawed with a **TANGENTIAL CUT**, the plane of the cut is at right angles to the radius of the log. The plane of this cut intersects the **GROWTH RINGS** in such a way that they appear in patterns that look like parabolas or portions of parabolas. The **RAYS** appear as segments of vertical lines. **Label** Figure 10.4B using the terms capitalized in the preceding description.

When wood is sawed with a **RADIAL CUT**, the plane of the cut is along a radius of the log. The **GROWTH RINGS** appear as more or less continuous vertical lines, fairly evenly spaced. **RAYS** appear as horizontal irregular bands with a shiny or satinlike surface. Obviously, not too many radial cuts can be made from a log, and furniture made from wood so cut is very expensive. To get the most lumber from a log, sawmills mostly cut logs tangentially. **Label** Figure 10.4C using the terms capitalized in the preceding description.

When wood is sawed with a TRANSVERSE CUT, the plane of the cut is at right angles to the long axis of the log. GROWTH RINGS appear as concentric circles. RAYS appear as straight segments of lines radiating from the center towards the edge of the face of the cut. Transverse cuts are not used in making lumber from a log.

Label Figure 10.4D using the terms capitalized in the preceding description.

ACTIVITY 6 **Examine** "Hough" slides, which are very thinly cut sections of wood, and identify the planes of cut.

ACTIVITY 7 **Examine** pieces of plywood. Plywood is made from several layers of thinly sliced wood (veneer) that are glued together. The grain of the wood in alternate veneer layers runs at right angles. This gives the plywood much greater strength than a single piece of wood of the same thickness. Because of the way it is cut, common fir or pine plywood always shows a tangential cut grain pattern. More expensive paneling may have the surface layer made from a hardwood such as birch or pecan. The pattern of this layer may be either tangential or radial. Cutting timber into plywood is a much more efficient use of the total log.

E. DATING THE PAST WITH TREE RINGS

Tree-ring dating is based on the general principle that a living tree forms a new layer or ring of wood each year. Rings made in wet years are wider than average; those made in dry years are narrower than average. The year-by-year sequence of varying ring widths forms a unique nonrepeating pattern. These rings are visible on pencil-thin cores of wood bored out of a tree trunk by means of a special instrument.

The same ring pattern occurs in trees growing within a few hundred kilometers of each other. Scientists have learned how to use ring patterns to tell climate as far back as 8400 years ago. Basically, the way it is done is as follows. First the scientist determines the dates of the rings on a wood core taken from a relatively young tree as, for example, *A* in Figure 10.5. Some of its earliest formed growth rings will match some of the later formed rings of an older tree, such as *B* in Figure 10.5. In turn, the earliest rings of *B* can be matched with the later rings of the older wood of sample *C*, and so on. Thus it becomes possible to put dates on older and older rings of older wood. The oldest woods that have thus been dated give us direct information on the past climate of a given area. We can see cycles of wet and dry years. By knowing the history of past cycles of precipitation, we can better analyze trends of change in our current climate. This is especially important for agriculture and watershed management.

Tree-ring dating, or dendrochronology, has proved to be the only tool that gives precise year-to-year history of past climate. It is more accurate than radiocarbon dating.

ACTIVITY 8 **Try** your hand at tree-ring dating. **Examine** cores that have been taken from trees of the same species growing in similar environments and preferably as near each other as possible. You will be given a pair of cores. One of the pairs will have a date marked on the mount. This represents the year that the tree was cored or cut down to obtain the sample. **Hold** the sample with the dated end to your right. Which end of the core is the inside of the tree, (a) the left end; (b) the right end? Circle (a) or

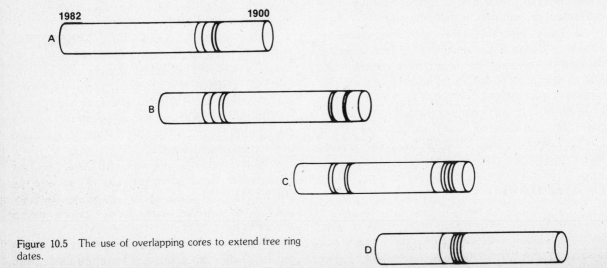

Figure 10.5 The use of overlapping cores to extend tree ring dates.

(b). Which is the outside of the tree, (a) the right; (b) the left end? Circle (a) or (b). Write down the year of the oldest (innermost) growth ring.

Remember that this oldest ring that you can *see* on the core may not represent the first year in which the tree was alive. Why not? Look at Figure 10.6. This shows that cores taken at levels *A*, *B*, and *C* will not give the true age of the tree, and that if 1976 is the first year of growth, you need to take a core from level *D* (near the ground) to get all the annual rings made by the tree.

Examine the second core of the pair. It will have a code number or other identification to show that it can be matched with the dated sample. Which end, (a) right or (b) left, of this core was nearer the outside of the tree? Circle (a) or (b). Which end represents the oldest wood, (a) right or (b) left? Circle (a) or (b). Count and write down the number of years' growth represented by this second core. _____. The second core will probably overlap the first one, that is, it will share some years in common with the first one, but it will also extend further back (or forward) in time.

Can you find the years that are shared in common? The simplest way to do this is to set down the cores alongside each other (being sure you have both cores with their most recent years kept oriented at the right). Slide one to the left or right until some of the ring series seem to match. Can you find such a match this way? With some types of wood this is possible, and is just as easy as indicated. If you have made a match, can you now date the youngest (that is, latest formed) ring of the second sample? What year was it formed? _____. What is the date of the oldest ring of the second sample? _____. Does this core have at either of its two ends rings that are older than the dated core? **Yes   No.**   Are there any rings that are younger than those in the dated core?   **Yes No.**

If you were unable to get a match in this simple way, you were probably in the majority. Even professionals don't use this method. For really old samples there are just too many rings to try to count. Dendrochronologists use techniques somewhat like the following procedure.

ACTIVITY 9   **Measure** the overall length of the dated core in millimeters _____. **Divide** this figure by the number of rings you have counted. The number is _____. The average ring width, in millimeters, is _____. **Enter** this average as the basic value on the upper graph in Figure 10.7. **Mark** the graph above and below this basic value so that each ring is represented as a dot above or below the average. **Plot** out the value (above or below average) for each year of the dated core. **Draw** a line connecting the dots of each year. **Repeat** this for the undated core of the pair on the lower graph, using a different numerical scale if necessary. Remember to *adjust* values on this second sample so that *its* average is at the midline, and *its* larger and smaller rings are graphed above or below the midline.

Now, **cut** the second graph off the bottom of the page along the marked line. **Slide** this graph, with heavy lines or dots for each year, along the upper graph. The patterns of larger and smaller rings should match up, approximately, at some point. You have then reached the same point as when the cores themselves matched. Is the second sample, overall, (a) older or (b) younger than the first? Circle (a) or (b). What years are represented by the second sample? _____ to _____.

It is by techniques similar to these that our knowledge of many past events can be dated precisely. Computers are commonly used for the scanning and matching. Suppose an archeologist or historian wishes to date the year that a village was destroyed by an earthquake. The wood of the beams in the houses, furniture, tool handles, and even firewood may be dated to within one or two years in many cases. The accuracy is far greater than that possible with carbon-14 dating, to the point that dendrochronology is often used to check the accuracy of refined carbon-14 techniques during their development.

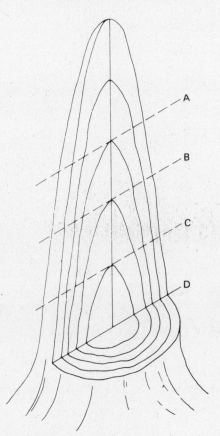

Figure 10.6   Rough sketch showing how each year's growth of wood is layered over the preceding year's. (Layers are exaggerated for the sake of clarity.)

EXERCISE 10   Student Name _____

QUESTIONS

1. Which is considered to be biologically the more successful way of life, the tree or the herb? _____

2. Which is slower to reproduce itself, the tree or the herb? _____

3. The photosynthate of the tree is chiefly used for what purpose(s)? _____

4. Why are herbs better as food crops for humans than are trees? _____

5. Which grower must wait longer to get a return on his investment in planting, the Nebraska wheat farmer or the Michigan apple grower? _____

6. What features on the twig would you count in order to determine how many years of growth are present on the twig? _____

7. Suppose an ecologist wanted to determine the previous five years' climatic conditions of a watershed (that is, the forested upland head) of the Columbia River. What twig feature(s) could he rely on in making some of his determinations? (a) _____ ; (b) _____

8. List three twig features that are useful for identifying woody plants during their dormant season when no leaves or flowers are present. (a) _____ ; (b) _____ ; (c) _____

9. Define a thorn. _____

10. What is the function of a lenticel? _____

11. Does a two-year-old twig have only primary growth, only secondary growth, or both? _____

12. List the structures that would be found in a bud. _____

13. In winter or in the dry season (in the tropics and subtropics) how could you determine the leaf arrangement of a tree since at this time it would be leafless? _____

14. The terminal bud inhibits the growth of lateral buds. How is this accomplished? _____

15. Which buds on a twig are more inhibited from growing: those nearer to, or those farther from, the terminal bud? _____

16. When pruning a fairly large branch off a tree or shrub, where should you make the cut? _____

17. How should cuts be made, straight across the stem or on a slant? _____

18. Many cultivated varieties (cultivars) of woody ornamentals are valued because they are sterile and don't produce any fruit or seeds to litter the ground. How then can the nurseryman continue to grow and offer such plants for sale? _____

19. Is heartwood physiologically active? _____

20. Boards cut from which type of wood (heartwood or sapwood) need a longer period of drying? _____ Explain why. _____

21. In general, which kind of tissue is more resistant to decay, living tissue or dead tissue? _____

22. Which part of a living tree, therefore, is more resistant to decay, (a) heartwood or (b) sapwood? Circle (a) or (b).

23. Why is lumber cut from heartwood more resistant to decay than lumber cut from sapwood?

    _____

    _____

24. Define bark. _____

    _____

    _____

25. Which is outermost, the earlywood of 1982 or the latewood of 1981? _____

    _____

26. In tree-ring dating methods, basically one tries to get a match in ring series of the younger part of a tree of known age with the older part of a tree of older and unknown dates. Is this statement true or false?

27. Figure 10.8 represents the six faces of a block of wood. Only one transverse (cross) section has had the ring and ray pattern drawn in. Cut out the figure along the dotted lines and fold the faces to make a six-sided block. Draw the ring and ray patterns as they would appear on each face. Glue or tape the faces together along the tabs.

EXERCISE 10   **Student Name** _____

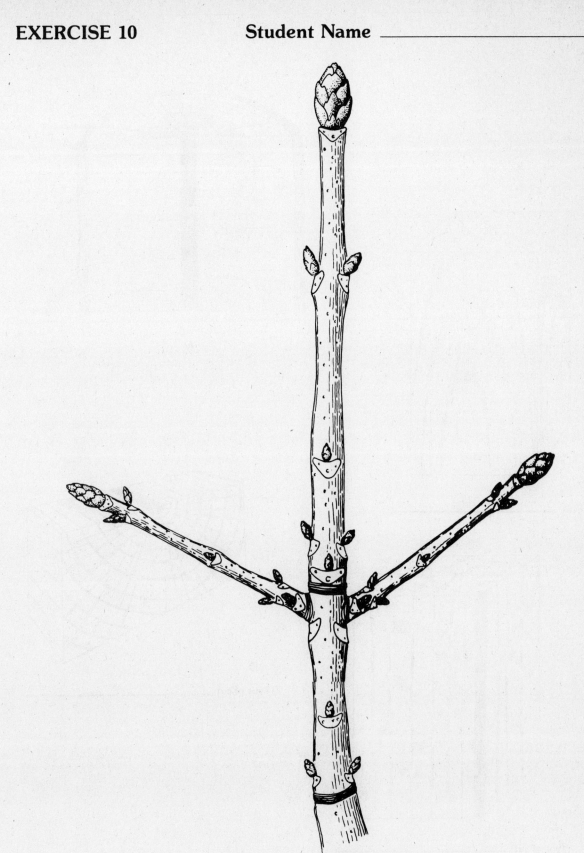

Figure 10.1   External features of woody twig.

A

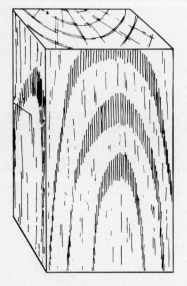

B

Type of cut _____

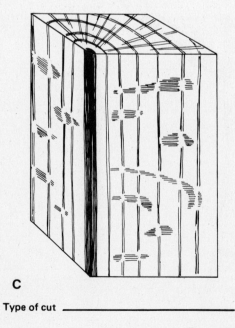

C

Type of cut _____

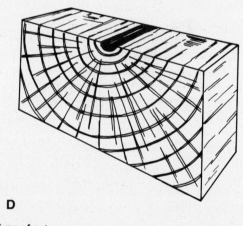

D

Type of cut _____

Figure 10.4   Wood sections.

EXERCISE 10

Student Name _____

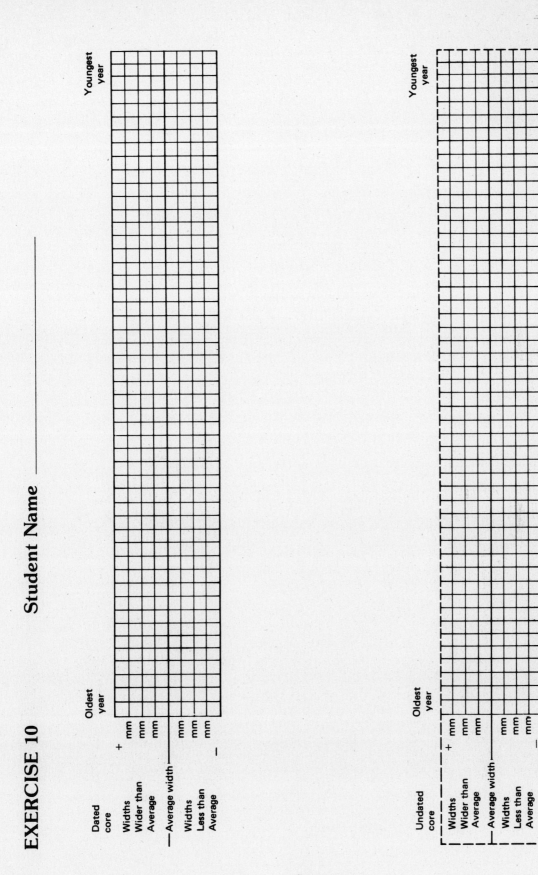

Figure 10.7

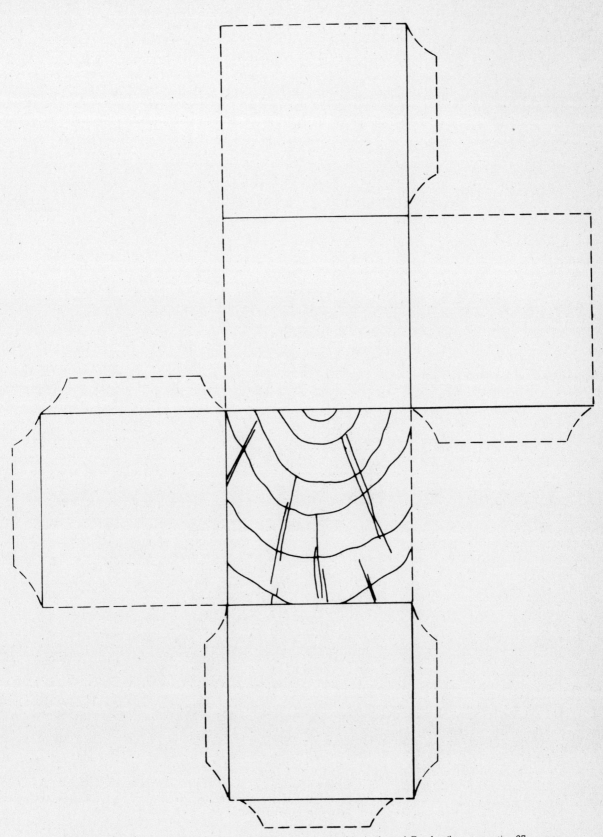

Figure 10.8 A figure to be cut out and folded to make a paper block of wood. For details, see question 27 on page 88.

LEAVES I: EXTERNAL FEATURES

EXERCISE 11

I. INTRODUCTION: THE RELATIONSHIP OF LEAF SIZE AND SHAPE TO HABITAT

The size and shape of a leaf are not completely arbitrary. In many cases (though not all) they are an adaptation to habitat. Just as many desert animals have big ears that dissipate heat, whereas their arctic cousins have small ones, so with many plants differences in leaf size and/or shape are adaptations to temperature and moisture stresses. This is best seen in plants living in hot dry, cold dry, or warm wet climates. In temperate moist regions there seems to be little correlation with climate, and no one leaf type is typical.

Plants are immobile and cannot move to a better site when there is a drastic rise or fall in air temperature or soil moisture. Plant parts most vulnerable to such changes are the leaves, flowers, and buds. In regions or in periods of intense sun heat, leaf temperature could soar drastically. But for many plants the size and shape of their leaves help prevent high leaf temperatures. The leaf can also prevent a rapid heat loss during cold periods.

METHODS OF TEMPERATURE CONTROL

One way that leaf temperature is lowered is by **evaporative cooling,** or **transpiration**—that is, the loss of water vapor from the leaf. This mechanism can be effective only when temperatures are only moderately high and there is abundant soil moisture, as in moist temperate and tropical climates. But in dry habitats, cooling by transpiration would waste the plant's water. Hence, in such places leaf temperature

is lowered in another way—by losing heat directly to the air without the loss of water. The leaf can also gain heat from the air if the air is warmer.

This exchange of heat with the air is by a process called **convection.** It works like this. When a leaf and the air are simultaneously warmed by the sun, the air always heats up more slowly and is thus always a few degrees cooler than the leaf. Excessive rise in leaf temperature is thus reduced by losing heat to the cooler air. Conversely, when a leaf and the air simultaneously lose heat (as at night), the air cools down more slowly and is thus always a few degrees warmer than the leaf. An excessive drop in leaf temperature can be prevented by gaining heat from the warmer air.

Convective flows between leaf and air are effective only near the edge (margin) of the leaf. Therefore, the greater the extent of the margin in relation to the area of the leaf blade, the greater the effect of convective heat exchange. The more irregular the margin, the greater the total length bordering any one section of the leaf blade. Compare the greater total length of the margins of the incised or lobed leaf blades with the margin of the entire leaf blade shown in Figure 11.1. In the same way, the length of the margin of a small leaf is proportionally much greater in relation to its blade area than is the length of the margin of a large leaf compared with its blade area. For this reason, the more irregular the margin, or the smaller and narrower the leaf, the greater the relative difference between margin length and blade area, and therefore, the more easily convective heat exchange occurs. Thus, plants living in deserts or droughty regions commonly have leaves that are narrow, small, or highly lobed or divided into smaller parts or leaflets.

The bigger and broader the leaf, the less readily does convective heat exchange occur. A big broad

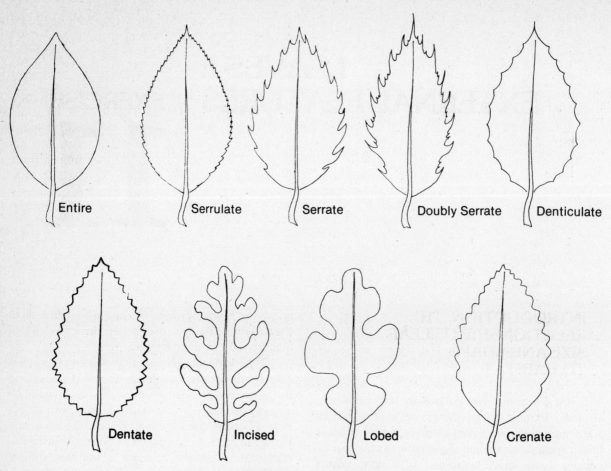

Figure 11.1   Leaf margins

leaf is more suited to evaporative cooling, or transpiration. Plants living in the shady understory of the wet tropical forest, where water is abundant, can "afford" the loss of water that occurs in the evaporative cooling process. Thus, in these habitats, we find plants with the biggest leaves in the world. Often they are not lobed or divided, and their margins are even.

In those tropical regions with seasonal dry periods, the taller forest trees often shed their leaves during the dry season. Without the leafy canopy of the trees, understory plants, especially the vines that climb the tall trees, are more exposed to the sun. For survival under these conditions, many of the plants with larger leaves must make use of convective as well as evaporative cooling. Such an adaptation often takes the form of highly lobed or perforated leaves—for example, the leaves of monstera and other *Philodendron* species.

A third way to control leaf (and plant) temperature is through **succulent leaves** and **stems**, which contain large amounts of water in the tissues. This massive, wet tissue slows the rate of daytime heat gain and nighttime heat loss.

People use leaf size and shape as a means of identifying many different plant species. Leaves,

along with flowers and buds, are referred to as key characters. Books called "keys" use these key characters arranged in an organized step-by-step sequence that leads the reader to the identification of a particular plant. Exercise 31 of this manual is a key for the identification of some common trees.

II. LEAVES OF PLANTS NATIVE TO HOT DRY AND/OR COLD DRY REGIONS

ACTIVITY 1   **Examine** in the laboratory or in the field the leaves of plants native to hot dry or cold dry regions. Examples to study are cone-bearing plants (conifers), such as pine (*Pinus* sp.), spruce (*Picea* sp.), fir (*Abies* sp.), or juniper (*Juniperus* sp.)

If available, **examine** sagebrush (*Artemisia tridentata*), creosote bush (*Larrea divaricata*), and tumbleweed (*Salsola* sp.). Note that in each case the leaf is small, with very little surface area, but by comparison has a considerable amount of edge (margin). The conifer leaf is needlelike (Figure 11.2A), scalelike (Figure 11.2B), or awllike (Figure 11.2C). These leaf types show an efficient design for

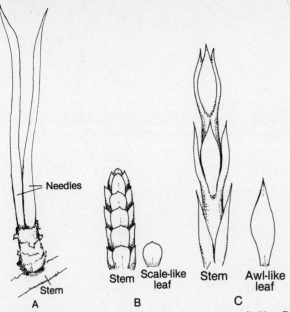

Figure 11.2 Examples of leaves of conifers: A, needle-like; B, scale-like; C, awl-like.

which method of temperature regulation—(a) convection or (b) transpiration? Circle (a) or (b). What other features of conifer leaves indicate their adaptation to dry regions? _____
_____ **Record** as many of the features as are applicable in Table 11.1.

ACTIVITY 2 Examine cactus plants such as species of *Mammillaria, Opuntia, Cleistocactus,* or *Cephalocereus senilis.* In cacti, all photosynthesis is carried out by the green, fleshy, watery stem. The leaves are reduced to spines that may discourage most animals from eating the watery stem.

Examine also representative plants known as the spurges, such as species of *Euphorbia.* Most of these are native to the deserts of Africa. Superficially they resemble the American cacti, but their thick, fleshy stems contain a milky sap that is unpalatable to animals.

Examine representative plants with succulent leaves such as species of *Kalanchoe, Aloe, Crassula, Sedum,* or *Haworthia.*

How do you explain the almost identical appearance of the American cacti and the African spurges, two groups that are unrelated and native to different continents? _____
_____. Why do you think that many terraria planted with a variety of small-leaved plants are unsuccessful? _____
_____
_____.

III. LEAVES OF PLANTS NATIVE TO WARM WET REGIONS

ACTIVITY 3 Examine plants native to the wet subtropics and tropics. **Record** your observations in Table 11.1. These plants are commonly grown as indoor plants in the temperate zone, but the original habitat of most is the moist, shady floor of a tropical forest. Examples of such plants are rubber plant (*Ficus elastica*), fiddle-leaf fig (*Ficus lyrata*), dumbcane (*Dieffenbachia* sp.), rex begonia (*Begonia rex*), pothos (*Scindapsus aureus*), arrowhead (*Syngonium* sp.), and others.

This type of leaf is an efficient design for which method of temperature regulation, (a) convection or (b) transpiration? Circle (a) or (b). Write a sentence or phrase defining this method. _____
_____ Give one reason why these plants are particularly suited for indoor plants? _____
_____ Most of these leaves have what kind of margin (edge), (a) even or (b) uneven? Circle (a) or (b). Is this the type of margin characteristic of large-leaved plants of the wet tropics? *Yes No.*

IV. LEAVES OF PLANTS NATIVE TO THE MOIST TEMPERATE ZONE

A. LEAVES OF DICOT PLANTS

ACTIVITY 4 Read the following sections and study Figures 11.1 and 11.3–11.7 as a guide. **Examine** the plants that have been set out for you and **record** the leaf characters in Table 11.1.

Leaf Parts

The leaf consists of a flattened laminar portion, called the **BLADE,** and a stalk, called the **PETIOLE,** which attaches the blade to the stem (Figure 11.3A, B). If the blade is attached directly to the stem, the leaf is described as **SESSILE.** If the leaf you are examining is still attached to its stem, look for the **AXILLARY BUD** (Figure 11.3B), a small bud situated in the angle (axil) between the base of the petiole and the stem. When this bud matures, it will produce a branch.

In many plants there are two small leaflike appendages at the junction of the petiole and the stem. These are **STIPULES** (Figure 11.3A). Their function varies and is not always clear-cut. They may serve to protect the leaf when it's young; they are

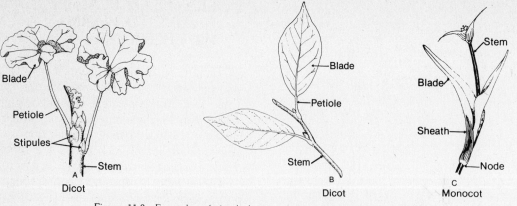

Figure 11.3 Examples of simple leaves of dicot and monocot plants.

usually photosynthetic, but sometimes they are modified into spines.

Leaf Types

Leaves are either **SIMPLE** or **COMPOUND**. A simple leaf has a blade consisting of one piece (Figure 11.3A–C). A compound leaf has a blade composed of a number of segments, called *LEAFLETS* (Figure 11.4A–C). There are two basic kinds of compound leaves, **PINNATELY COMPOUND** and **PALMATELY COMPOUND**. In a pinnately compound leaf the leaflets occur in a linear sequence lined up along both sides of a central axis, called the **RACHIS**. If the rachis is unbranched, the leaf is described as **PINNATE** (Figure 11.4A). If the rachis is branched one or more times, the leaf is described as **DOUBLE PINNATE** (Figure 11.4B) or **TRIPLE PINNATE**. The honey locust tree (*Gleditsia triacanthos*) commonly has both single and doubly pinnate leaves.

A compound leaf that is **PALMATE** (Figure 11.4C) is one in which three, five, seven, or more leaflets are all attached at one point near the tip of the petiole, and they radiate out from this tip. A palmate leaf with three leaflets is commonly referred to as

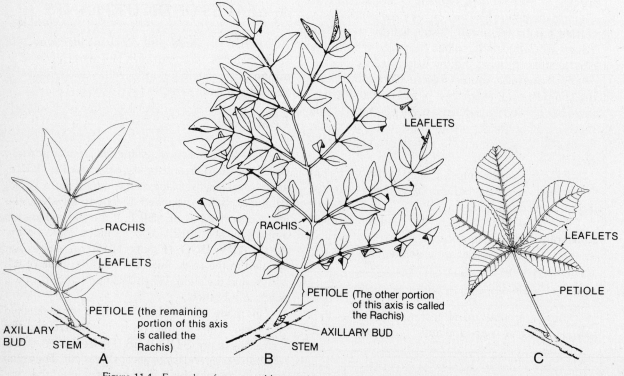

Figure 11.4 Examples of compound leaves: *A*, pinnate; *B*, doubly pinnate; *C*, palmate.

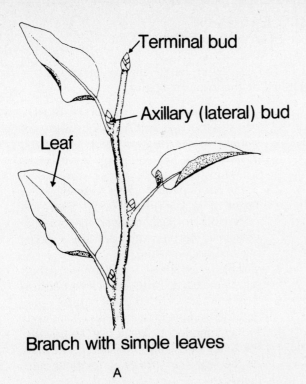

Branch with simple leaves

A

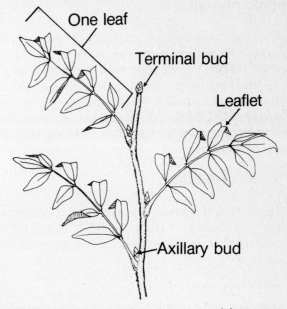

Branch with pinnately compound leaves

B

TRIFOLIOLATE. The botanical name (genus) for the common field clover (*Trifolium*) really describes the trifoliolate leaf of the plant. When you find the proverbial "four-leaf clover," you really have a four-leaflet clover. Both simple and compound leaves may occasionally occur on the same plant, as in Boston ivy (*Parthenocissus tricuspidata*).

Some Distinctions Between Simple and Compound Leaves and Between Branches and Pinnately Compound Leaves   It is sometimes difficult to distinguish between a branch and a compound leaf. Figure 11.5B and the following explanations will help in this matter.

1. Buds occur in the axils of leaves but not in the axils of leaflets.
2. All the leaflets of a compound leaf occur in the same plane, whereas the blades of adjacent simple leaves are not usually all oriented in the same plane.
3. Very large pinnately compound leaves sometimes resemble an entire branch. The branch will have a terminal bud, the leaf won't.

Leaf Arrangement

Leaves are attached to the stem in different arrangements (Figure 11.6). In the **ALTERNATE** (or **SPIRAL**) arrangement, one leaf occurs at each node (Figure 11.6A). This is the most common arrangement. The **OPPOSITE** arrangement with

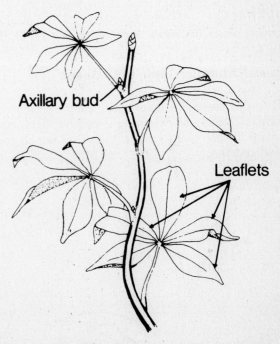

Branch with palmately compound leaves

C

Figure 11.5 Compound leaves (*B* and *C*), like simple leaves (*A*), always have an axillary bud in the angle between the leaf petiole and the stem.

two leaves at a node (Figure 11.6B) is somewhat less common, and the **WHORLED** arrangement (Figure 11.6C) with three or more leaves at one node is the least common, particularly in woody plants. It is, however, not uncommon in herbaceous plants.

Leaf Margin

The leaf margin is the leaf edge. There is a great variation in margins. Some are shown in Figure 11.1. An **ENTIRE** margin is an even or regular edge. A **LOBED** margin is a deeply indented edge, making the blade lobed. Indentations between lobes are called **SINUSES**. All other margin types are different degrees of dissection, ranging between entire and lobed.

Leaf Venation

Venation refers to the arrangement of veins in the leaf (Figure 11.7). In the **NET VENATION** characteristic of dicots, there are one or more large main veins from which smaller veins branch and interconnect, forming a meshlike network. The **PINNATE VENATION** pattern (Figure 11.7A) has one main vein extending from the base of the leaf to its tip and bearing many lateral, parallel branch veins. The **PALMATE VENATION** pattern (Figure 11.7B) has several large main veins that radiate out from a common point at the base of the leaf. **PARALLEL** venation (Figure 11.7C) is typical of many monocots and is discussed in Section III. B.

Leaf Shapes, Tips, and Bases

The shape, the tip or apex, and the base of the leaf or leaflet are extremely varied but fairly constant for any given species. These will not be considered here.

B. LEAVES OF MONOCOT PLANTS

The leaves of monocots are usually distinguished by the following features.

1. They usually consist of a narrow **BLADE** whose base is a **SHEATH** that wholly or partly encloses the stem (Figures 11.3C and 11.7C). Very few monocots have leaves with a distinct petiole.
2. They have parallel veins extending the length of the leaf (Figure 11.7C). There are exceptions to this, however; many monocots have net venation, and some dicots have parallel venation.
3. The blade and sheath are considered to be a highly modified, flattened petiole because they are parallel-sided, have parallel veins, and the blade is rarely lobed.
4. In many monocots, especially the grasses, there is a persistent meristem at the base of the leaf that allows it to grow indefinitely.
5. They rarely show true opposite arrangement and are normally alternate or scattered.

ACTIVITY 5 **Examine** the leaves of representative monocot plants and **record** the leaf characters in Table 11.1. Suggested plants are any grass, sedge (*Carex* sp.), or rush; gladiolus (*Gladiolus* sp.); lily

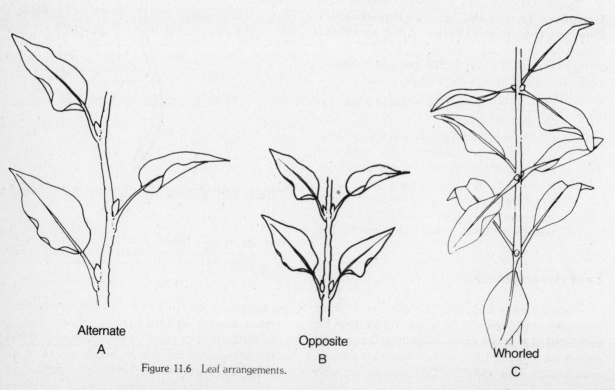

Alternate

A

Opposite

B

Whorled

C

Figure 11.6 Leaf arrangements.

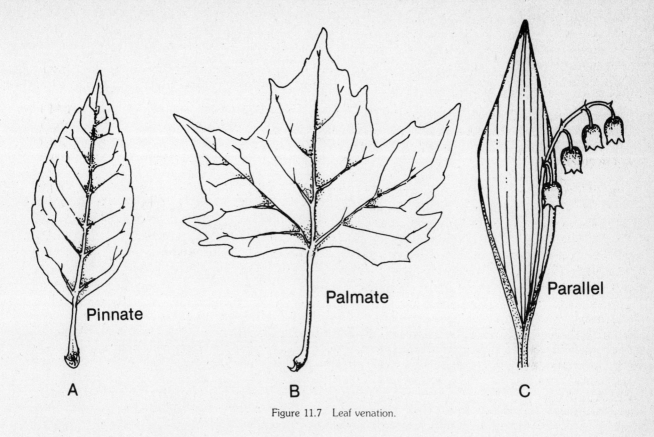

Pinnate

Palmate

Parallel

A

B

C

Figure 11.7 Leaf venation.

(*Lilium* sp.); or tulip (*Tulipa* sp.). Common indoor plant suggestions for study are spiderwort (*Tradescantia* sp.), wandering Jew (*Zebrina* sp.), waterweed (*Anacharis canadensis*), *Sansevieria* sp., *Dracaena* sp., *Rhoeo discolor,* and pineapple (*Ananas comosus*).

Most of the grasses of the temperate zone are basically adapted to open areas of full sunlight and occasionally to regular periods of drought. What structural design (shape, dimensions, and so on) does the grass leaf show as an adaptation to hot, drought conditions? _____

_____

_____

_____

_____

Look at examples of monocot plants whose leaves have net pinnate venation rather than the parallel venation typical of most monocots. These might be dumbcane (*Dieffenbachia* sp.), pothos (*Scindapsus aureus*), and species of *Philodendron* and *Maranta*.

EXERCISE 11   **Student Name** _____

QUESTIONS

1. The leaves (needles) of conifers are adapted to which type(s) of climate, hot and dry, hot and wet, cold and dry, or cold and wet? _____

2. In regions with dry climates, which method of temperature regulation is predominant in leaves of plants, convection or transpiration? _____

3. List the structural features of a leaf adapted to dry climate _____
_____
_____

4. Plants with very large leaves and even margins are common to what type(s) of climate, hot and dry, hot and wet, cold and dry, or cold and wet? _____
_____

5. Such leaves are adapted to which method of temperature regulation? _____
_____

6. Can a leaf be both cooled and warmed by convection? _____

7. When a leaf is warmed by convection, does the heat come from the sun or the air? _____

8. Does there seem to be any correlation between climate and leaf type among plants native to the moist temperate zones? _____

9. What leaf shape or type predominates in the moist temperate zone? _____
_____

10. Describe the attachment of the typical monocot blade to the stem. _____
_____

11. What is a "key"? _____
_____
_____

12. What is one way to distinguish between a large pinnately compound leaf and a branch?
_____
_____

13. How does a simple leaf differ from a compound leaf? _____
_____

14. Does a four-leaf clover really have four leaves? _____ What does it have? _____
_____

15. The cells of cacti have special proteins that are resistant to high temperature. The cells also store abundant water. What function does the water serve? _____
_____

Table 11.1 EXTERNAL FEATURES OF LEAVES

NAME OF PLANT	TYPE OF CLIMATIC CORRELA-TION (IF ANY)	IF COM-POUND, PINNATE OR PALMATE	IF SIMPLE, SHAPE OF BLADE OR NEEDLE	ARRANGE-MENT	TYPE OF MARGIN OF LEAF OR LEAFLET	VENATION

LEAVES II:
INTERNAL STRUCTURE

EXERCISE

12

INTRODUCTION

Even though land plants are well adapted to life out of water, photosynthetic reactions demand aqueous conditions. This is a holdover from ancestral green algae in which photosynthetic cells were in direct contact with water. The terrestrial plant is not surrounded by water, but its photosynthesizing cells still require an aquatic medium. This is because the carbon dioxide necessary for photosynthesis is found in the cell not as a free gas but dissolved in water. Thus you will see that the leaf of the land plant is really an aquaculture—a life-support system of photosynthesizing cells called chlorenchyma. This system is sandwiched between two protective layers, the upper epidermis and the lower epidermis. The epidermis is porous, allowing diffusion of carbon dioxide and oxygen. Water is unavoidably lost through these pores, and the potential exists, therefore, for the system to dry out. That this does not happen as readily as you might expect is due to many factors that will be considered in this and later exercises.

You have seen that the external features of the leaf can, in certain cases, help to abate temperature stress. In this exercise you will see how the internal features and the epidermis of the leaf help to abate moisture stress. You will study the photosynthetic tissues and the ways they are modified to receive light and to maintain a saturated internal atmosphere.

I. THE MESOMORPHIC LEAF

The majority of flowering plants are mesophytes, that is, plants requiring moderate amounts of water regularly throughout the growing season. Their leaves are described as **mesomorphic**.

A. *THE DICOT LEAF*

ACTIVITY 1 **Examine** a prepared slide of the cross section of a leaf of a representative mesophytic dicot such as lilac (*Syringa* sp.) or privet (*Ligustrum* sp.). Using the lowest magnification on the microscope, orient the slide so that the section appears as a flat, ribbonlike mass extending horizontally across your field of view. If the midrib or main vein "bulges" downward, then you have the leaf correctly oriented. **Look** at Figure 12.1 as a guide.

In this section you are seeing all the cell layers that occur from the top to the bottom of the leaf. The outermost layers at the top and at the bottom are the UPPER EPIDERMIS and the LOWER EPIDERMIS. All the cells between them comprise the middle of the leaf, the MESOPHYLL, that is, the wet aquaculture of chlorenchyma and veins. **Change** to higher magnification and **examine** these layers and cell types in detail.

Upper Epidermis

The UPPER EPIDERMIS and the lower epidermis form a continuous layer around the leaf, but the two layers differ in certain ways. The upper epidermis is a layer of closely packed cells. This compact fit helps prevent excessive water loss and provides mechanical support. The outer walls of the cells are covered by a waxy, noncellular layer, the CUTICLE. This further retards water loss. Do you see any openings or pores in the upper epidermis? *Yes   No.*   Beneath the upper epidermis are the cells of the mesophyll containing abundant small bodies, the chloroplasts. Do the epidermal cells also contain chloroplasts? *Yes   No.*   **Label** the upper epidermis and cuticle in Figure 12.1C.

Mesophyll

MESOPHYLL is all of the portion of leaf between the upper and lower epidermis. It is the part that can be likened to an aquaculture system since it is usually saturated with water and water vapor. It contains veins, a certain amount of fibrous tissue, a labyrinth of free spaces, and chlorenchyma cells. The chlorenchyma in the upper part of the mesophyll is known as PALISADE TISSUE; in the lower part it is known as SPONGY TISSUE. Chlorenchyma are thin-walled parenchyma cells containing abundant chloroplasts. The colorless walls and glass-clear protoplasm of these cells reflect and refract the incoming light rays, scattering them in zigzag paths. This ricocheting of each ray lengthens its path and multiplies its chances of striking a chloroplast before it passes out of the leaf. **Bracket** the mesophyll of Figure 12.1.

Palisade Tissue   This is a layer of evenly spaced, cylindrical cells whose long axes are at right angles to the upper epidermis. This arrangement puts the cells at an acute angle to the incoming sun rays and reduces the effect of their intensity. Full, intense sunlight during the brightest part of the day actually reduces the rate and the efficiency of photosynthesis. Therefore, light intensity that is less than maximum (as achieved through the orientation of the palisade cells) results in faster rates of photosynthesis and also an overall greater efficiency of the photosynthetic reaction in the leaf.

Most photosynthesis of the leaf occurs in these palisade cells, which have more chloroplasts than spongy tissue cells. Although they appear to be closely packed, palisade cells are slightly set apart so that the surface of each is exposed to the wet atmosphere of the mesophyll. In what practical way does this contribute to the reaction of photosynthesis? _____
_____.

Usually only one layer of palisade cells is present in the mesomorphic leaf. However, leaves that are continually exposed to direct sunlight (as opposed to those partially or periodically shaded) tend to have two or more layers of palisade. This makes the leaf thicker. **Label** the palisade tissue in Figure 12.1C.

Spongy Tissue   This is the lower portion of the mesophyll. It is a more open zone of irregularly shaped cells and large intercellular spaces. Bumps or lumpy extensions of each cell touch those of other cells. The limited surface contact between cells leaves much of the cell surface free or exposed. What is the advantage of this? _____
_____. In growing conditions where the leaf is shaded this random arrangement of cells and their more or less cubical shape favor maximum light absorption. The large air spaces in the SPONGY TISSUE allow for rapid, easy flow of carbon dioxide, oxygen, and water vapor throughout the mesophyll. **Label** the spongy tissue in Figure 12.1C.

Veins   VEINS are cylindrical strands of vascular tissue and occur mainly in the mesophyll. They run in all directions, so you will see some in cross section (appearing circular in outline) and some in profile view (appearing as a more or less complete or incomplete horizontal band). **Label** the small vein shown cut in two planes in Figure 12.1C.

Examine the large MIDVEIN (midrib or main vein) in the center of the leaf.

The XYLEM CELLS in the vein are easily recognized by their thick, angular walls (stained bright red in the slide-making process) and by an absence of cell contents. Xylem brings water into the mesophyll. PHLOEM CELLS are much less distinct; they are thin-walled, usually of smaller diameter than xylem cells, and located below the xylem, that is, on the side between the xylem and the lower epidermis.

Study the ring of large, thin-walled parenchyma cells that encircle and may radiate out from the veins. These cells are the BUNDLE SHEATH CELLS. **Note** that they are also in close contact with both the palisade and spongy tissue. Bundle sheath cells are important because they are thought to serve as flow channels between veins and chlorenchyma. This increases conduction in the leaf. Sheath cells envelop the vein over most of its length and close over and around its tip. In many veins these bundle sheaths extend to the upper and lower epidermis and help provide mechanical support for the blade. In large veins SCLERENCHYMA FIBER cells are part of the sheath. If you look at the midrib you will see above and below it a mass of thick-walled sclerenchyma cells. Such a "reinforced" vein serves as a flexible yet strong girderlike support for the leaf.

Label the midvein and its parts in Figure 12.1C, using the terms capitalized in the preceding description.

Lower Epidermis

The **LOWER EPIDERMIS** is similar to the upper epidermis except that it has many small pores or **STOMATES** that control the passage of gases into and out of the leaf and the escape of water vapor. Each stomate is flanked by two sausage-shaped **GUARD CELLS**. Guard cells are smaller than the adjacent epidermal cells and have a few chloroplasts and, most noticeably, a very prominent nucleus. **Find** the guard cells and stomates. Guard cells and stomates are best seen in an epidermal peel, which you will make in Activity 2. Changes in the shape of the guard cells result in the opening and closing of the lens-shaped pore between them. Such shape changes in a guard cell are due to changes in its water content. Stomates are opened as a result of the guard cell's response to light, among other things. They are usually closed at night or if it is windy, owing to a loss of turgor pressure in the guard cells. When light and heat intensities are high, the leaf can stabilize its temperature somewhat by evaporating water from its mesophyll out through the stomates. **Label** the lower epidermis in Figure 12.1C, using the terms capitalized in the preceding description.

ACTIVITY 2 **Examine** stomates. **Using** a razor blade, peel off a strip of the lower epidermis from a leaf of wandering Jew (*Zebrina* sp.) or any species of *Sedum*, German ivy (*Senecio mikanioides*), corn (*Zea mays*), or other suitable plants, and make a wet mount of the tissue. Use the technique demonstrated by your instructor. Make sure that the outer surface of the epidermis is up when you place the tissue on the slide. Note the large epidermal cells, which will have different shapes and patterns or arrangement, depending on the species. In *Sedum* the cells have "wavy" contoured walls and are arranged like the pieces of a jigsaw puzzle; in corn the cells are rectangular and arranged in parallel rows. Do the epidermal cells have chloroplasts? *Yes   No.* **Find** the guard cell pairs and the pore between them. Do the guard cells have chloroplasts?   *Yes   No.*

B. ESTIMATING THE NUMBER OF STOMATES OF A LEAF

Different species of plants have different numbers of stomates per unit area of leaf surface and on different parts of the same leaf.

ACTIVITY 3 **Using** any number of different plant species, **count** the stomates seen in an epidermal peel. **Compute** the average number of stomates per square centimeter of the leaf. **Divide** the counting among the members of the class and **record** the averages in Table 12.1. Proceed as follows. Count the number of stomates visible in the field of view using the 40X objective and the 10X ocular of the microscope.

First, measure the diameter of the field of view of your microscope at this power by viewing a thin plastic ruler with millimeter markings, Your lab instructor may have done so already, using a more accurate micrometer slide, and the field diameter will be posted in the laboratory.

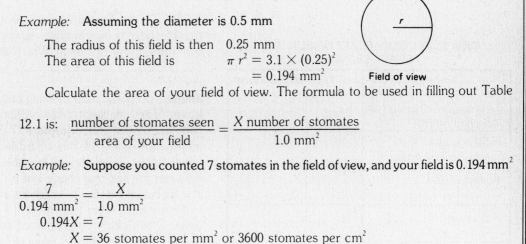

Example: Assuming the diameter is 0.5 mm

The radius of this field is then   0.25 mm
The area of this field is $\pi r^2 = 3.1 \times (0.25)^2$
                                   $= 0.194 \text{ mm}^2$ **Field of view**

Calculate the area of your field of view. The formula to be used in filling out Table

12.1 is: $\dfrac{\text{number of stomates seen}}{\text{area of your field}} = \dfrac{X \text{ number of stomates}}{1.0 \text{ mm}^2}$

Example: Suppose you counted 7 stomates in the field of view, and your field is 0.194 mm^2

$$\frac{7}{0.194 \text{ mm}^2} = \frac{X}{1.0 \text{ mm}^2}$$
$$0.194X = 7$$
$$X = 36 \text{ stomates per mm}^2 \text{ or } 3600 \text{ stomates per cm}^2$$
$$(\text{since } 100 \text{ mm}^2 = 1 \text{ cm}^2)$$

Table 12.1 EXAMPLES OF AVERAGE NUMBERS OF STOMATES ON
MESOMORPHIC LEAVES

NAME OF PLANT	AVERAGE NUMBER OF STOMATES IN 400× FIELD	AVERAGE NUMBER OF STOMATES PER CM² OF LEAF SURFACE	HEAT LOSS (CAL/MIN)

Under conditions of high temperature stress, the leaf can keep its temperature lowered by evaporative cooling. A typical water loss rate from a leaf is 0.0005 gram of water lost per square centimeter per minute. This is an equivalent loss of about 0.3 calorie lost each minute. For fun, you might want to compute the figures and fill in the last column in Table 12.1.

C. SUN AND SHADE LEAVES OF MESOPHYTIC PLANTS

The leaves of a plant, particularly those on a tree or shrub, are not all equally exposed to the sun. In a tree, for instance, the leaves in the outer part of the crown (sun leaves) receive much more sunlight than those in the lower or deeper recesses of the crown (shade leaves). More water is lost by sun leaves than by shade leaves because of the exposure to the sun's heat. The quantity of light influences the development of the palisade tissue, and the quantity of water influences the growth of all the cells of the leaf: the more water available, the larger the cells, and consequently, the larger (in breadth and length) the overall size of the leaf.

ACTIVITY 4 **Examine** the leaves of *Zebrina* sp. that has been grown in bright sunlight and compare with those grown in shade. Which leaves are larger? _____. Which are thicker? _____. **Examine** a prepared slide showing both sun and shade leaves of any species in cross section. **Compare** the two leaf types and **complete** Table 12.2. From information given in this exercise and your own observations, you should be able to distinguish between the sun and shade leaf on the slide.

Tobacco plants whose leaves are to be used to make the outer layer (wrapper) of a cigar are grown under a cloth canopy supported over the crop by an elaborate wooden framework. Based on your knowledge of sun and shade leaves, explain why these expensive, shade-grown leaves are better

Table 12.2 COMPARISON OF SUN AND SHADE LEAVES

FEATURE	SUN LEAF	SHADE LEAF
Comparative thickness of each (comparative width of each)		
Number of layers of palisade parenchyma		
Number of veins (abundant or few)		
Size of veins (large or small)		
Cuticle (thick or thin)		

suited for this use. _____
_____
_____
_____
_____

D. THE HIGH-EFFICIENCY LEAF

The monocots most valuable to man are those known as the grasses. Chief among these are sugar cane and the cereals—corn, wheat, rice, sorghum, and so on. It is now known that many grasses, notably sugar cane, corn, and sorghum, have evolved more efficient photosynthetic reactions, particularly those trapping carbon dioxide. This explains their extreme value as a food source. Other plants with these high rates include many aggressive weeds, notably pigweed (*Amaranthus* sp.), lambs-quarters (*Chenopodium* sp.), tumbleweed (*Amaranthus albus*), and purslane (*Portulaca oleracea*). The more **COMPACT MESOPHYLL** of the leaves of these grass crops and weeds maximizes conduction between the cells and also minimizes any possibilities of drying. Numerous veins provide each cell with more direct access to water. Big **SHEATH CELLS** around veins actively accumulate starch in **SPECIALIZED CHLOROPLASTS**, and their close contact with veins aids in rapid removal of the photosynthate. All this adds up to higher rates of photosynthesis, which results in greater yield and/or more vigorous growth. The corn leaf is a good example of a high-efficiency leaf.

ACTIVITY 5 **Study** a prepared slide of a cross section of the leaf of corn (*Zea mays*). Is the mesophyll differentiated into palisade and spongy tissue? *Yes No.* The vertical position of the corn leaf reduces the need for palisade cells. Explain how this is so. _____
_____ Are there as many large air spaces in the mesophyll as in the dicot leaf previously studied? *Yes No.* The veins are (a) more numerous or (b) less numerous than in the dicot leaf previously studied. Circle (a) or (b). Would you say that the mesophyll is more compact? *Yes No.* Are there bundle sheaths around each vein? *Yes No.* These sheaths are (a) large and prominent or (b) small and inconspicuous. Circle (a) or (b). The chloroplasts of the sheath cells are (a) larger or (b) smaller than those of other mesophyll cells. Circle (a) or (b). These chloroplasts are specialized types, very active in accumulating starch.

Another feature worth noting in the corn leaf is the large, bubblelike **BULLIFORM CELLS**. These have thin walls and occur in the epidermis in rows parallel to the veins. During the hottest part of the afternoon or in droughts, when water losses from the leaf are high, bulliform cells lose water and turgor more rapidly than other epidermal cells. As a result, they shrivel, causing the leaf to roll up or fold. This reduces evaporation from those stomates on the leaf surface now enclosed in the roll. Bulliform cells occur in all grasses and most monocots.

Locate bulliform cells on the corn leaf and compare their appearance in an expanded leaf versus a rolled or folded leaf.

Label the corn leaf cross section in Figure 12.2, using all the terms capitalized in the preceding description of features of a high-efficiency leaf.

THE XEROMORPHIC LEAF

The xeromorphic leaf is one that is designed to minimize its water losses. Plants that grow in dry, droughty places or in the actual desert (including the arctic cold desert) have xeromorphic leaves. Other plants that live in the wet tropics also have xeromorphic leaves. Surprising? No. In this case it is an adaptation for survival during the annual dry season.

The special tissue modifications for conserving water in the xeromorphic leaf are:

1. Thick cuticle.
2. Epidermal cells with thick walls.
3. Several layers of sclerenchyma below the epidermis and in other regions.
4. Sunken stomates—that is, stomates inside grooves or cavities recessed down from the leaf surface.
5. Stomate cavities often lined with hairs.
6. In some plants, hairs on the leaf surface.
7. A more compact and uniform mesophyll with very few air spaces.
8. In some plants, palisade tissue on both sides of the leaf.
9. In some plants, a reduced number of stomates per unit of leaf area.

A. THE XEROMORPHIC LEAF OF PINE

Pines are adapted to both hot dry and cold dry habitats. Besides having most of the above listed modifications, the leaf of pine is shaped like a prism. This makes for a low ratio of surface to volume, which automatically reduces water loss. Think about it. A wet towel will dry out less quickly if it is (a) rolled up or (b) laid out flat. Circle (a) or (b). Besides water regulation, why (or how) can temperature be regulated by this narrow prism shape? _____
_____

ACTIVITY 6 **Examine** a prepared slide of the leaf (needle) of pine. **Look** at Figure 12.3 as a guide. Most pines bear their needles in clusters (bundles) of two, three, or five. Depending on the species, therefore, your slide will show the needles in groups of two, three, or five. One or two VEINS occupy the center of each needle. The veins are embedded in a mass of thin-walled cells thought to conduct water from the vein to the surrounding MESOPHYLL. **Note** how the mesophyll is compact. The cells are filled with resins and their walls have many indentations, which give them a wavy outline. What effect does this have on the rigidity of the leaf? _____

_____

RESIN CANALS occur here and there in the mesophyll. **Find** the SUNKEN STOMATES. **Review** the list of xeromorphic features, then reexamine the pine needle cross section and list those xeromorphic features that you can recognize.

_____
_____
_____
_____
_____
_____

Label Figure 12.3, using the terms capitalized in the preceding description.

B. THE XEROMORPHIC LEAF OF THE UNDERSTORY PLANTS OF THE WET TROPICS

The large-leaved plants growing in the moist shade of the tall trees of the wet tropics make up much of the understory. During the dry season many of the tall trees are able to conserve water by temporarily shedding their leaves. This opens up the canopy and exposes the understory plants to the sun's heat. In the preceding exercise, we noted modifications of leaf shape allowing for convective cooling of understory plants. But many of these same species also have internal xeromorphic features that reduce their water losses during dry periods.

ACTIVITY 7 **Examine** living plants native to the tropics, such as rubber plant (*Ficus elastica*), fiddle-leaf fig (*Ficus lyrata*), and the large-leaved *Philodendron* sp. What tissue is responsible for making these leaves rather stiff? _____
_____. Without this tissue, what would happen to the leaf when its moisture content was low? _____.
In temperate climates, these plants are successfully

grown as indoor plants. How do you explain the adaptation of these plants to this unnatural environment?

_____

_____

If prepared slides are available, **examine** the tissues of the xeromorphic leaves of these tropical plants.

III. THE HYDROMORPHIC LEAF

The hydromorphic leaf is found in plants that grow either submerged in water or with their leaves floating on the surface of the water and their roots anchored in the soil at the bottom.

ACTIVITY 8 **Review** the list of modifications given for the xeromorphic leaf. Think about the different conditions impinging on the leaf surrounded by water or floating on it, and for each of the six features listed in Table 12.3, write a one-word or one-phrase descriptive modification that you would expect to find in a hydromorphic leaf.

Table 12.3

FEATURE	MODIFICATIONS EXPECTED TO OCCUR IN A HYDROMORPHIC LEAF
Cuticle	
Walls of the epidermal cells	
Position of stomates	
Amount of air spaces in the mesophyll	
Amount of sclerenchyma	
Number or veins (few or many)	

After you have completed Table 12.3, **examine** a prepared slide showing the cross section of the leaf of a hydrophytic plant, such as water lily (*Nymphaea* sp.) or water poppy (*Castalia* sp.). Do your observations agree with your descriptions?   *Yes*   *No*.

IV. COMPARISON OF THE MESOMORPHIC, XEROMORPHIC, AND HYDROMORPHIC LEAVES

ACTIVITY 9. **Fill** in Table 12.4 to show the major distinctions between leaves of these three types.

EXERCISE 12   Student Name _____

QUESTIONS

1. Why must photosynthetic cells (chlorenchyma) be bathed in a watery film? _____
_____

2. The leaf is primarily involved in photosynthesis. What other major role does the internal
structure of the leaf play? _____
_____

3. How is the path of a light ray inside the leaf increased? _____
_____

4. What is the advantage of lengthening the path of light? _____
_____

5. What internal design helps to reduce the intensity of light rays entering the leaf? _____
_____
_____

6. What is the value in reducing the intensity of light? _____
_____

7. In which part of the leaf do air and gases move freely about and enter and exit from the leaf?
_____

8. What is the role of sheath cells in a mesomorphic leaf? _____
_____

9. Give three features of the "high-efficiency" leaf that allow for higher rates of photosynthesis.
(a) _____ (b) _____;
(c) _____.

10. What two general kinds of plants have been identified as having high-efficiency leaves?
(a) _____; (b) _____

11. Why are sun leaves thicker than shade leaves? _____
_____

12. Why are shade leaves broader and longer than sun leaves? _____
_____

13. How does a grass leaf cut down on its water losses on a hot day? _____
_____

14. Why do understory plants of wet tropical forests have xeromorphic leaves? _____
_____
_____

15. Why are understory plants of the wet tropical forests so successfully grown as indoor
plants in the temperate parts of the world? _____
_____

16. Many of the understory herbs (e.g., spring wild flowers) of temperate deciduous forests
have thin, soft leaves with little cuticle. Explain. _____
_____
_____

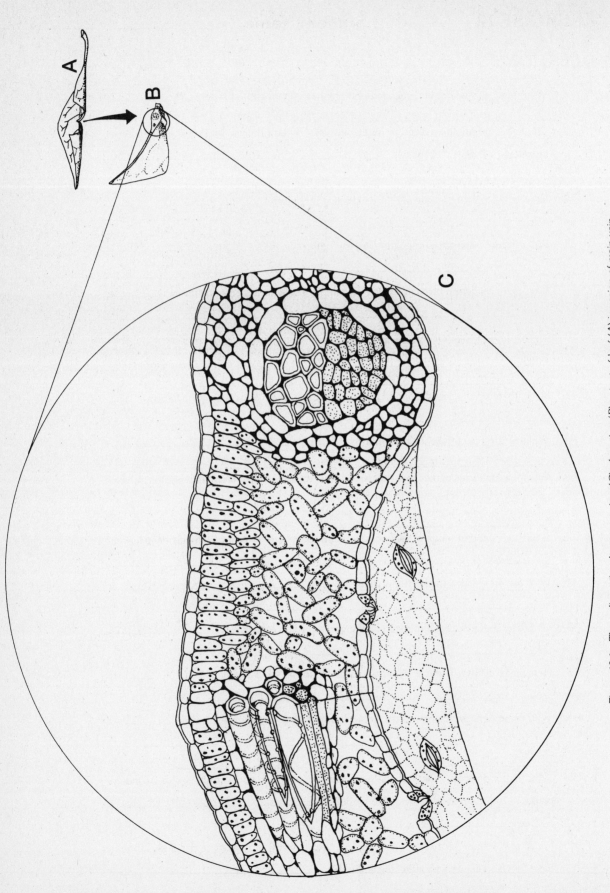

Figure 12.1   Three dimensional cutaway view (C) of section (B) cutout of leaf (A) in region of main vein.

EXERCISE 12   Student Name _____

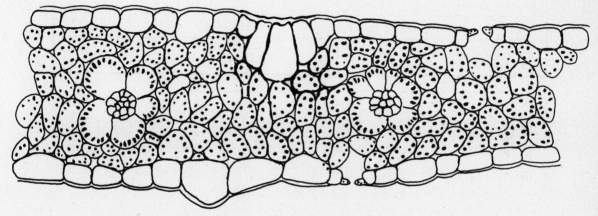

Figure 12.2   Cross section of an area of the high-efficiency leaf of corn (*Zea mays*).

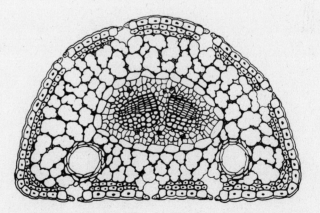

Figure 12.3 Cross section of the xero-morphic leaf (needle) of pine (*Pinus* sp.) showing its special cells and features for conserving water.

Table 12.4 COMPARISON OF MESOMORPHIC, XEROMORPHIC, AND
HYDROMORPHIC LEAVES

FEATURE	MESOMORPHIC LEAF	XEROMORPHIC LEAF	HYDROMORPHIC LEAF
Cuticle (thick or thin)			
Walls of epidermal cells (thick or thin)			
Amount of sclerenchyma (fibrous tissues) (little or much)			
Amount of air spaces in mesophyll (little or much)			
Nature of the mesophyll (thick or thin)			
Position of stomates (sunken or at surface)			
Overall size of xeromorphic versus mesomorphic leaf (large or small)			

FLOWERS

INTRODUCTION

Mention the word flower and right away most people think of something colorful, beautiful, and perhaps fragrant. Bees, birds, and many other animals "agree" on this point. Why is there this attraction of animals to flowers? The explanation is simple, but the phenomenon is complex.

The flower is the reproductive organ of the plant. More important, it is the organ for sexual reproduction, that is, reproduction involving fertilization of an egg by a sperm. Although many plants are self-fertile (i.e., a sperm cell can fertilize an egg cell of the same plant), it is usually preferable for the uniting sperm and egg to be from different plants. The genetic variety achieved through this cross breeding between individuals offers many advantages over self-fertilization.

So for plants, rooted and thus immobilized, it is necessary for some agent to bring the sexes together. For plants with colorful flowers, the agents are bees, butterflies, certain other insects, certain birds, and bats.

These agents (pollinators) transfer pollen from one plant to another. The transferred pollen grains later germinate into a pollen tube. This tube grows through the tissues of the female part of the flower and delivers its sperm nuclei to the vicinity of the egg, and fertilization takes place.

The real trick, of course, in achieving this pollen transfer is first to get the attention of the transfer agents so they will come to the flower; second, to accommodate the agent physically so that its body or head and mouth parts can fit into the flower; and third, to reward the visiting animal so that it will continue to visit other flowers of the same type.

How is this done? You know the answer already! Color and scent attract the agent's atten-tion. Often the shape does too. Sugary nectar is the edible reward in most cases. In many cases pollen is also eaten or collected (by bees making beebread, for example). In some flowers a special kind of pollen is produced that serves only as food for insects.

The shape and design of the different flower types are variously modified and thus accommodate the visit of a certain type of agent (with its particular body or head and mouth parts) while at the same time excluding others. For instance, many insects like to sit down and crawl over the flower, and certain petals form "landing platforms." Flowers pollinated by hummingbirds have no such platforms, and the bird hovers while it sucks nectar from the flower. Absence of the petal platform in the bird-pollinated flower is effective in discouraging insect pollinators.

We assume that structural floral modification that encourages one type of agent to visit while excluding others is perhaps nature's way of assuring that the food (reward) reserves remain in abundance for one type of pollinator. Such a reliable food supply "guarantees" a reward most of the time or for a longer period, and thus leads to a greater constancy in the visits of that pollinator.

Floral structure is related also to the population levels (size) of the pollinator. A hectare of clover or alfalfa might have a million plants in bloom at any one time. Such a field could never "rely" on hummingbirds as pollinators because hummingbird populations are not large over any one area. Bees, though, typically occur in populations of very large numbers of individuals—more than adequate for pollinating large numbers of plants all in bloom at once.

By no means, however, does even a majority of flowering plants have flowers that show special structural adaptations for specific pollinators. Many have no specialized structural design and can be visited by a wide variety of pollinators. These flowers are called the "generalists." The generalists are probably more common, and the more highly specialized types are in the minority.

Wind-pollinated flowers cannot be overlooked. Hundreds of species of plants rely on the wind as the agent of pollen transfer, and for this they are marvelously modified (as you will see).

Finally, only a very few (aquatic) plants utilize water for pollination.

I. STUDY OF FLOWERS POLLINATED BY INSECTS, BIRDS, AND WIND

A. INSECT-POLLINATED FLOWERS AND BIRD-POLLINATED FLOWERS

ACTIVITY 1 Rather than robbing you of the fun of discovery, we ask that you **try to determine** which of the flowers provided for study are pollinated by insects and which by birds. You need to know, of course, something about the pollinators. So **study** Table 13.1. **Using** the information given, **examine** the flowers provided (or color slides) or the diagrams of flowers (Figures 13.1 to 13.5) and **list** the features you see in the flower that would attract and accommodate either a hummingbird or an insect or a particular kind of insect.

Based on the features you see in the flower, **identify** the animal pollinator of the flower. Bear in mind that these are only a few of the more common examples. The most important pollinators are bees, butterflies, birds, hawkmoths, flies, beetles, and bats. Not all can be considered at this time, so the table contains only five of these.

B. WIND-POLLINATED FLOWERS

ACTIVITY 2 If plants are available, **examine** flowers (or look at color slides or herbarium specimens) of wind-pollinated plants.

Answer these questions

	Circle one
Are the flowers large and conspicuous?	Yes No
Are the flowers colorful?	Yes No
Do the flowers have a scent?	Yes No

For the answer to this last question, you may have to rely on your past experience (if any) if no fresh flowers are available.

C. POLLEN OF ANIMAL-POLLINATED PLANTS VERSUS POLLEN OF WIND-POLLINATED PLANTS

ACTIVITY 3 **Examine** pollen taken from bird- or insect-pollinated flowers and pollen from wind-pollinated flowers. Prepare a standard wet mount of the pollen. Identify and examine them on separate slides first. Then combine them on one slide so you can compare the grains for differences in size and surface features.

Answer these questions.

Which grain type is larger?

Underline one
Animal-pollinated   Wind-pollinated

Which grain type has a smooth surface or a surface with little pattern to it?

Underline one
Animal-pollinated   Wind-pollinated

Which grain type has a rough or sculptured surface?

Underline one
Animal-pollinated   Wind-pollinated

Explain the differences you have seen in these pollen types.

_____
_____
_____
_____
_____

Table 13.1 SOME CHARACTERISTICS OF FIVE COMMON ANIMAL POLLINATORS

POLLINATOR	SENSE OF SMELL	COLORS RESPONDED TO	FEEDING BEHAVIOR	TYPE AND LENGTH OF MOUTH PART
Honey bees and bumblebees	Good	Yellow, blue, violet, orange	Alights on flowers and crawls over it.	Short tongue
Butterflies	Not strong	Variety	Alights on flower and crawls over it.	Long tongue
Hawkmoth	Good	White	Hovers	Very long tongue
Hummingbird	Poor	Variety, but especially red	Hovers	Very long bill and tongue

Floral Features That Accommodate the Pollinator

_____
_____
_____
_____

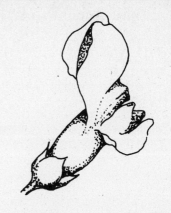

Fragrant yellow

Type of Pollinator: _____

Figure 13.1   Flower of snapdragon (*Antirrhinum majus*).

Floral Features That Accommodate the Pollinator

_____
_____
_____
_____

Red. No fragrance
Brightly colored
Numerous Stamens

Type of Pollinator:

_____

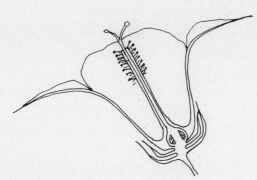

Figure 13.2   Flower of hibiscus (*Hibiscus* sp.)

Floral Features That Accommodate the Pollinator

Type of Pollinator: _____

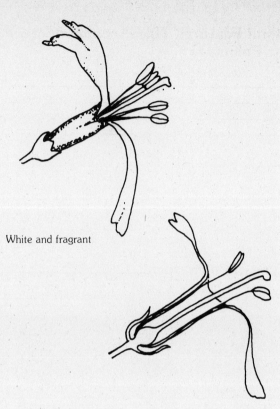

White and fragrant

Figure 13.3   Flower of honeysuckle (*Lonicera* sp.)

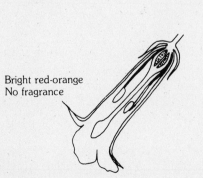

Bright red-orange
No fragrance

Figure 13.4   Flower of trumpet creeper/(*Campsis radicans*).

Floral Features That Accommodate the Pollinator

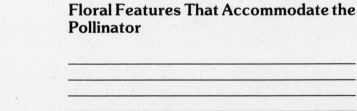

Type of Pollinator:

_____

Floral Features That Accommodate the Pollinator

_____
_____
_____
_____

Type of Pollinator: _____

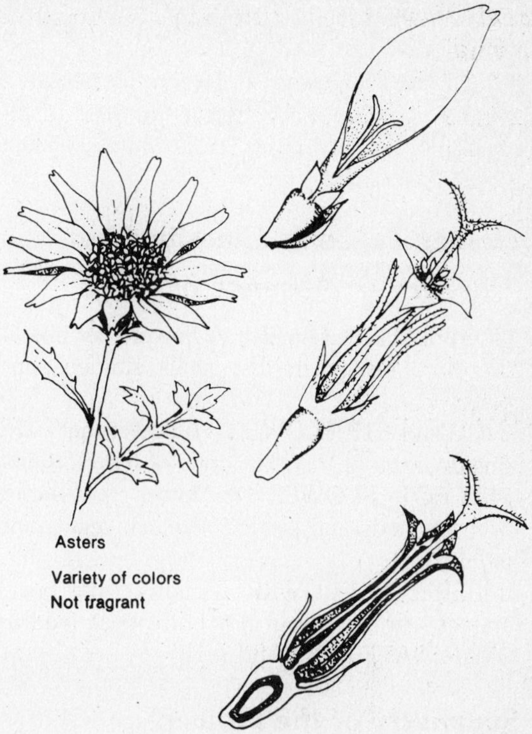

Asters

Variety of colors
Not fragrant

Figure 13.5   Aster (_Aster_ sp.).

II. THE BASIC PARTS OF FLOWERS

From the work and study of the previous activities in this exercise, it should be obvious to you that the flower is an extremely functional structure. Now let's look at the basic parts of the flower and learn some of the terminology.

SOME TERMS USED FOR SIMPLE FLOWERS

A. Flower Parts

PEDICEL   The stalk that bears the flower

RECEPTACLE   The somewhat enlarged tip of the pedicel from which the floral parts arise

SEPAL   One of several leaflike structures that make up the outermost (lowermost) circlet of floral parts. It is usually green, but may be brown and scalelike. It protects the flower when it is in bud stage, and either remains green as flower matures or becomes another color, often like that of the petals.

CALYX   Collective term for all the sepals

PETAL   One of several colored or white leaflike structures that occur in one or more circlets above the sepals

COROLLA   Collective term for all the petals

TEPAL   Term applied to petals and sepals that are identical. Tepals commonly occur in flowers of monocot plants.

PISTIL   The seed-bearing organ. Commonly a pear-shaped or bowling-pin-shaped body found in the center of the flower, it is composed of an ovary, style, and stigma.

OVARY   The fat, basal part of the pistil. It has one or more internal cavities (locules) containing egg-bearing bodies called ovules.

STYLE   Usually a long and slender necklike part extending from the ovary. It serves as a passageway for pollen tube growth into the ovary.

STIGMA   The tip of the style. It is sticky and serves as a simple device by which pollen, which is blown or carried to the flower by pollinators, stays stuck to the flower. It aids in germination of the pollen grain.

OVULE   Small round or oval body occurring singly or in groups inside the cavity or cavities of the ovary. It contains the egg cell and other related cells. After fertilization, it ripens into the seed.

LOCULE   The name applied to each cavity within the ovary

STAMEN   A pollen-producing structure, consisting of an anther and a filament

ANTHER   The pollen-bearing part of the stamen. It is supported by a slender filament and usually consists of four pollen sacs.

FILAMENT Slender stalk that supports the anther

NECTARY A gland that secretes nectar, a sugary liquid. It is located at the base of the pistils or stamens or in special hornlike cups called spurs.

B. Flower Types as Classified by:

1. Presence or Absence of Parts

COMPLETE FLOWER A flower that has all four floral organs; sepals, petals, stamens, and pistils

INCOMPLETE FLOWER A flower that lacks one or more of the four kinds of floral organs

PERFECT FLOWER A flower that bears both stamens and pistils (may lack sepals and petals)

IMPERFECT FLOWER A flower that bears stamens or pistils but not both; described as STAMINATE or PISTILLATE

2. Symmetry of the Flower

RADIAL SYMMETRY Symmetry based upon a wheel plan—that is, the flower is divisible on more than one axis into two equal halves that are mirror images of each other. Also known as REGULAR or ACTINOMORPHIC SYMMETRY (Figure 13.6A)

BILATERAL SYMMETRY A symmetry in which the flower is distinctly divisible into right and left sides; that is, it is divisible into mirror images on only one axis. Also known as IRREGULAR or ZYGOMORPHIC SYMMETRY (Figure 13.6B)

3. Position of the Ovary

HYPOGYNOUS FLOWER A flower in which the ovary is superior because the stamens, petals, and sepals arise from a level below the base of the ovary (Figure 13.7A)

PERIGYNOUS FLOWER A flower in which the ovary is superior but the bases of the stamens, petals, and sepals develop as a floral cup around the pistil (Figure 13.7B)

EPIGYNOUS FLOWER A flower in which the ovary is inferior because the stamens, petals, and sepals arise from a level that is above the base of the ovary (Figure 13.7C)

III. SIMPLE, SINGLE, COMPLETE, AND PERFECT FLOWER WITH RADIAL SYMMETRY

ACTIVITY 4 **Examine** some simple, single, regular flowers that are commercially available or in season such as petunia (*Petunia hybrida*), tulip (*Tulipa gasneriana*), gladiola (*Gladiolus* sp.), lily (*Lilium* sp.), primrose (*Primula* sp.), poppy (*Papaver* sp. or *Eschscholtzia* sp.), *Browallia* sp, *Amaryllis* sp. *Hoya* sp., *Clivia* sp., *Azalea* sp., or *Magnolia* sp.

Using the terms given in Section II.A, **locate** as many of the parts described as possible. **Note** also which parts are absent. **Count** the number of petals and sepals. If sepals and petals each total three or some multiple of three, then your flower is from a monocot plant. If the sepals and petals each total

Divisible along several axes into two mirror images

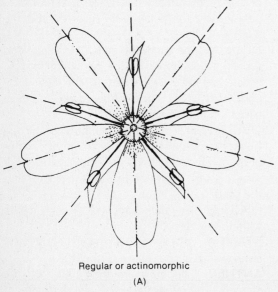

Regular or actinomorphic

(A)

Divisible along only one axis into two mirror images

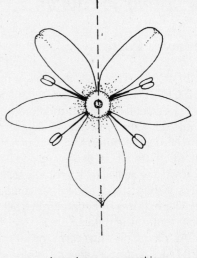

Irregular or zygomorphic

(B)

Figure 13.6 Symmetry in flowers.

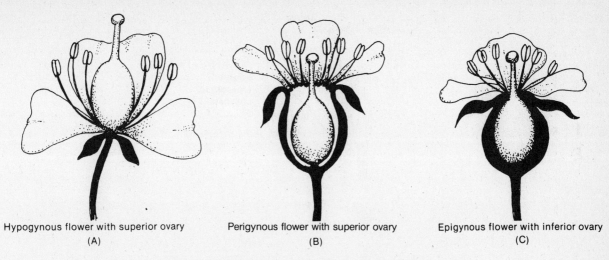

Hypogynous flower with superior ovary
(A)

Perigynous flower with superior ovary
(B)

Epigynous flower with inferior ovary
(C)

Figure 13.7   Flower types based on the position of the ovary.

four or five or some multiple of these, then your flower is from a dicot plant. **Record** your findings in Table 13.2.

Determine the symmetry of each flower and the type of flower, using the information given in Section II.B.1 and 2. **Record** your findings in Table 13.2. If possible, try to **determine** the position of the ovary in relation to the other parts of the flower. Based on the information given in Section II.B.3, **record** the flower type in Table 13.2. **Label** Figure 13.8, using as many of the defined terms as possible.

IV. THE FLOWERS OF COMPOSITES

Daisies, sunflowers, asters, and dandelions are members of the composite family of plants. What you see is not really a single flower with numerous petals, but many tiny separate flowers clustered together on a common receptacle. This composite of many miniature flowers is called a **HEAD**. It superficially resembles a single flower.

The miniature flowers of composite plants are of two kinds, (1) the **RAY FLOWER** (Figure 13.9), in which the corolla looks like one petal, tubular at its base but mostly long and strap-shaped, and (2) the **DISK FLOWER** (Figure 13.10) in which the corolla forms a tube of fused petals. In some species, such as the sunflower (*Helianthus* sp.), daisy (*Chrysanthemum leucanthemum*), and aster (*Aster* sp.), the head is composed of both ray and disk flowers. In these plants the ray flowers occur at the periphery (perimeter) of the head. In daisies (Figure 13.11), for instance, the ray flowers form the white "petals." Disk flowers occupy the central portion of the head. In other species, such as dandelion (*Taraxacum officinale*) (Figure 13.12), the head consists only of ray flowers.

ACTIVITY 5 **Examine** the head of a dandelion flower. Using tweezers, **pull off** some of the tiny flowers. This is best done by thrusting the tweezers deep down around the base of the small flowers and pulling out several at once. **Use** Figure 13.9 as a guide and **study** one flower. **Use** your hand-lens. Note the strap-shaped corolla. The single strap is composed of several fused petals. It is tubular at its base. **Find** the pistil, stamens, and ovary. These ray flowers are (a) perfect or (b) imperfect? Circle (a) or (b).

Examine the head of a sunflower, daisy, aster, *Cineraria* sp., or other plant that has both ray and disk flowers in the head. Use tweezers to extract some of the small flowers from the head, as you did with the dandelion. **Use** Figure 13.10 as a guide. **Examine** a disk flower. It has a slender tubular corolla of fused petals surrounding the pistil and stamens. **Slit** open the corolla with a razor blade and expose the pistil and stamens. **Note** that the anthers enclose the style and that the stigma is bifurcate (two-parted). The disk flower is (a) perfect or (b) imperfect? Circle (a) or (b). **Examine** a ray flower from this head. Does it generally resemble the ray flowers of the dandelion?   *Yes   No.*   Are anthers present?   *Yes   No.*   This ray flower is (a) pistillate, (b) staminate, or (c) perfect? Circle (a), (b) or (c).

There are several advantages to the way the composite type flower is built: (1) the head is usually more conspicuous to pollinators than is a single modified body or head parts to get to the nectar in flower; (2) it is probably less dependent on one particular kind of pollinator; (3) pollen is often exposed to all comers; (4) insects need no highly the flowers; and (5) the construction of the head allows several flowers to be pollinated by a single insect during one visit. This head type, a composite of several miniature flowers, is unique to the

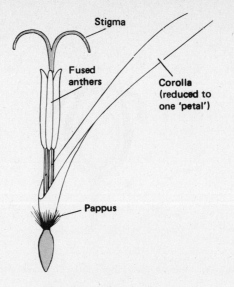

Figure 13.9   One ray flower.

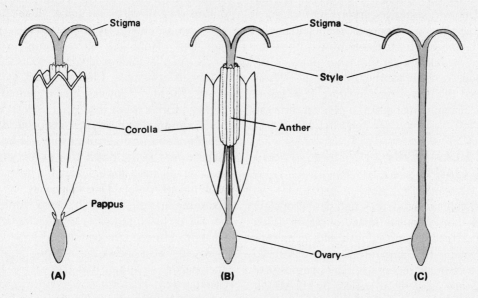

Figure 13.10   Disk flower in three stages of dissection.

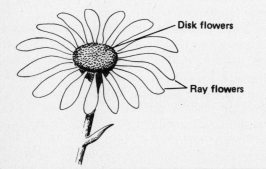

Figure 13.11   Daisy head composed of both ray and disk flowers.

Figure 13.12   Dandelion head composed of ray flowers only.

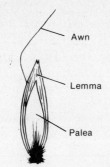

Figure 13.13   Closed floret.

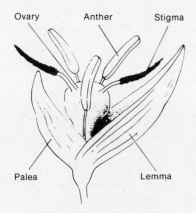

Figure 13.14   Open floret.

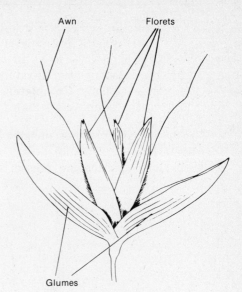

Figure 13.15   One spikelet.

composite family (hence its name). It, along with other characteristics of the group, makes the family the most highly evolved dicots among flowering plants.

V. THE GRASS FLOWER

Species in the grass family (Poaceae) have flowers that are highly modified in comparison with the ones you have examined so far. The individual grass flower, the **FLORET** (Figures 13.13 and 13.14), is enveloped by a pair of modified leaves, the **BRACTS**. The outer and larger bract is termed the **LEMMA,** and the inner and smaller one is termed the **PALEA**. The palea encloses the stamens and pistil.

A group of florets is termed a **SPIKELET** (Figure 13.15). At the base of each spikelet is a pair of modified leaves called **GLUMES**. The number of florets in a spikelet varies with the species. A cluster of spikelets makes up an **INFLORESCENCE** (Figure 13.16). The arrangement of spikelets in the inflorescence varies with the species.

Figure 13.16   Inflorescence of oats.

ACTIVITY 6 **Examine** a grass inflorescence, preferably oats (*Avena sativa*). With the aid of the drawings provided (Figures 13.13 to 13.16) and the use of a hand-lens, dissecting needle, and forceps, **identify** a spikelet and a floret. **Completely open** the floret and determine the answers to the following:

1. Are petals and sepals present in the floret? *Yes   No.*
2. How many stamens are present in the floret? _____
3. How many florets are there per spikelet? _____.
4. Describe the stigma part of the pistil. _____.
5. Is the flower (floret) colorful and aromatic? *Yes   No.*
6. On the basis of this flower's morphology, identify the pollinating agent for grass. _____

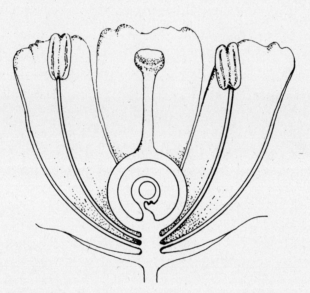

Figure 13.8 Generalized representation of a simple, single, regular flower.

EXERCISE 13   **Student Name** _____

QUESTIONS

1. Is a perfect flower also a complete flower?   Yes   No.   Give a reason for your

   answer. _____

2. If you found a flower with six petals and three sepals, would you identify it as a monocot

   or a dicot? _____

3. Commonly, flowering crabapple trees have flowers with five petals. Are these trees mono-

   cots or dicots? _____

4. In general, which type of flower would you expect to produce the greater amount of pollen:
   wind-pollinated flowers or insect-pollinated flowers? Circle one and explain your choice.

   _____

   _____

5. The flowers of many plants are bisexual and self-fertile. What then is the advantage of

   cross fertilization between two individual plants? _____

   _____

6. Dependence on birds, bats, bees, and other insects as pollinators is common in flowering

   plants. What features of the flower serve to: (a) attract the pollinator? _____

   _____ (b) physically accommodate the

   pollinator? _____

   (c) "reward" the pollinator? _____

   _____

7. Studies show that bees don't respond to pure red colors (and this is probably true for
   certain other insects). Birds, however, including the hummingbird, do respond to red
   color. Asia, Africa, and the Americas have many red flowering plants, whereas Europe
   has almost none. On the basis of this information: (a) which would you expect to be the

   chief pollinating agents in Europe, birds or bees (and other insects)? _____

   _____ (b) would you expect to find many hummingbirds in Europe?

   _____

8. What is meant by the "generalist" type of flower, as it relates to pollination? _____

   _____

9. In most cases, if flower structure encourages one particular type of pollinator, what is

   the effect on other possible pollinators? _____

10. If the major function of the flower is the successful achievement of pollination, what

    features of the composite flower can you cite that support this function? _____

    _____

    _____

Table 13.2 VARIATION IN SIMPLE FLOWERS

FEATURE	PLANT NAMES									
parts present										
parts absent										
type of flower based on parts present										
symmetry										
no. of petals										
no. of sepals										
plant type based on number of petals										
ovary position										
type of plant based on ovary position										

FRUITS

INTRODUCTION

For almost everyone, fruit as a food is something sweet such as apples, pears, and oranges. Vegetables aren't sweet. A grocery list of them might include string beans, corn, green pepper, tomato, squash, cucumber, and eggplant.

But the truth is, these particular vegetables **are** fruits, just as are apples, pears, and other sweet fruits. They are all fruits because they have developed from one or more flowers. Each part of a fruit has developed from some part of the flower, usually the ovary or cluster of ovaries and in some cases, from other floral parts. With some exceptions, fruits contain seeds. A seed develops from an ovule of the flower.

Vegetables that are not derived from the flower but from plant parts such as leaves, stems, petioles, or roots include such foods as lettuce, cabbage, asparagus, celery, radish, and carrot. The term vegetable for them is all right because they are vegetative parts.

As you learned in the previous exercise, the flower serves to achieve sexual reproduction. Later, it is "remodeled" into a new organ, the fruit (Figure 14.1), which serves to protect the product of reproduction, the seed (the embryonic plant). In some ways too, the fruit is modified in such a way that it is dispersed by wind and animals (or, in a few cases, by water), thus bringing about dispersal of the plants. (In many plants, it is the seed that is modified and dispersed by these agents.)

I. FRUIT DEVELOPMENT

When the fruit develops, the matured ovary wall, the **PERICARP**, may thicken and become differentiated into three layers. These layers may or may not

be easy to distinguish, depending on the species. These three layers are the **EXOCARP** (outer epidermis), the **MESOCARP** (middle layer), and the **ENDOCARP** (inner layer). The peach is a good example of a fruit with three distinct layers. The peach skin is the exocarp, the fleshy part is the mesocarp, and the stony pit is the endocarp. The seed is inside the pit (Figure 14.1).

ACTIVITY 1 **Examine** different preserved or fresh specimens illustrating fruit development from the flower. These show the gradual stages of transition of the floral parts into the fruit. Some suggested examples are chilli pepper (*Capsicum frutescens* v. *longum*), horse chestnut (*Aesculus hippocastanum*), apple (*Malus* sp.), strawberry (*Fragaria* sp.), bean (*Phaseolus vulgaris*); maple (*Acer* sp.), parsnip (*Pastinaca sativa*), or others that have been assembled.

II. RELATIONSHIP BETWEEN THE PISTIL STRUCTURE OF THE FLOWER AND THE STRUCTURE OF CERTAIN SIMPLE FRUITS

In the previous exercise, you learned that the pistil of the flower typically has a round or oval part, the ovary, from which there extends a tubular part, the style, whose tip is called the stigma. This general shape is derived from the shape of the basic building block of the pistil—namely, the carpel, a highly modified leaf.

Depending on the species, a pistil may be composed of one carpel (Figure 14.2A) or of two carpels (Figure 14.2B–E). In many cases, there are three, four, five or more carpels making up one pistil. If two or more carpels compose a pistil, they are

127

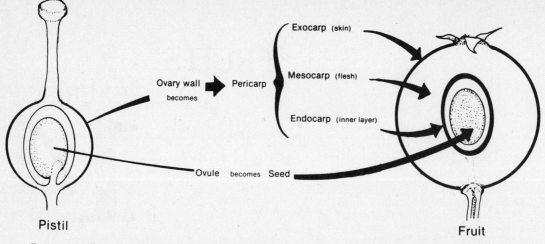

Figure 14.1   Example of the development of parts of the fruit from the pistil of the flower, in this case using the peach (*Prunus persica*).

usually fused together. This fusion can occur in a variety of ways, some of which are shown in Figure 14.2B–E. Any pistil composed of two or more carpels is described as a compound pistil.

The interior chamber(s) of the ovary of a pistil is called the locule (see Figure 14.3C). Seeds (matured ovules) are found inside the locules. Often the locule may also have one or more internal partitions (Figure 14.4C).

The number of carpels making up the pistil and the degree of fusion between them is often most apparent in the mature fruit, especially in plants with dry, dehiscent fruits such as follicles, capsules, and so forth. Figures 14.3 and 14.4 illustrate how the pistil construction determines the character of such fruits.

III. THE STRUCTURE AND IDENTIFICATION OF SOME COMMON FRUITS

CLASSIFICATION OF FRUITS

"Package design" is really an appropriate phrase for the description of fruit types. This design is related first and foremost to the protection of the seed(s) inside it.

Second, the design may be related to the release of the seed from the protective fruit. Third, in many plants the design plays an important role in seed dispersal by various agents, though in many other plants the seed itself is modified for dispersal.

One of the chief reasons why the flowering plants are the dominant, most diverse group of land plants is their capacity for wide dispersal. The ability of a plant to expand its habitat plays a major role in the origin of species.

Botanists have looked at the great variety of design in fruits and have established a classification system, which is given in brief outline in Table 14.1.

ACTIVITY 2   **Dissect** and **examine** specimens of fruits that have been made available for you in the laboratory.

Using the information from Table 14.1 as an aid (where necessary), key out as many of the fruits as possible. **Use** the *Key to Fruit Types* (pp. 129–131). **Record** your findings in Table 14.2.

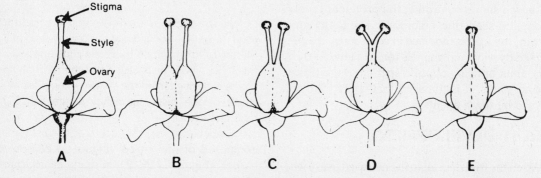

Figure 14.2   Pistils of flowers showing examples of the number of carpels and the degrees of fusion of carpels. *A,* Pistil consisting of one carpel. *B–E,* Pistils composed of two carpels and showing different degrees of fusion.

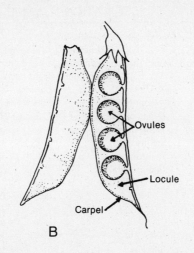
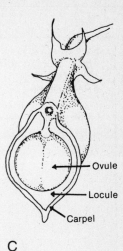

Ovules

Locule

Carpel

Ovule

Locule

Carpel

Figure 14.3 Legume fruit type formed from the unicarpellate pistil of the pea flower (*Lathyrus* sp.) *A*, Pea flower. *B*, Legume-type fruit formed from the single carpel composing the pistil of the flower. *C*, Cross section of the legume.

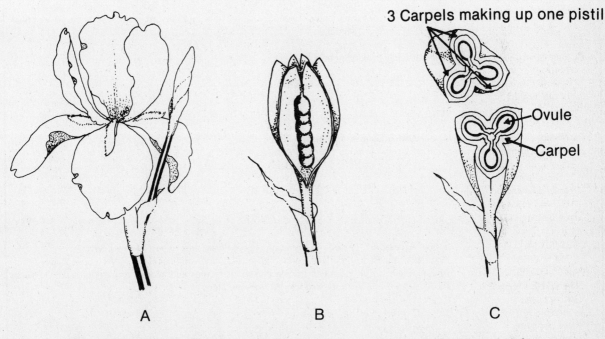

3 Carpels making up one pistil

Ovule

Carpel

Figure 14.4 Capsule-type fruit formed from the tricarpellate pistil of the iris flower (*Iris* sp.). *A*, Iris flower. *B*, Mature capsule-type fruit split open and showing the tricarpellate construction of the flower. *C*, Cross section of the capsule.

IV. A KEY TO FRUIT TYPES

Directions for Using Key: This key consists of pairs or couplets of choices, 1a–1b, 2a–2b, and so on. Compare your plant specimen with the descriptions given in couplet 1a–1b. Choose whichever statement (1a or 1b) is more applicable to your specimen and go on to the next couplet as indicated by the number at the extreme right. Continue through the key, each time choosing between a pair of choices until you terminate at the name of a plant.

1a.	Fruit produced from several flowers crowded on the same inflorescence . **Multiple fruit** (*examples:* pineapple, mulberry, Osage orange)	
1b.	Fruit produced from a single flower 2	
2a.	Fruit derived from more than one pistil . . . . **Aggregate fruit** (*examples:* buttercup, tulip tree)	
2b.	Fruit derived from a single pistil, **simple fruits** 3	
3a.	Fruit fleshy, usually indehiscent 4	
3b.	Fruit dry at maturity, dehiscent or indehiscent 7	

Table 14.1 CLASSIFICATION OF FRUITS

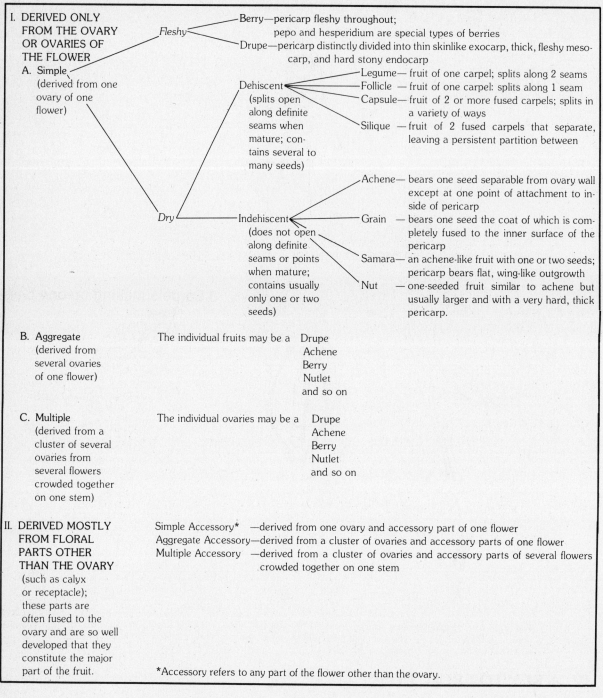

B. Aggregate
(derived from
several ovaries
of one flower)

The individual fruits may be a   Drupe
Achene
Berry
Nutlet
and so on

C. Multiple
(derived from a
cluster of several
ovaries from
several flowers
crowded together
on one stem)

The individual ovaries may be a   Drupe
Achene
Berry
Nutlet
and so on

II. DERIVED MOSTLY
FROM FLORAL
PARTS OTHER
THAN THE OVARY
(such as calyx
or receptacle);
these parts are
often fused to the
ovary and are so well
developed that they
constitute the major
part of the fruit.

Simple Accessory*   —derived from one ovary and accessory part of one flower
Aggregate Accessory—derived from a cluster of ovaries and accessory parts of one flower
Multiple Accessory   —derived from a cluster of ovaries and accessory parts of several flowers
crowded together on one stem

*Accessory refers to any part of the flower other than the ovary.

4a.   Flesh of fruit consisting of accessory tissue that surrounds the papery carpels, giving the
appearance of an inferior ovary . . . . **Pome**
(*examples*: apple, pear, hawthorn)
4b.   Flesh of fruit derived from ovary wall 5
5a.   Pericarp with a fleshy mesocarp and a stony endocarp . . . . **Drupe**
(*examples*: cherry, plum, coconut)
5b.   Pericarp without a stony endocarp, more or less fleshy throughout . . . . **Berry**
(*examples*: tomato, banana)
See also 6 for modifications of the berry type:
6a.   Septa (partitions) evident in cross section; the thick leathery exocarp separable from the
inner layers of the pericarp . . . . **Hesperidium**
(*examples*: orange, lemon)

6b.   Septa not evident; exocarp leathery to hard or woody, but not separable from the inner layers of the pericarp . . . . **Pepo**
(*examples:* cucumber, gourd)

7a.   Fruit indehiscent 8
7b.   Fruit dehiscent 13

8a.   With one or more wings . . . . **Samara**
(*examples:* elm, ash, maple)

8b.   Without wings 9

9a.   Pericarp clearly differentiated into exocarp, mesocarp, and stony endocarp . . . .
. . . . **"Drupaceous" nuts**
(*example:* hickory)

9b.   Pericarp not clearly differentiated into three layers, but fruit sometimes partly or completely enclosed by a papery or leathery husk or cup as in the acorn 10

10a.   Fruit two-carpellate, the indehiscent carpels separating at maturity but remaining attached to a common axis . . . . **Schizocarp**
(*examples:* carrot, dill)

10b.   Fruit one-carpellate, or of two or more united carpels that do not separate at maturity
. 11

11a.   Fruits relatively large; pericarp thick and stony, seed at maturity separated from ovary wall
. . . . **Nut**
(*examples:* hazel, acorn, chestnut)

11b.   Fruits relatively small, often very minute; pericarp thin, not stony, seed not completely separate from ovary wall 12

12a.   Pericarp completely fused to seed coat . . . . **Grain or Caryopsis**
(*example:* all members of the grass family)

12b.   Pericarp and seed coat united at only one point . . . . **Achene**
(*examples:* dandelion, beggar-ticks, sedges, sunflower)

13a.   Fruit derived from a simple pistil; one carpel 14
13b.   Fruit derived from a compound pistil; carpels two or more, united 15

14a.   Dehiscent along two sutures (seams) . . . . **Legume**
(*examples:* pea, sweet pea; some legumes are not readily dehiscent such as peanut and honey locust)

14b.   Dehiscent along one suture . . . . **Follicle**
(*examples:* milkweed, larkspur)

15a.   Fruit two-loculed (with two seed cavities), the two halves splitting away from a persistent (remaining attached to the stalk) central septum (partition) . . . . **Silique**
(*examples:* shepherd's purse, mustard)

15b.   Fruit one- or several-loculed, dehiscing in various ways but without persistent central septum if the fruit is two-loculed . . . . **Capsule**
(*examples:* poppy, cotton, tulip, lily)

V. SEED AND FRUIT DISPERSAL

ACTIVITY 3   **Observe** demonstration specimens of seeds and fruits that exhibit specialized features that aid in dispersal. Fruits, and even many seeds, exhibit an immense variety of mechanisms for dispersal by wind, water, birds, mammals, and even ants. Examples suggested for study are as follows:

1.   **WINGED FRUITS** Elm (*Ulmus* sp.), maple (*Acer* sp.), ash (*Fraxinus* sp.), dock (*Rumex* sp.), birch (*Betula* sp.), tulip tree (*Liriodendron tulipifera*), and so on.

2.   **WINGED SEEDS** Trumpet vine (*Campsis radicans*) and catalpa (*Catalpa* sp.).

3.   **STICKY FRUITS** Mistletoe (*Phoradendron* sp.). Bits of fruit with seeds stick to the bird's beak. Seeds are transported quite some way by the bird and lodge in bark crevices when the bird cleans its beak.

4.   **PLUMED FRUITS** Dandelion (*Taraxacum officinale*) and goatsbeard (*Tragopogon* sp.).

5.   **PLUMED SEEDS** Milkweed (*Asclepias* sp.) and cottonwood (*Populus* sp.).

6.   **SPINY FRUITS** Cocklebur (*Xanthium* sp.), burdock (*Arctium* sp.), beggar tick (*Bidens* sp.), and sandbur (*Cenchrus* sp.).

EXERCISE 14   **Student Name** _____

QUESTIONS

A. Check (✔) each statement as true or false.   **True | False**

1. A peanut is a nut.
2. All legumes readily dehisce.
3. A fruit that is a true nut has a thick, stony pericarp.
4. The inner core of the apple is the ovary wall.
5. Animal-dispersed fruits are necessarily colorful and attractive.
6. Many seeds have structural adaptations for dispersal.
7. A grain and an achene are approximately the same in their makeup.
8. The fleshy part of the apple is the mesocarp.
9. A cherry is a berry.
10. All seeds are dispersed by fruits.

B. Complete the following by choosing the correct word or phrase. Check (✔) either column a or b.   **a | b**

1. Peaches and almonds are both drupes. The hard shell enclosing the edible almond seed is (a) an endocarp; (b) a stony pericarp.
2. If each of the tiny black, gritty things on the surface of a strawberry is an achene (see the key), then the fruit type of the strawberry is (a) an aggregate; (b) an accessory aggregate.
3. The seed is the matured (a) ovary; (b) ovule.
4. The pit of a drupe is (a) the hard seed coat; (b) the endocarp.
5. The coconut that is usually sold in the produce department in a grocery store is (a) the whole fruit; (b) just the endocarp and seed.
6. Choose the term that more accurately completes the statement: A simple fruit is derived from (a) one ovary; (b) one flower.
7. A fruit that is formed from other floral parts besides the ovary is classified as (a) an accessory fruit; (b) a multiple fruit.
8. The apple is (a) an accessory fruit; (b) a berry.
9. Spiny fruits show particular adaptation for dispersal by (a) mammals; (b) birds.

Table 14.2 CHARACTERISTICS OF SOME COMMON FRUITS

NAME OF PLANT	NAME OF FRUIT	DRY OR FLESHY	DEHISCENT OR INDEHISCENT	STRUCTURE(S) OTHER THAN OVARY INVOLVED IN FRUIT FORMATION	FRUIT TYPE

WATER RELATIONS: OSMOSIS AND DIFFUSION

EXERCISE

15

INTRODUCTION

Water relations experiments are interesting, easy to do, and of great practical value. But they are time-consuming. This exercise may be completed in one or two three-hour class periods.

If done in one class period,

1. OSMOSIS, Sections II–IV, should be started early in the period.
2. DIFFUSION, Sections V–VIII, can be done while the osmosis experiments are running.

If done in two class periods,

1. DIFFUSION, Sections V–VIII, could be done in the first period (with reference to the terms given in the Introduction).
2. OSMOSIS, Sections II–IV, could be done in the second period.

In one sense, plants need water more than animals. Animals grow mainly by the synthesis of protein building units. Plants grow by cell expansion from water pressure inside the cell. This occurs when the cell is young. Continued growth depends on the formation of additional young cells through the process of cell division. For this, water is also needed. Foods, hormones, nutrients, fertilizers, herbicides—everything that moves through the plant moves in water. Plants also lose large amounts of water vapor through their leaves in a process known as transpiration.

Water moves into, through, and out of the plant in two ways, by diffusion and by osmosis (which is a special type of diffusion). Today's exercise consists of a few experiments and demonstrations that illustrate how plants relate to water, in both their structure and their function.

A. SOME TERMS AND DEFINITIONS

1. DIFFUSION   The net movement of a substance from a region where it is more highly concentrated to a region where it has lower or zero concentration. This movement is influenced by temperature and a number of other conditions.
2. OSMOSIS   The diffusion of a solvent (usually water) through a differentially permeable membrane.
3. DIFFERENTIALLY PERMEABLE MEMBRANE (SEMIPERMEABLE MEMBRANE)   A membrane that allows some molecules to pass through it more readily than other molecules.
4. OSMOMETER   An instrument or apparatus that is used to demonstrate and measure osmosis.
5. OSMOTIC PRESSURE   The pressure that theoretically could develop in a cell if it were under the perfect conditions of an unlimited supply of pure water available to flow into the cell.
6. TURGOR PRESSURE   Pressure that develops in a cell as a result of the osmotic uptake of water. This pressure is exerted against the cell wall and is equal to, but in the opposite direction of, the wall pressure exerted against the cell contents.
7. TURGOR   The normal inflated condition of a living cell caused by internal water pressure. Tissues and organs develop rigidity or turgor because of the turgor pressure within the cells pushing outward against the cell wall.

8. FLACCID   The limp, flabby condition of cells, tissues, and organs due to the loss of water.
9. PLASMOLYSIS   The shrinkage of the protoplast of a living plant cell away from the cell wall as a result of diffusion of water out of the cell. Because the water diffuses through the cell's membrane, the passage occurs by osmosis.
10. TRANSPIRATION   The diffusion of water from plant tissues into the atmosphere in the form of water vapor.
11. POTOMETER   An apparatus that is used to demonstrate and measure transpiration.
12. IMBIBITION   The diffusion of water into colloidal material, followed by swelling of the material.
13. GUTTATION   The diffusion of liquid water out of plant leaves.

I. OSMOSIS

A. OSMOSIS IN A NONLIVING SYSTEM

Osmosis is a special kind of diffusion, the movement of a solvent (usually water) through a differentially permeable membrane.

The instructor will have ready for the class some "sausage" osmometers, or students can make them. These are tubes made of cellulose acetate film. This film is readily permeable to water but less so to other substances. In this way it is analogous to a living cell membrane. What term is used to describe this type of membrane? _____

_____

All the sausages contain a 10 percent sugar solution plus a dye for color. Those labeled as set (A) have for some time before class been immersed in a 20 percent sugar solution. Those labeled as set (B) have been immersed for the same length of time in pure water.

ACTIVITY 1   **Work** in teams of two or more and **proceed** as follows.

1. Each student or group of students will work with one sausage from set (A) and one from set (B).
2. Remove sausage (A) and sausage (B) from the pan or beaker in which they have been soaking.
3. Wash off with water the one that was in the sugar solution.
4. Note whether each sausage is flaccid or turgid. Record this in Table 15.1.
5. Dry and weigh each sausage.
6. Compile the weights for the class on the chalkboard and record in Table 15.1.
7. Immerse sausage (A) in a beaker or pan containing pure (tap) water.
8. Immerse sausage (B) in a beaker or pan containing a 20 percent sugar solution.
9. After one hour, remove the sausages, note their condition, dry, and weigh them.
10. Compile the class data on the chalkboard.
11. Record these data in Table 15.1.
12. Calculate the average net loss or gain in weight and record in Table 15.1.

OBSERVATIONS

1. Which set of sausages initially contained the greater concentration of water? Circle one: (A) (B).
2. After immersion of set (B) sausages in the sugar solution, (a) water moved out of the sausage; (b) sugar moved out of the sausage. Circle (a) or (b).
3. When taken from the sugar solution, set (A)

Table 15.1   CHANGES IN WEIGHT AND TURGOR PRESSURE IN SAUSAGE OSMOMETERS

	SET (A)		SET (B)	
	When Taken From Sugar Solution	After One-Hour Immersion in Water	When Taken From Water	After One-Hour Immersion in Sugar Solution
Condition of sausages (flaccid or turgid)				
Weight (g)				
Average weight change after one hour (g)		_____ weight (g) _____ gain or loss		_____ weight (g) _____ gain or loss

Table 15.2   WEIGHT, VOLUME, AND TISSUE CONDITION CHANGES IN PLANT TISSUE DUE TO OSMOSIS

	TIME	LENGTH (MM) Plug 1	Plug 2	DIAMETER (MM) Plug 1	Plug 2	WEIGHT (G) Plug 1	Plug 2	TISSUE CONDITION (FLACCID OR TURGID)
Osmotic changes in tap water	Start of experiment							
	End of experiment							
	Change: loss or gain							
Osmotic changes in a strong salt solution	Start of experiment							
	End of experiment							
	Change: loss or gain							
		Length (mm)			Diameter (mm)			Weight (g)
Comparison of average osmotic changes in tap water and strong salt solution	In tap water							
	In strong salt solution							

sausages had a (a) low; (b) high turgor pressure. Circle (a) or (b).

4. Weight changes in both sets (A) and (B) are due primarily to gains and losses of (a) sugar; (b) water. Circle (a) or (b).

B. OSMOSIS IN A LIVING SYSTEM

ACTIVITY 2   **Work** in teams of four students and **proceed** as follows.

1. Obtain a cork borer about 7–10 mm in diameter (or use a "french-fry" cutter) and cut four plugs of tissue about 3–5 cm long from potatoes, turnips, or rutabagas.
2. Make certain that all four plugs are from the same potato, turnip, or rutabaga, and are the same in length and diameter.
3. Measure the length and diameter of each plug to the nearest millimeter.
4. Weigh each plug to the nearest 0.1 g.
5. Squeeze and bend (gently) each plug to note if it is turgid or flaccid.
6. Record these data in Table 15.2.
7. Place two plugs in tap water and the other two in a strong salt solution.
8. After 60–90 minutes remove the plugs from the liquids and note their degree of turgor.
9. Measure and weigh each plug.
10. Record all data for your group in Table 15.2.

Complete Table 15.2 with the comparison of the average changes in tap water versus the changes in the salt solution.

OBSERVATIONS

1. Are the results of this experiment comparable to the results obtained with the sausage osmometers?   *Yes   No.*
2. If you wish to retain more of the water-soluble vitamins when you boil potatoes, should you cook them (a) in salted water; (b) in unsalted water? Circle (a) or (b). Why is this so? _____

II. DIFFERENTIALLY PERMEABLE MEMBRANES DEMONSTRATING OSMOTIC PRESSURE

Now we will examine the role of the cell membrane during osmosis. We will also see how the cell's starch content affects the osmotic pressure differently from that of sugar.

Although water may freely diffuse across cell membranes, substances dissolved or suspended in the water may or may not do so. The membrane is selective: some substances pass through, others do not. The film, cellulose acetate, can be used to simulate the living cell membrane.

ACTIVITY 3   **Observe** the two sausage osmometers that were set up sometime before or during the class period. Diagrams of the setups are shown in Figure 15.1. **Draw** in and **label** the approximate fluid levels in each tube as you observe them.

The sausages are made of differentially permeable cellulose acetate. Sausage (A) contains a glucose solution; sausage (B) contains an equal volume of starch solution. Each was weighed at the

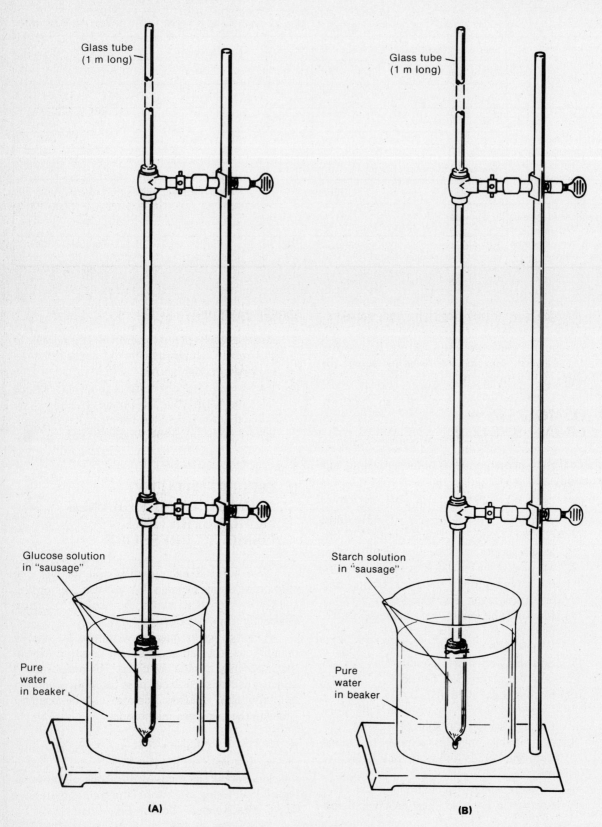

Glass tube
(1 m long)

Glass tube
(1 m long)

Glucose solution
in "sausage"

Starch solution
in "sausage"

Pure
water
in beaker

Pure
water
in beaker

(A)

(B)

Figure 15.1   Osmometer demonstration.

start of the experiment. A glass tube about 1 meter long is inserted in each sausage and securely tied to it. The beakers were filled with pure water at the start of the experiment. If, at the start of the experiment, any liquid moved up to the glass tube, the height was noted and recorded.

Observe the instructor perform certain tests. We will now see what substances have or have not passed through these membranes.

OBSERVATIONS

1. Has starch passed through the membrane into the beaker of water in (B)?

    Test for starch: A sample (about 15 ml) of liquid is removed from the beaker and put in a test tube. A few drops of iodine–potassium iodide (IKI) are added. A blue color indicates the presence of starch (a blue-colored starch–iodine complex is formed).

    Result: (a) Starch present; (b) no starch. Circle (a) or (b).

    Conclusion: The membrane (a) is; (b) is not permeable to starch. Circle (a) or (b).

2. Has glucose passed through the membrane into the beaker of water in (A)?

    Tests for glucose:

    a. A strip of glucose indicator paper (trade name Tes-tape) is inserted into the liquid in the beaker. The new color in the paper compared with the sample color chart indicates the presence of glucose.

    b. An alternative test: add a few drops of Benedict's solution to a test tube containing some of the liquid from the other beaker. The tube is heated in a beaker of boiling water. If sugar is present, a yellowish to reddish brown precipitate will form.

    Result: (a) Glucose present; (b) no glucose. Circle (a) or (b).

    Conclusion: The membrane (a) is; (b) is not permeable to glucose. Circle (a) or (b).

3. Which builds up osmotic pressures, starch or sugar? _____

    Examine the osmometers one or more times during the class period and **make** notations of any changes on Figure 15.1. **Fill in** Table 15.3 with the information as indicated.

4. Has water moved into the sugar solution in (A)? _____; into the starch solution in (B)? _____

5. Both membranes (that of (A) and (B)) are equally permeable to water. Do you agree? *Yes   No.*

6. The differentially permeable membranes are equally permeable to starch and sugar molecules. Do you agree? *Yes   No.* What reason can you give? _____
_____.

7. Is there much tendency for the water to move into the starch solution osmometer *Yes No.*

8. Can the sausage in (B), therefore, ever build up much osmotic pressure? *Yes No;* become turgid? *Yes   No.*

9. In order for the plant to keep osmotic pressures low and within a manageable range for long-term storage situations, which, therefore, is the preferable storage form of carbohydrate, starch or sugar? _____

III. PLASMOLYSIS AND WILTING

A. CELL CHANGES IN PLASMOLYSIS

ACTIVITY 4 **Make** a wet mount of elodea (*Anacharis canadensis*) or species of *Rhoeo* or *Zebrina*. If using *Rhoeo* or *Zebrina*, strip off the epidermis from over the midrib or from the lower surface of the leaf

Table 15.3   COMPARISON OF SUGAR AND STARCH IN REGARD TO OSMOTIC PRESSURE BUILDUP

	AT START OF EXPERIMENT	AFTER _____ MINUTES	AFTER _____ MINUTES
Height of liquid in tube (mm) in (A) (sugar solution)	mm	mm	mm
in (B) (starch solution)	mm	mm	mm
Condition of sausage (flaccid or turgid) in (A) (sugar solution)			
in (B) (starch solution)			
Weight of sausage (g) in (A) (sugar solution)	g	g	g
in (B) (starch solution)	g	g	g

and mount it. If using elodea, mount an entire leaf.

Examine the cells with the microscope. **Note** that in *Rhoeo* and *Zebrina* colored cell sap appears to fill the cells. A cell in this condition with the CYTO-PLASM and CELL MEMBRANE pressed against the WALLS is described as a TURGID CELL. One large VACUOLE or several smaller vacuoles (not visible to you) occupy the cell and most of the cell sap is contained within them. A turgid cell has no EMPTY SPACE. In elodea the vacuole is usually obscured by the numerous CHLOROPLASTS in the CYTO-PLASM.

Without moving the slide, **draw** off the water with a piece of paper towel and replace the water with a 5 percent salt solution. **Note** the change in any one cell.

OBSERVATIONS

1. Is water entering or leaving the vacuole? _____
    _____

2. Immediately after you added the salt solution, which had the greater concentration of water, the cell sap or the salt solution? _____
    _____

3. Is the protoplasm still pressed against the cell walls?   *Yes*   *No.*
4. The change that you see in the cell was caused by (a) loss of water; (b) influx of toxic salt. Circle (a) or (b).
5. A cell in this condition is described as _____
    _____

Fill in the contents of the cells in Figure 15.2. Let (A) represent a turgid cell and (B) a plasmolyzed cell. Label your drawings with the terms capitalized in the preceding description.

B. PLANT CHANGES (WILTING) RESULTING FROM CELL PLASMOLYSIS

ACTIVITY 5   **Observe** herbaceous plants (or cuttings) such as geranium (*Pelargonium hortorum*); tomato (*Lycopersicon* sp.), coleus (*Coleus blumei*), and so on, that have been placed in solutions of different salt concentrations (0.0, 0.5, 3.0 percent). Alternatively, or in addition, **observe** plants that have received different concentrations of fertilizer. **Record** your observations in Table 15.4.

OBSERVATIONS

1. The cells of the wilted plants have (a) lost water; (b) lost turgor pressure; (c) gained salt; (d)

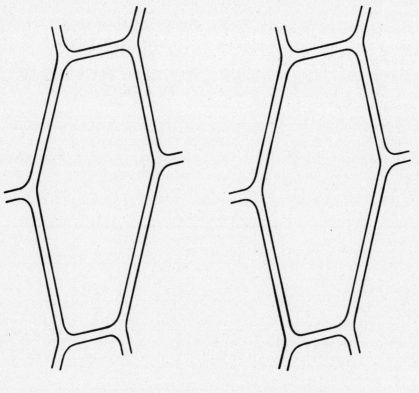

(A)   (B)

Figure 15.2   *A*, Turgid cell. *B*, Plasmolyzed cell.

Table 15.4 PLANT CHANGES DUE TO CELL PLASMOLYSIS

PLANT NAME OR PLANT TISSUE	SALT CONCENTRATION			FERTILIZER CONCENTRATION		
	0.0%	0.5%	3.0%	___ %	___ %	___ %

Record for degree of wilting: none = 0, moderate = +, severe = + +

been plasmolyzed. Circle as many as are correct.

2. The stems and leaves of the wilted plants have (a) become flaccid; (b) lost turgor pressure; (c) taken up salts. Circle as many as are correct.

3. If you were to return these plants to containers with pure water, many of the plants would regain their turgor because (a) water would move into the cells by osmosis; (b) salt would wash out of the cells. Circle (a) or (b).

4. This movement of water occurs because (a) the plant cells have increased in salt content; (b) the concentration of water in the plasmolyzed cells is less than that in the container. Circle (a) or (b).

IV. DIFFUSION OF SOLIDS

A. *DIFFUSION OF SOLIDS IN WATER AND ITS APPLICATION TO FERTILIZERS*

ACTIVITY 6 **Work** alone or in groups, and **proceed** as follows.

1. Using forceps, place one crystal of potassium permanganate ($KMnO_4$) on the moist surface of agar in a petri dish (near the center of the dish). If agar plates are unavailable, use dishes with tap water instead.
2. Do not joggle or otherwise disturb the plate.
3. After 30–60 minutes, note the distance of diffusion of the permanganate.

OBSERVATIONS

1. What is the radius of the area of diffusion? _____ mm

2. How much time elapsed since the $KMnO_4$ was set on the agar (water)? _____ min

3. The rate of diffusion of the $KMnO_4$ through the agar (or water) was _____ mm/h

4. Would you say then that $KMnO_4$ is water soluble? **Yes No.** (agar gel is a colloid containing water)

5. Is there still a fairly high concentration of $KMnO_4$ in the center of the dish? **Yes No.**

6. If instead of $KMnO_4$, we were considering a soil fertilizer, the rate of diffusion would become very important. There are a number of fertilizers called "slow-release fertilizers." In these, the fertilizer is either coated with or combined with slowly soluble materials such as resins, plastic, glass, and so on. The main effect of these on the fertilizer is to limit what? _____
_____

7. Plant roots will suffer "fertilizer burn" when too much fertilizer is applied at any one time. How do you explain fertilizer burn? Is it a matter of (a) osmosis; (b) concentration; (c) both (a) and (b)? Circle (a), (b), or (c).

B. *WATER MOVEMENT IN STEMS*

Water moves up stems mostly because of the evaporative pull of transpiration. Any material in the water is carried up the stem. The water itself must remain as an unbroken fine column. Xylem elements provide the conduit for this column.

ACTIVITY 7 **Observe** stalks (petioles) of celery (*Apium graveolens*), the shoots of jewelweed or touch-me-not (*Impatiens*), the flowering stems of white varieties of carnation (*Dianthus* sp.), or *Chrysanthemum* species that have been set upright in beakers containing a water-soluble dye.

OBSERVATIONS

1. Describe the leaves and the petals. _____
_____
_____

2. Through what tissue of the stem does the dye move? _____

3. This movement up stems into the flowers is put to good practical use by florists preparing for certain holiday sales. Can you think of an example, and for what holiday(s)? _____
_____
_____

   If time permits, **examine** thin cross sections cut from the midpoint of the celery petiole (several centimeters above the liquid level).

4. How would you describe the distribution of the dye, (a) spread about; (b) localized? Circle (a) or (b).

5. Name the structure in which the dye is visible. _____

6. Name the tissue (within the petiole) through which the dye is moving. _____
_____

V. DIFFUSION OF GASES

Oxygen and carbon dioxide are gases that **DIFFUSE** into and out of plants. Their concentration in the air around the leaves influences the rates of respiration and photosynthesis.

Commercial greenhouse operators in climates where the greenhouses remain closed much of the winter may experience reduced crop growth due to low levels of carbon dioxide. Addition of carbon dioxide to the greenhouse atmosphere is good practice and has become common among growers of roses, carnations, mums, foliage plants, and so on. Large carbon dioxide generators produce the carbon dioxide, which is then metered and allowed to diffuse through the house. Crop response is so good that many growers now consider adding carbon dioxide to greenhouse atmosphere in winter just as essential as fertilizing or watering the plants.

A. DIFFUSION OF GASES

ACTIVITY 8 It is not as easy to demonstrate diffusion of CO_2 or O_2 as certain other gases, so the instructor will demonstrate *diffusion of hydrochloric acid gas* and *ammonia gas*. **Observe** the instructor's demonstration.

1. A large glass tube (50–70 cm long and 5 cm in diameter) is plugged at both ends with rubber or cork stoppers. To each stopper is attached a wad of cotton.

2. A few drops of one normal (1N) ammonium hydroxide (NH_4OH) are added to one wad of cotton and the same amount of 1N hydrochloric acid (HCl) to the other wad.

3. The ends of the tubes are marked for reference.

4. The stoppers are placed one in each end of the glass cylinder and the cylinder is held horizontal.

5. **Note** the time on your watch to the second when the tube is plugged at both ends.

6. **Observe** that a white band or cloud soon forms in the tube.

7. **Note** the time (to the second) that the white band first formed.

8. **Note** where the band appeared: in the center or closer to one end, and which end.

OBSERVATIONS

1. The elapsed time for the diffusion was _____ _____ seconds.

2. The band appeared where? _____
_____

3. The speed of diffusion of the two gasses is calculated by

$$\frac{\text{tube length (cm)}}{\text{time (seconds)}} = \_\_\_\_\_ = _____ \text{cm/sec (speed)}$$

4. How can you tell which gas moved faster? _____
_____

5. The HCl solution is HCl dissolved in water. NH_4OH is NH_3 (ammonia) dissolved in water. When vaporizing out of their water solutions, they diffuse as HCl and NH_3.
   a. What is the molecular weight of HCl? _____ _____; of NH_3? _____
   b. Which is lighter in weight? _____
   c. Which, therefore, should move faster? _____ _____

6. Does this correspond to your answer to question 4? _____

7. The white band is a precipitate formed between the two gases. What is it? _____
_____

8. Which would diffuse faster through a greenhouse, CO_2 or water vapor? _____
_____ Why? _____
_____

B. TRANSPIRATION AND COMPARISON WITH EVAPORATION

ACTIVITY 9 **Observe** the demonstration of transpiration and evaporation. Three bell jars are on glass plates. When originally set up, they were dry.

Under one jar is a plant rooted in a sealed pot or peat cube; under the second is a wet sponge. The third jar is empty and serves as a control. All three jars have been under lights for at least 12 hours. **Compare** the relative amounts of condensed water on the inside of the bell jars.

OBSERVATIONS

1. The bell jar with the most condensed water is the one covering the _____

2. Is there water condensed under the jar covering the wet sponge? *Yes   No;* under the control? *Yes   No*

3. Water moved from the plant and the sponge as a gas, called _____

4. This loss of water from the sponge is called _____

5. The loss of water from the plant is called _____ _____

6. Is the loss in both instances basically the same phenomenon? _____

7. Define transpiration and use the term you supplied for question 4 _____ _____ _____

8. The sponge has many big pores; the leaf has many tiny pores (stomates). In this demonstration, which pore size is more effective in diffusing moisture? _____

9. Do you think a larger sponge would have lost more water than the leaves of the plant? *Yes   No*

C. DEMONSTRATING AND MEASURING THE RATE OF TRANSPIRATION

ACTIVITY 10   **Work** in teams of two and set up a potometer as shown in Figure 15.3. This is a simple (and even portable) device by which you can see and measure the uptake of water due to the pulling force developed by transpiration. **Proceed** as follows.

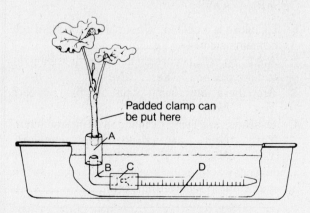

Figure 15.3 Potometer after assembly.

1. **Assemble** all the parts of the potometer and put them together in a pan as shown in Figure 15.3 under 5 cm of water. Figure 15.3 shows a micropipette or capillary tube (D) connected to a glass or plastic elbow bend (B) by a piece of tubing (C). The elbow is also connected to a 5-cm long piece of heavy gum rubber tubing (A). The inside diameter of the tube A should be the same as the diameter of the stem of the plant that will be used.

2. **Force** all air out of the system by squeezing the tubing A and holding all the parts under water.

3. **Obtain** a leafy stem of some plant such as geranium (*Pelargonium* sp.), pine (*Pinus* sp.), tomato (*Lycopersicon* sp.), or potato (*Solanum tuberosum*). Quickly put the cut end below the surface of the water in the pan and with a razor blade remove about 1 cm from the plant's stem.

4. Fit the stem into the tubing A. Keep everything under water while you do this.

5. If any air bubbles appear inside the apparatus, expel them by *gently* squeezing on the rubber tube A.

6. Lift the assembly (potometer) out of the pan. It should look like Figure 15.3.

7. The potometer can be carried around the laboratory and exposed to different conditions of light and air movement.

8. If desired, a padded burette clamp may be used at the point shown to hold the potometer upright to a ring stand after removing it from the pan of water.

9. After removing the potometer from the water, **observe** the water being taken up by the plant. You will see this by looking carefully at the micropipette or capillary tube. As water is taken up by the plant, air will move into the tip of the pipette.

10. **Note** the time required for the air column to move toward the plant. (This may be expressed as mm/min if you use a capillary tube and ruler, or 0.01 ml/min if you are using a micropipette.)

11. **Record** at least three successive minutes to see if the rate changes under normal classroom conditions.

12. Repeat for bright light or sunlight; repeat for dark; repeat for moderate air movement (not violent motion).

13. **Record** results in Table 15.5.

OBSERVATIONS

1. What conditions produced the fastest rate of transpiration? _____

Table 15.5 TRANSPIRATION RATE UNDER VARIOUS CONDITIONS OF LIGHT AND AIR MOVEMENT

AIR	LIGHT	TRANSPIRATION RATE
Still	Classroom	
	Bright light	
	Dark	
Moving	Classroom	
	Bright light	
	Dark	

2. What conditions produced the slowest rate? _____

3. Does gentle moving air speed up or slow the rate? _____

4. What could excessive wind do to the water content of a plant? _____

5. Which type of plant is more prone to death by desiccation on windy winter days, deciduous plants or evergreens? _____

6. What is the name of the pores in the leaf through which water is transpired? _____

7. During transpiration, does water exit from the plant as a liquid or a vapor? _____

provided in these questions and statements, list two environmental conditions and a plant factor that are responsible for guttation:
a) _____; (b) _____; (c) _____

5. In plants, is guttation just another name for dew? _____

6. Root pressure is very low and practically non-existent in some plants. It is of minor importance in moving water up the plant. In temperate regions of the world, less than 1 percent of water loss is due to guttation. In what part of the world would you expect guttation to be very common? _____

VI. GUTTATION

A. LOSS OF WATER FROM LEAVES

ACTIVITY 11 **Examine** seedlings of barley (*Hordeum vulgare*), wheat (*Triticum* sp.), or tomato (*Lycopersicon* sp.) set under a bell jar. The soil or growing medium was well watered when the demonstration was first set up several hours ago.

OBSERVATIONS

1. Localized at the _____ of the leaves are a few _____

2. These have been forced out of the leaves through pores, called hydathodes. Is this then a type of transpiration? _____

3. This phenomenon is guttation. It occurs when the rate at which water enters the plant is more rapid than the rate of water loss.
   a. What soil conditions would induce this? _____

   b. What atmospheric conditions? _____

4. The water in the xylem elements of roots is under pressure, called root pressure. Root pressure can often be observed when the rate of water loss is slow or nil. With the information

VII. IMBIBITION

A. DIFFUSION OF WATER INTO COLLOIDAL SUBSTANCES FOLLOWED BY THEIR SWELLING

ACTIVITY 12 **Examine** samples of material in the dry state and again after they have imbibed water. Suggested material is seeds, old mature pine cones, wood veneer, and so on. (Pine cones float and will need to be held under water with a rock, for example.)

OBSERVATIONS

1. Cellulose is a colloid. What part of the plant cell contains cellulose? _____

2. Are living cells necessary in order for imbibition to take place? **Yes No**

3. Are there any living cells in the veneer? **Yes No**

4. In seeds, are living cells involved in imbibition **Yes No;** are dead cells involved? **Yes No**

5. In what way(s) is the swelling effect of imbibition important to seed germination? _____

EXERCISE 15

Student Name _____

QUESTIONS

1. When you plant a young (for example, five-year-old) tree in the spring, which is the greater threat to its survival (a) desiccation; (b) insufficient photosynthesis? Circle (a) or (b).

2. One summer day you go to a nursery and select a tree for purchase, but the nurseryman prefers to wait until late October to dig it up and then plant it in your yard. Give at least two

   advantages to fall planting: (a) _____

   (b) _____

3. Certain fertilizers with controlled rates of diffusion are called _____

4. Many herbicides can kill the entire plant because they move through the vascular tissue. How do they move? By (a) osmosis; (b) diffusion. Circle (a) or (b).

5. When trees are transplanted during the growing season, they are subject to transplant shock. This shock is brought on because the root–shoot balance has been upset. What

   plant process is the chief cause of desiccation (transplant shock)? _____

   _____

6. A common practice when transplanting trees in the springtime is to spray the leaves with antidesiccants. These allow the diffusion of oxygen and carbon dioxide but prevent diffus-

   sion of what? _____

7. Before the advent of chemical antidesiccants, what do you suppose was (and still is) a common method of bringing the mass of leaves, twigs, and branches back into balance

   with the reduced root mass? _____

8. Many people in regions of cold climates start their vegetable gardens early by germinating and growing the seedlings indoors. When these are later transplanted outside in the ground,

   the recommendation is to do this on cloudy days or in the evening. Explain why. _____

   _____

9. If imbibition did not take place when a dry starchy seed is wetted, could water enter the

   seed by osmosis? _____ Sometime in a test you may be asked to *explain* this.

10. The common kitchen practice of soaking wilted vegetables in cold water to crisp them

    demonstrates the cook's confidence in the process of _____

11. Many shrubs are grown from seedlings in a container of some sort and sold directly to the customer for planting in the ground. Other shrubs are grown in the field, then put temporarily in a container for shipping and sale.
    a. Which shrub has the better chance of survival after planting in the ground? (a) field-grown plant; (b) container-grown plant. Circle (a) or (b).
    b. Which one should be top-pruned after planting in the ground? (a) the field-grown plant; (b) the container-grown plant. Circle (a) or (b).

12. Explain the basis of your answers to Questions 11.a and 11.b. _____

    _____

    _____

    _____

13. Aside from extra lighting (which is not usually practiced), what is one way that commercial

    growers of greenhouse tomatoes increase the yield during the winter months? _____

    _____

    _____

14. A plant wilts or becomes flaccid because some or many of its cells have become _____

    _____

15. Most herbaceous plants have very little wood or fibers. What then gives them structural strength and form? _____

16. The first hard frost of autumn kills all the garden vegetables and flowers. This can be explained as due to death of the _____ of the cells with the subsequent loss of _____ from the plant.

17. Osmosis occurs only in living tissue. (a) True; (b) false. Circle (a) or (b).

18. Plasmolysis of living cells is caused by the influx of salt. (a) True; (b) false. Circle (a) or (b).

19. In plants that can tolerate desiccation or lower temperatures in the fall and winter, would you expect that their cell sap contains (a) little or (b) much sugar. Circle (a) or (b). Can you explain why this is true? _____

PHOTOSYNTHESIS

INTRODUCTION

The basic framework of organic matter on earth is made primarily of carbon, and green plants are the only ongoing means of capturing carbon from the atmosphere.

Through photosynthesis the green plant:

1. Constructs organic compounds from water, minerals, gases, and so on
2. Captures the energy of the sun's light waves and converts it into a biologically usable chemical form of energy
3. Stores this chemical energy in the organic compounds
4. Replenishes the atmosphere of the earth with oxygen

Thus, photosynthesis is the ultimate source of structural materials, as well as energy, for all life. Even the photosynthesis of plants of the geological past provides us with energy today in the form of oil, coal, and gas.

I. THE PIGMENTS OF PHOTOSYNTHESIS

Much of plant breeding today is focused on producing plants that are more efficient in photosynthesis. To do this, scientists study, analyze, and experiment with the pigments of photosynthesis. The easiest way to analyze the individual pigments is to separate them from each other. A technique called chromatography may be used to separate the pigments. It has been known for many years and can be done in a variety of ways. Some ways are technologically quite sophisticated.

In today's work you perform the most basic technique of paper chromatography. Paper chromatography works on the principle that different pigments, when dissolved in a solvent (such as water or ether), will move (travel) through a piece of paper at different rates of speed. The rate is determined by (1) how soluble the pigment is in the solvent and (2) the degree of adhesion of the pigment to the surface of the paper.

A. SEPARATION OF THE PIGMENTS IN INK BY PAPER CHROMATOGRAPHY

ACTIVITY 1   **Proceed** as follows.

1. Obtain a piece of filter paper long and narrow enough to fit inside a test tube (see Figure 16.1A.
2. Crease the paper lengthwise to make it more rigid.
3. Ink a straight line across the paper about 2–3 cm from one end and let it dry. (Use the ink provided.)
4. Put a little water (about 1 cm or so in depth) in a test tube.
5. Insert the paper in the tube. The inked line should be above the water level. Do not wet the line.
6. Set the tube in a rack or wooden block.
7. After a few minutes, examine the pigment separation on the paper strip.

OBSERVATIONS

1. What color ink did you use to make the line on the paper? _____
2. How many ink colors are visible after separation of the pigments? _____
3. What colors are there? _____

_____

4. The combination of all the pigments leads to some masking of individual colors. Do you agree? **Yes   No**

5. Is the distance traveled through the paper the same for all pigments? **Yes   No**

6. Are all the pigments, therefore, equally soluble in the water? **Yes   No.** Are they equally adsorbed by the paper? **Yes   No**

7. The differential adsorption to the paper and the differential solubility of these pigments in water has effectively done what to them? _____

B. SEPARATION OF LEAF PIGMENTS BY PAPER CHROMATOGRAPHY

A leaf extract has been prepared by macerating leaves in a small volume of acetone. This can be done either with a mortar and pestle or in an electric blender. By paper chromatography, you can separate the different pigments out of the leaf extract just as you separated the different pigments out of the ink.

ACTIVITY 2   **Proceed** as follows.

1. Get a test tube and add to it a small amount of solvent (about 1 cm or so in depth). The solvent is a mixture of 95 parts ether and 5 parts acetone. Plug the tube and set it aside.

2. Obtain a strip of filter paper long and narrow enough to fit inside a test tube (Figure 16.1A) and crease it lengthwise.

3. With a medicine dropper, apply a few drops of the leaf extract to the filter paper strip about 2–3 cm from one end. Allow the spot to dry. Two or three repeated applications to the same spot may be needed. The spot must be dry before you proceed to step 4.

4. Insert the paper in the test tube. Do not permit the extract on the paper to be directly wet by the solvent. Don't "buckle" the paper. Plug the tube and set it aside.

5. In a few minutes you will see the pigments begin to separate. As the solvent passes through the paper, the various pigments are dissolved in it and move up the paper at different rates.

Examine the chromatograph at frequent intervals because if the pigment separation continues for too long, some of the pigments will be superimposed on each other near the top of the strip.

The four pigments that separate out into their respective colors are

1. Chlorophyll a—blue green color
2. Chlorophyll b—olive green color
3. Xanthophyll—pale yellow
4. Carotene—orange yellow

Draw the lines on the diagram of the paper strip (Figure 16.1B) to show the position and general appearance of the pigments. **Label** each pigment line.

OBSERVATIONS

1. Chlorophyll b is the least soluble. Which position, therefore, should it occupy on the strip?

   _____

2. Carotene is the most soluble. Where should it be located? _____

II. THE NECESSITY OF CHLOROPHYLL IN PHOTOSYNTHESIS

The leaves of many plants are naturally variegated, that is, they have margins or patches that are not green but are usually white or some shade of yellow or pink. These patches lack the green pigment, chlorophyll. Variegated leaves are considered an attractive feature and are of horticultural value.

For a botany class they are of value because they can be used to demonstrate that where there is no chlorophyll in a leaf, there is no production and storage of the photosynthate, starch.

ACTIVITY 3   **Test** for starch in the variegated leaf. **Proceed** as follows.

1. Obtain a leaf from some variegated variety of plant, such as *Coleus blumei*, Algerian ivy (*Hedera canariensis* v. *variegata*), geranium (*Pelargonium hortorum* v. *marginatum*), and so on.

2. **Diagram** and **label** the leaf with the green and nongreen zones in the space provided as Figure 16.2 on page 154.

3. Boil the leaf in water to kill the tissue and to remove the water-soluble red and blue pigments that mask chlorophyll in some of these plants.

4. Boil the leaf in alcohol for a few minutes to extract the pigments. *Caution: Do not heat the alcohol over an open flame.* Use an enclosed-element electric hot plate or use the beaker of hot water (from step 3) as a means of heating a small beaker of alcohol.

5. Set the leaf in a petri dish and add a few drops of dilute iodine-potassium-iodide (IKI).

6. A brown or dark purple coloration in the leaf indicates the presence of starch.

7. **Label** the leaf diagram you drew, indicating which area gave a positive starch test.

OBSERVATIONS

1. Did the white or nongreen zones of the leaf give a positive test for starch?   *Yes   No*
2. Did the green zones of the leaf give a positive test for starch?   *Yes   No*
3. Is chlorophyll present throughout the leaf? *Yes   No*
4. Is the presence of starch a fairly good indication of chlorophyllous tissue in a leaf? *Yes   No*

III. THE LIGHT SPECTRUM AND PHOTOSYNTHESIS

As you probably know, light from the sun (or a light bulb) is a mixture of many different wave lengths. All the colors of the rainbow are present in this spectrum. As a matter of fact, a rainbow *is* a display of the visible part of the spectrum. Not all colors (wave lengths) in the spectrum are equally useful in photosynthesis, and the pigments of the leaf do not absorb all wave lengths equally.

ACTIVITY 4 **Observe** the spectrum of normal sunlight [or of white artificial light (incandescent, not fluorescent)] by means of a spectroscope, prism, or diffraction grating. This may be set up as a demonstration, or small, hand-held units may be available for each group.

What colors do you see? _____ _____ Is blue/violet at one end?   *Yes   No*.   Which end? _____ What color is at the other extreme? _____

If the spectroscope has numbers along with the colors, **note** the number associated with each color band. These numbers are the wavelength of each color, and they are expressed in nanometers (nm) or one thousand millionths of a meter.

Place a test tube containing a solution of leaf pigments in front of the light source. This may be a diluted solution of the leaf extract used in Activity 2 or a separate, small bottle already prepared for this activity. **Note** that the light intensity will be lower. If you can't see any light, dilute the strength of your pigment extract, or move the spectroscope toward a brighter light. Shift the leaf extract back and forth several times so that you see first the full spectrum and then the darkened one.

Are all wave lengths (colors) changed equally? _____ Are some blocked more than others? _____ Which are dimmed the most? _____ _____ Which are changed the least? _____ _____ Are any wave lengths (colors) almost the same? _____ Which one(s)? _____ _____ Which

wave lengths are probably most impo_____ plant in carrying out photosynthesis? _____ _____

Which are the least important? _____ _____ Based on your findings, why do most leaves appear green? _____ _____ _____

Repeat this observation with a thin, intact leaf instead of the leaf-extract solution. A young leaf of German ivy (*Senecio mikanioides*) or other relatively thin leaf from almost any species is satisfactory. Do the results differ from the previous observation? _____ In what way(s)? _____ _____

IV. THE NECESSITY OF LIGHT IN PHOTOSYNTHESIS

ACTIVITY 4 **Observe** bean plants (*Phaseolus vulgaris*) or geranium (*Pelargonium* sp.) in which several leaves have squares (about 4 cm^2 in size) of heavy black paper attached to them (by paper clips). The black paper forms a light-tight screen. The plants were kept in the dark for 72 hours. Then they were exposed to light for 8 to 12 hours just before class.

Remove one of the leaves; remove the black paper and test for starch as follows.

1. Boil the leaf in water.
2. Boil the leaf in alcohol. *Caution: Do not heat the alcohol over an open flame. Take precautions as noted earlier* in Activity 3, Step 4.
3. Set the leaf in a petri dish and pour some dilute IKI over it.
4. If starch is present, the tissue will turn brown or dark purple.

OBSERVATIONS

1. Did the tissue that had been kept under the light screen give a positive test for starch?   *Yes   No*
2. Did the remainder of the leaf show the presence of starch?   *Yes   No*
3. What is the conclusion? _____ _____
4. Why were the plants kept in the dark for 72 hours? _____ _____ _____
5. What would happen to the plants if they were kept in the dark for several weeks at room temperature? _____ _____
6. Commercial greenhouse growers of ornamental

...ants in the dark for
...sh to delay growth for
... plants suffer no set-
...environmental factor
...he grower to allow for

V. THE NECESSITY OF CARBON DIOXIDE IN PHOTOSYNTHESIS

ACTIVITY 5   **Observe** the following demonstration.

1. Three test tubes numbered 1 to 3 are each partly filled with water that has previously been heated to remove any dissolved air, then cooled to room temperature.
2. A few drops of phenol red indicator are added to each tube. Phenol red is an indicator here for the presence of carbon dioxide. Phenol red turns yellow in an acid such as carbonic acid (H_2CO_3) formed when CO_2 combines with water. When no CO_2 is present, the indicator is red.
3. Using a soda straw, the instructor will briefly blow his breath into the liquid in tubes 1 and 2 just until the liquid turns yellow. The third tube will remain as the control with no CO_2 in it.
4. A sprig of elodea (*Anacharis canadensis*) or parrots-feather (*Myriophyllum brasiliense*) is placed in tube 1.
5. All three tubes are put under very bright light for a half hour or so.

OBSERVATIONS

1. Have any color changes taken place?   *Yes*   *No.*   (The plants may be removed to aid in color comparisons.)
2. Has the control tube 3 changed color?   *Yes*   *No*
3. Has tube 2 changed color?   *Yes*   *No*
4. What does this indicate?_____
_____
5. Has tube 1 with the plant changed color?   *Yes*   *No*
6. What is the color of the liquid in tube 1? _____
_____
7. Is there any CO_2 in the liquid of tube 1 now?   *Yes*   *No*
8. What can you conclude has happened to the CO_2?_____
_____
_____

VI. OXYGEN AS A BY-PRODUCT OF PHOTOSYNTHESIS

ACTIVITY 6   The equation below is a very simplified summary of the process of photosynthesis. The oxygen produced by photosynthesis, which replenishes the earth's atmosphere, is derived from the water used in the process.

$$\text{carbon dioxide} + \text{water} \xrightarrow[\text{chlorophyll}]{\text{light (energy)}}$$

organic compounds + oxygen

Observe the demonstration that shows the release of oxygen by a photosynthesizing aquatic plant and the influence of light on the rate of photosynthesis.

A few sprigs of elodea (*Anacharis canadensis*) or parrots-feather (*Myriophyllum brasiliense*) have been placed in two jars or beakers with clean tap water. In order to enrich the carbon dioxide content of the water, several drops of 1 percent $NaHCO_3$ have been added to the water, or the gas has simply been bubbled in by means of blowing through a soda straw.

A funnel trap has been placed over the plants (as shown in Figure 16.3) and a test tube inverted over the stem of the funnel. The gas given off by the plant will be collected in the test tube.

One beaker of plants has been kept in relatively dim light (usually normal classroom light); the other has been kept under fairly bright light.

Note the difference in the amount of gas that has accumulated in each tube. **Watch** what happens when the instructor inserts a glowing splinter in the tubes.

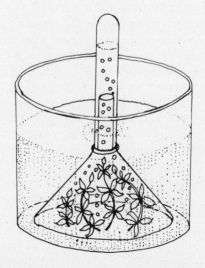

Figure 16.3.   Oxygen release during photosynthesis.

OBSERVATIONS

1. Which plant has produced more gas, (a) the well lighted plant; (b) the dimly lit one? Circle (a) or (b).
2. What happened to the glowing splinter when it was inserted in the test tube? _____ _____
3. What is the gas that accumulated in the tube? _____

VII. STARCH—A STORAGE PRODUCT OF PHOTOSYNTHESIS

Sugars are an early product of photosynthesis, but they are generally converted to starch for storage purposes. Why is this? The answer is that sugars are osmotically active compounds and therefore are not readily "isolated" from metabolism. One way to keep storage products intact and isolated from metabolism is to make them osmotically inactive. Starch is an osmotically inactive substance—thus its advantage as a storage product. Starch is more exten-

sively stored than proteins or fats. Some plants do store the sugar sucrose, for example, sugar cane (*Saccharum officinarum*) and sugar beets (*Beta vulgaris*).

ACTIVITY 7 **Prepare** a wet mount of starch grains of potato (*Solanum tuberosum*) by lightly scraping the cut surface of a raw white potato with a scalpel. Mount the scrapings in water and cover with a cover glass. Then allow a drop of IKI to run under the cover glass while you hold a piece of paper toweling at the opposite edge of the cover glass to draw off some of the water.

Starch grains often have a laminated appearance much like the growth lines on a clam shell. Starch grains can be used to some extent to identify the plant because they are characteristic of the species.

Prepare other wet mounts using flour made from the starch of any of the following plants: sweet potato (*Ipomoea batatas*), rice (*Oryza sativa*), oats (*Avena sativa*), or others. **Examine** these for differences in the starch grains.

EXERCISE 16   **Student Name** _____

QUESTIONS

1. Radiant energy is not directly usable by biological systems. What form of energy is usable?

   _____

2. Can radiant energy be stored for future use (a) by man; (b) by animals; (c) by plants? Circle as many as apply.

3. Can chemical energy be stored? _____

4. Where is chemical energy stored? Be specific, give the three basic organic storehouses:

   (a) _____; (b) _____; (c) _____

5. What process converts radiant energy into chemical energy? _____

   _____

6. The biological and technological world is like an upside-down pyramid; everything rests on

   one point. What biological reaction occupies that point? _____

   _____

7. Although you saw four pigments separate out in the leaf chromatography experiment, only one is the primary pigment of photosynthesis. The others are accessory pigments. Which

   pigment is it? _____

8. In today's exercise you saw that light influences the rate of photosynthesis. Name at least

   five physical environmental factors that can alter the rate of photosynthesis: (a) _____

   _____; (b) _____; (c) _____;

   (d) _____; (e) _____

9. Oxygen is a by-product of photosynthesis. From which of the original compounds is it

   derived? _____

10. The most common storage compound in plants is _____

11. Do you think it would be wasteful or beneficial if a grower in northern Michigan gave extra

    applications of fertilizer to his greenhouse crops during December and January? _____

    _____

12. Explain your answer to Question 11. _____

    _____

13. It is usually recommended that plants grown indoors all winter be placed outdoors in a

    semishady area during the summer. Why in a semishady place? _____

    _____

    _____

14. What element makes up the basic framework of organic matter? _____

15. Can anything, other than green plants, capture or "fix" this element? _____

    _____

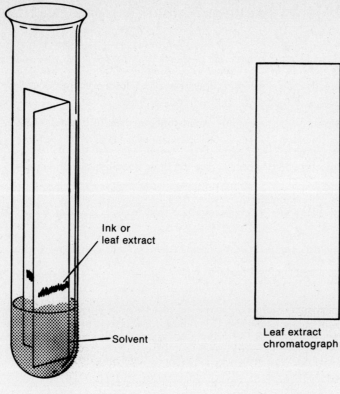

Ink or
leaf extract

Solvent

Figure 16.1   (A)   (B)

Leaf extract
chromatograph

Figure 16.2 Diagram of variegated leaf (student drawing).

DIGESTION AND RESPIRATION

EXERCISE 17

INTRODUCTION

Green plants capture the energy of light and put it away for storage in organic molecules. These molecules are mostly starches and fats and (to a lesser degree) proteins. The energy is retrieved through the process of *respiration*.

Respiration does not retrieve the energy directly from the storage molecules. These must first be converted to sugars, and it is from the sugars that respiration then retrieves the energy. The conversion of storage molecules to sugar and other substances is *digestion*. A more up-to-date term for this process might be "mobilization of reserve compounds." So, digestion and respiration are two closely related functions.

I. DIGESTION

When water is present (and other suitable conditions exist), living systems can convert starch to sugar. The conversion is mediated by enzymes secreted by the living system. Essentially, digestion makes an insoluble substance into a soluble one.

A. *PRODUCTION OF STARCH-DIGESTING ENZYMES BY A LIVING SYSTEM*

Germinating seeds are excellent sources of starch-digesting enzymes and are easily used to demonstrate the effects of the enzymes.

Starch-Digesting Ability of Germinating Seed

ACTIVITY 1 **Observe** the starch-digesting ability of a germinated seed.

1. Two days before this class, a germinating corn seed was cut lengthwise and placed (with the cut surface down) on a starch-agar plate (petri dish).
2. The instructor now removes the seed from the plate.
3. Dilute iodine-potassium-iodide (IKI), a starch indicator, is poured over the plate and left on for one or two minutes. Areas of the plate containing starch will react with the IKI and turn a blue-black. Those without starch will not turn color.
4. The IKI is poured off and the plate rinsed with water.

OBSERVATIONS

1. Did all the agar plate turn blue?   *Yes   No*
2. Describe the agar zone in the vicinity occupied by the seed. _____
3. What is your conclusion? Is starch present in the vicinity of the seed?   *Yes   No*
4. Is starch present in the remainder of the agar plate?   *Yes   No*
5. Briefly, what can you conclude about materials exuding from the cut seed? _____
_____

Extraction and Testing of Reducing (Digestive) Enzymes from Germinating Seeds

Extraction of the Enzyme

ACTIVITY 2 **Observe** the extraction.

1. One cup of five- to six-day-old germinating wheat seeds is mixed with one to two cups of a very weak buffer solution (0.001 percent sodium acetate). (This controls the acidity of the solution and allows the en-

zymes to work better than in just plain water.)

2. The seeds and buffered water are mixed in an electric blender for a few minutes.
3. The mixture is filtered with cheesecloth and the filtrate allowed to stand for 15–20 minutes to permit the solid residue to settle out.
4. The supernatant should contain reducing enzymes.
5. The supernatant will be referred to as the *extract* in the subsequent tests.

Testing the Extract for the Presence of Reducing Enzymes We need to show if starch reducing enzymes are present in the extract. If starch is added to the extract and if enzymes are present, the starch should be converted to sugar. If enzymes are not present, the starch will remain starch.

The presence of sugar can be demonstrated by means of an indicator called Benedict's solution. **Observe** a demonstration of this sugar test. The instructor will add a few drops of Benedict's solution to a test tube containing a sugar solution. The tube is heated to boiling in a water bath. If sugar is present, a yellow to reddish brown precipitate will form.

The presence of starch can be demonstrated by means of the indicator IKI. You have already used this indicator on several occasions. Material that contains starch turns dark when IKI is added to it.

We need to test how long it takes for the enzyme to take effect and to continue being effective. Therefore, sugar and starch tests will be made at the start and at 15-minute intervals for a period of 45 minutes.

ACTIVITY 3 **Look** at Figure 17.1 as a guide. **Proceed** as follows.

1. Obtain a 125-ml flask and fill it about one quarter full with the extract (about 50 ml).
2. Add an equal volume of a 1 percent starch solution to the flask.
3. Shake the flask well.
4. Fill two small test tubes each with about 2–3 cm of the starch-extract mixture.
5. Immediately test one tube for starch, using IKI.
6. Immediately test the second tube for sugar, using Benedict's solution as outlined above.
7. Repeat steps 3 through 6 at 15-minute intervals.
8. Record the results and your interpretations and conclusions in the spaces provided in Figure 17.1.

B. THE EFFECT OF HEAT ON DIGESTIVE ENZYMES

Many factors alter or inhibit enzyme activity. These include such things as dehydrated tissues, extremes of acidity or alkalinity, and extremes of temperature. In the following test we will see the effect of a temperature extreme.

ACTIVITY 4 **Observe** or **proceed** with the following tests. **Look** at Figure 17.2 as a guide.

1. Fill five test tubes one quarter full with water.
2. Add 15 drops of 1 percent colloidal solution of starch to each tube.
3. To tubes 3, 4, and 5 add 15 drops of amylase (a starch-digesting enzyme).
4. Boil tube 3 in a water bath for 15–20 minutes.
5. Set all tubes aside for 45 minutes to allow any reactions to proceed.
6. Test each of the tubes as shown in Figure 17.2. Test tubes 1 and 5 for starch with IKI and test tubes 2, 3, and 4 for sugar with Benedict's solution.
7. **Record** the test results and your interpretations and conclusions in the spaces provided in Figure 17.2.

II. RESPIRATION

Respiration releases the energy stored in glucose (or some glucose product). This energy is then used to drive all the reactions needed to sustain life. Below is a summary equation of the chemical changes that occur in **aerobic respiration**.

glucose + oxygen⟶
carbon dioxide + water + energy

A. OXYGEN AS A FACTOR IN SEED GERMINATION

Seed germination is a period of rapid cell division and high respiration rates. Energy is released, and a great deal of oxygen is consumed in the process.

ACTIVITY 5 **Observe** the demonstration.

1. The demonstration was started five to six days prior to class.
2. Two wide-mouthed bottles contain water, but one (bottle 2) also has a small vial containing pyrogallate, which absorbs the oxygen that is inside the bottle.
3. At the start of the experiment, some thoroughly soaked (but not germinating) bean or corn seeds were placed in cheesecloth bags and suspended in the bottles over the water.

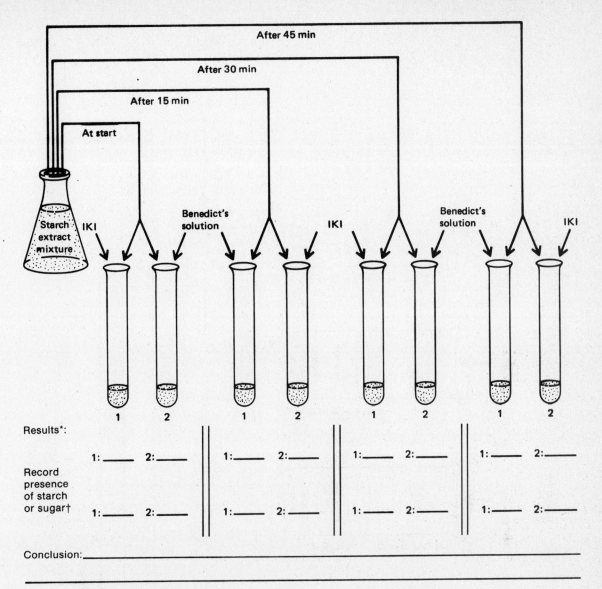

Figure 17.1 Presence and extent of activity of starch-reducing enzymes from germinating seeds.

*Use the following symbols: o = no change, +, ++, and +++ = intensity of positive test.

†Code: St = Starch; Su = Sugar

OBSERVATIONS

1. Did both batches of seed germinate? *Yes No.* Equally so? *Yes   No*
2. Which bottle shows more evidence of germination of seeds, (a) the bottle with oxygen or (b) the bottle without oxygen? Circle (a) or (b)
3. Did both bottles contain the same amount of water? *Yes   No*
4. What accounts for the difference in the germination of the seeds? _____
_____
5. **Complete** this statement by filling in the blanks. For most plants, seed germination will not occur in the absence of _____, even though ample amounts of _____ are available.

B. OXYGEN AS A FACTOR IN ROOT GROWTH

Root respiration is extremely critical. Oxygen must be available to roots or the plants will die. People and animals destroy air spaces in soil by trampling the ground. Compacted soil like this cannot support plant growth. Witness the grassless shortcut path across a lawn, and the dead or dying trees encircled by paved parking lots.

It's hard to demonstrate soil aeration directly, but root aeration in general can be demonstrated.

ACTIVITY 6 **Compare** the two sets of cuttings that have been rooted in water for a few weeks.

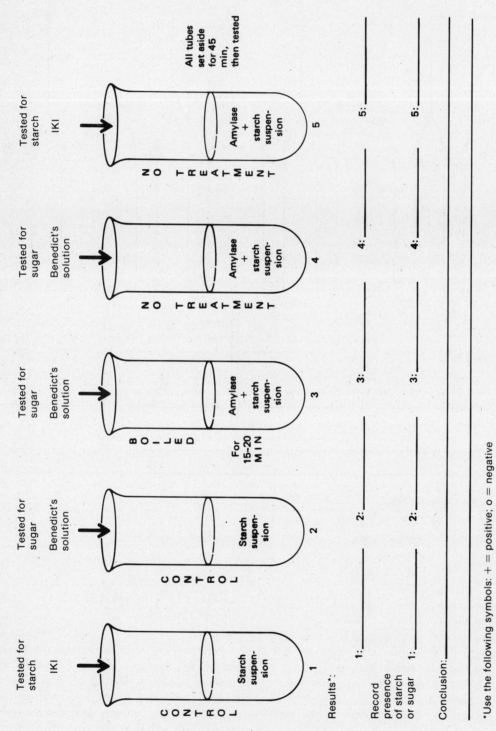

Tube 1: CONTROL — Starch suspension — Tested for starch — IKI

Tube 2: CONTROL — Starch suspension — Tested for sugar — Benedict's solution

Tube 3: BOILED For 15–20 MIN — Amylase + starch suspension — Tested for sugar — Benedict's solution

Tube 4: NO TREATMENT — Amylase + starch suspension — Tested for sugar — Benedict's solution

Tube 5: NO TREATMENT — Amylase + starch suspension — Tested for starch — IKI

All tubes set aside for 45 min, then tested

Results*:

Record presence of starch or sugar

1: _____ 2: _____ 3: _____ 4: _____ 5: _____

1: _____ 2: _____ 3: _____ 4: _____ 5: _____

Conclusion: _____

*Use the following symbols: + = positive; o = negative

Figure 17.2 The effect of heat on heat-reducing enzymes.

These cuttings may be from coleus (*Coleus blumei*), tomato (*Lycopersicon* sp.), wax begonia (*Begonia semperflorens*), geranium (*Pelargonium* sp.), wandering Jew (*Zebrina pendula*), purple heart (*Setcreasea purpurea*), or willow (*Salix* sp.).

Set *A* has had no oxygen added to the water other than what diffuses in from the air above the water or from splashing bubbles when the water is replenished.

Set *B* has been growing with the fish tank aerator bubbling air into the water in the container.

OBSERVATIONS

1. Which set has the better root and shoot growths, set *A* or *B*? (Circle one.)
2. If these plants survive growing in water, why won't plants survive in water-logged soil? _____

_____

_____

3. Some plants (but very few) thrive in flooded soil. Can you name one important food crop that does so? _____

C. RESPIRATION RATES AT DIFFERENT TEMPERATURES

Balancing the rate of respiration with the rate of photosynthesis is a predicament for the plant. To keep living cells functioning, respiration must supply them with energy at all times, day and night. Respiration uses the foods made by photosynthesis. There is a direct positive correlation between respiration and temperature. When temperatures are high, respiration rates are high too (rapidly using up the stored foods).

But photosynthetic rates depend on both temperature *and* light. If temperature is high but light is low, then the rate of photosynthesis is slow. In this situation, food production may not keep pace with the food losses (of respiration). If this kind of imbalance occurs too often or for too long, not enough food reserves remain for normal growth and development. For this reason, plants kept in heated offices and homes, with usually far less than optimum light for photosynthesis, do not grow as well as they do in their natural environment.

ACTIVITY 7 **Observe** the apparatus that has been set up and **look** at Figure 17.3 as a guide. An easy way to perceive differences in respiration rates is to measure the quantity of carbon dioxide given off by plants exposed to different temperatures. The apparatus in this demonstration is used to collect the carbon dioxide produced by seeds respiring at different temperatures. The apparatus (Figure 17.3) and demonstration consist of the following.

Apparatus

1. Four bottles are connected in a series to a hand aspirator (air displacement pump).
2. The system is set up to "wash" air (that is, pump it through) from one bottle to the next. Any CO_2 in the air in these bottles is either absorbed (in bottle 1) or precipitated out as a white precipitate (in bottles 2 and 4).
3. Three different bottles of bottle 3 will be put into the series, one at a time. These bottles contain wheat seeds that have been germinating for several days at different tem-

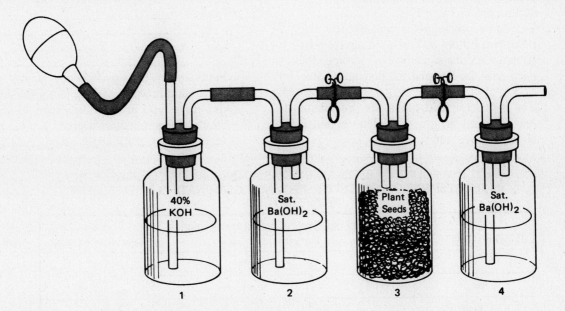

Figure 17.3 Apparatus for demonstrating the production of carbon dioxide in respiration.

peratures: one in the refrigerator at 5°C, one at room temperature of 20°C, and one in the incubator kept at 30°C. (Because other classes may have already used these bottles earlier in the day, your demonstration may show only the amount of CO_2 that has accumulated since then.)

4. Bottle 1 contains potassium or sodium hydroxide (KOH or NaOH). When the room air is pumped through the system, the CO_2 in it (the room air) should be removed by the KOH (or NaOH).

5. Bottles 2 and 4 contain barium hydroxide ($Ba(OH)_2$). Bottle 2 will precipitate out any remaining CO_2 that washes over from bottle 1.

6. During the germination test period, bottles 3 should be stored with tubing and pinch clamps as shown. After the bottle is connected into the gas-washing series, the clamps are removed.

7. When the aspirator is squeezed, CO_2 given off by the seeds will be "air washed" into bottle 4, where it will precipitate out as a white precipitate, barium carbonate ($BaCO_3$).

8. For each bottle 3 used, a new bottle 4 will be used.

Demonstration

Observe the procedure by the instructor.

1. One bottle 3 is put into the series.
2. The hand pump is squeezed four to five times to force air through the system.
3. **Note** the precipitate that forms in bottle 4 and record in Table 17.1.
4. The same procedure is repeated for the other two bottles 3. The hand pump must be squeezed the same number of times for each analysis. **Record** the results in Table 17.1.
5. The instructor next filters the precipitate and will oven-dry it with the filter paper at 80°C for one hour.

6. Each sample will be weighed to the nearest 0.1 g on a torsion balance.
7. **Record** the weights in Table 17.1.

OBSERVATIONS

1. Does temperature influence the rate of respiration?   *Yes   No*
2. The temperature difference between 5°C and 20°C is fourfold (assuming that no plant activity can take place at 0°C or below). Was there a fourfold difference in the respiration rate between these two?   *Yes No.*
3. How do you explain the result as answered in question 2? _____
_____
_____

4. List at least three other factors or conditions that influence the rate of respiration.
   (a) _____ ;
   (b) _____ ;
   (c) _____
5. The seeds of peas, turnips, and leaf lettuce can be planted outdoors five to six weeks *before* the last frost. If we had used seeds of these plants, would the results have agreed with those obtained with wheat? *Yes   No.* Explain. _____
_____
_____
_____

6. Green peppers and watermelon seed should not be planted until two weeks *after* the last frost. If we had used seeds of these plants, would the results have agreed with those obtained with wheat?   *Yes   No.* Explain. _____
_____
_____

7. In view of the facts given in questions 5 and 6, can you now list another factor (besides the three you listed in question 4) that relates to germination rates? _____
_____

Table 17.1   AMOUNT OF $BaCO_3$ FORMED AS A MEASURE OF CO_2 PRODUCTION IN THE RESPIRATION OF WHEAT SEEDLINGS GERMINATING AT DIFFERENT TEMPERATURES

TEMPERATURE (°C)	RELATIVE AMOUNT OF PRECIPITATE VISIBLE (0, +, ++, etc.)	$BaCO_3$ PRODUCTION (g)
5		
20		
30		

8. How can we be sure that the precipitate in bottle 4 came from the seeds? Couldn't some CO_2 have been in the air that was pumped from the room into the system?

_____

_____

_____

D. ANAEROBIC RESPIRATION

So far, you have been observing and studying various aspects of respiration that involve oxygen—in other words, aerobic respiration. Many bacteria and yeasts normally respire without oxygen. This is anaerobic respiration. Carbon dioxide is again one of the end products, and measuring the amount of it produced is again a method of observing the effect of temperature—this time upon anaerobic respiration.

Below is a simplified summary of the chemical changes that occur in anaerobic respiration.

glucose $\xrightarrow{\text{enzymes}}$

carbon dioxide + ethyl alcohol + energy
(or lactic acid)

Fermentation

The conversion of sugar to ethyl alcohol is called fermentation and is one of the most common types of anaerobic respiration. It is of great commercial value.

ACTIVITY 8 **Work** in groups of two or four or as directed by the instructor and **proceed** as follows.

1. The instructor will have available a solution in which yeast cells have been actively fermenting sugar for at least a half four beforehand.
2. Obtain from the instructor two empty fermentation tubes (Figure 17.4), and fill each with the fermenting yeast solution. Tilt the tubes (as demonstrated by the instructor) to fill the upright arm with the solution.
3. Set one tube aside on your desk where the room temperature is approximately 20°C.
4. Immediately cool down the second fermentation tube with an ice pack or other procedure as directed by the instructor.
5. Place the cooled tube in a refrigerator kept at approximately 5°C.
6. After at least one hour, remove the tube from the refrigerator and compare with the one on your desk.

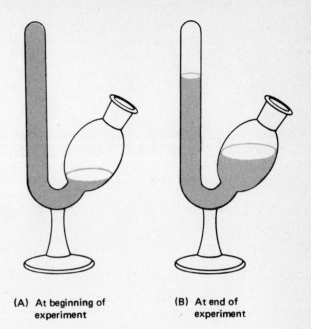

(A) At beginning of experiment

(B) At end of experiment

Figure 17.4   Fermentation tube.

OBSERVATIONS AND ADDITIONAL TESTS

1. Does the solution still completely fill the arm of both tubes? **Yes   No**
2. Which tube shows the greater displacement of the solution from the arm, (a) the 5°C one; (b) the 20°C one? Circle (a) or (b).
3. Measure (with a ruler) the empty part of the arm and record here. (If the tube arm is calibrated, read the amount in milliliters.)
   Refrigerated solution:
   gas accumulation _____ ml; or _____ mm
   Room temperature solution:
   gas accumulation _____ ml; or _____ mm
4. Add four to five drops of phenol red indicator to the fermentation tube.
5. Without disturbing the gas in the arm, shake gently to distribute the indicator.
6. What is the color of the solution after adding the phenol red? _____
7. You have used this indicator before in previous exercises. This color change is an indication of the presence of what? _____

CONCLUSIONS

1. Under which conditions did respiration proceed at the faster rate, (a) 5°C; (b) 20°C? Circle (a) or (b).
2. What was your evidence for this? _____

   _____
3. Is CO_2 produced during anaerobic respiration?
   Yes   No
4. What was your evidence for this? _____

   _____

5. If you had kept one of the fermentation tubes at 50°C, the amount of gas in the arm would have been (a) greater than the 20°C tube; (b) greater than the 5°C tube; (c) less than the 5°C tube; (d) greater than the 5°C tube but less than the 20°C tube. Circle (a), (b), (c), or (d).

6. Explain your answer to question 5. _____
_____
_____

EXERCISE 17

Student Name _____

QUESTIONS

1. The depth of the feeder roots of trees is really quite shallow (20 to 40 cm). Do you think this is a response to the availability of moisture or of oxygen? _____

2. Weed seeds can lie dormant far below the soil surface for many years. If a farmer deep plows the ground, in a few days the area has many weeds germinating. What two factors can you deduce that the weed seed required before germination could occur? (a) _____ ; (b) _____

3. People who grow indoor plants know that during the winter, night temperatures should be about 10–15 degrees cooler than day temperatures. What immediate effect does this have on what process in the plant? (a) _____

_____

_____

4. What is the purpose of this practice? _____

_____

_____

5. Two amateur gardeners, Jane and Zane, disagree on whether or not they should soak their seeds before planting them. Jane planted most of her seeds without soaking but did soak the few hard-coated varieties overnight. Zane soaked all of his seeds for two or three days just to make sure they had plenty of moisture. Who will have the smaller germination percentage, Jane or Zane? _____ Explain. _____

_____

6. Jane and Zane are also house-plant enthusiasts, but again disagree on cultural practices. Zane's plants are healthier than Jane's, even though the soil in his pots gets dry between waterings. Jane waters her plants well every day. What is Jane systematically withholding from her plants? _____

7. One of the main problems that commercial greenhouse operators have is maintaining even heating throughout the house. What two plant processes particularly are affected by temperature differences? (a) _____ ; (b) _____

_____

8. A grower with severe problems of uneven heating in his greenhouses can't sell all his chrysanthemums for the same price. Why not? _____

_____

9. Commercial greenhouse growers of ornamental plants face a financial crisis with ever-rising fuel costs. What do you suppose is the focus of many plant-breeding programs as a result of this problem? _____

_____

10. Do dormant seeds respire? _____

11. Plants that are used for interior design in offices, lobbies, and similar places are often maintained by professional plant care services. The plants are kept in good condition by having matched sets of two or three of each kind of plant. They are used in rotation; one is on display while the other two are in a greenhouse. In light of what you have learned in this exercise, explain why this is the best way to manage these display plants. _____

_____

_____

12. It's been known for 50 years or more that storing apples at a temperature of 3° to 4°C keeps them firm and sweet for five or six months. What is the physiological basis for this practice? _____

13. More recently, the storage life and quality has been doubled to almost a year by flooding the cold storage vault with CO_2. How is this practice similar to that described in question 12? _____ How is it different?

    _____

14. Without herbicides, modern agriculture wouldn't be modern agriculture. Without in-depth knowledge of the normal plant processes of photosynthesis, growth, respiration, and digestion, chemists could not have developed herbicides. From the previous statement, what do you conclude to be the basis of action of herbicides in general? _____

    _____

15. Sodium chlorate is an herbicide. If its main *effect* is the depletion of root reserves, what would you conclude to be its *mode of action* on the physiology of the root tissues? _____

    _____

PLANT MOVEMENT AND GROWTH RESPONSES TO STIMULI

EXERCISE 18

INTRODUCTION

Many people think that animals, but not plants, can respond to stimuli from their environment. This idea is wrong. Plants respond to many of the same stimuli that affect animals. It is just that plant response is usually slower and may bring about permanent or semipermanent changes in the growth form of the plant.

Plant growth is controlled by the interplay and balance between several hormones (growth-promoting and growth-inhibiting). We will study only the auxins, a class of growth-promoting hormones, which were for many years thought to be the chief cause of all plant growth response. Modern studies show, however, that there isn't a simple cause-and-effect relationship between auxins and plant growth response in the living plant.

In this exercise we will consider just two types of plant responses to stimuli: (1) tropisms and (2) turgor movements. A tropism is an irreversible "bending" or growth curvature induced by auxins. A turgor movement is a reversible, relatively fast movement induced by water pressure changes in the cells.

I. GROWTH IN RESPONSE TO AUXIN

A. THE EFFECT OF AUXIN ON STEM GROWTH CURVATURE

You learned in previous exercises that plant cells grow larger because of the expanding force of internal water pressure. Auxin promotes this en- largement by making the cell wall more plastic, and it then becomes extended by the turgor pressure.

Whether the differential response of the shoot in the following activity reproduces the natural tropic response of a plant or merely simulates it is not clear.

ACTIVITY 1 **Work** alone or in groups as directed by the instructor. **Proceed** as follows.

1. **Obtain** from the instructor a small pot or peat cube containing seedlings of oats (*Avena sativa*) or barley (*Hordeum vulgare*) (or other suitable grasses) that are several days old.
2. Examine the tips of the plant and identify the **COLEOPTILE**, a colorless sheath enclosing the green stem. The coleoptile is easy for you to work with because of its simplicity of structure.
3. Obtain several toothpicks and a small quantity of lanolin paste in which auxin is mixed (in the proportion of 1 part auxin per 10,000 parts lanolin). This auxin is indole acetic acid (IAA).
4. Using a toothpick, apply the auxin paste to the upper 6–10 mm of one side of the coleoptile of each seedling. Some seedlings could be left untreated to serve as controls.
5. For a reference marker for later identification, poke a clean toothpick in the soil or peat on the same side of the seedling where the auxin is applied.
6. Set the pots in a warm dark chamber for one or two hours, then remove.
7. **Note** any changes in the stem tips.

OBSERVATIONS

1. Are the auxin-treated stem tips still growing straight up?   *Yes   No*

165

2. How do they look? (a) tip curved; (b) entire stem curved. Circle (a) or (b).

3. Is the curvature toward or away from the side on which the auxin was applied? _____

4. Note the sketch at the right. Which is greater, the distance A–A' or B–B'? _____

5. How do you account for the greater distance in B–B'? _____

6. Look at your specimens. Which side (A–A' or B–B') corresponds to the side on which you put the auxin paste? _____

7. Why were the seedlings placed in the dark during the experiment? _____
_____

8. Do the untreated seedlings show any curvature? _____

9. Is there a correlation between auxin concentration and differential growth (more growth on one side than on another)?   *Yes*   *No*

10. Did the higher auxin concentration (a) inhibit or (b) promote growth on the side to which it was applied? Circle (a) or (b).

11. In the stems of the untreated seedlings, would you describe the auxin distribution as (a) equal throughout or (b) unequal? Circle (a) or (b).

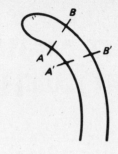

II. PHOTOTROPISM— A GROWTH CURVATURE (BENDING) IN RESPONSE TO LIGHT COMING FROM ONE SIDE

A. POSTURE, PIGMENTA- TION, AND GROWTH OF PLANTS IN RESPONSE TO LIGHT AND DARKNESS

In the preceding section you have caused the stem to "bend" in the dark simply by applying auxin to one side. You saw also that the bend was in a direction opposite to the side receiving the extra auxin.

This next demonstration shows you the response of plants to light coming from one direction, and the role auxin plays in this response. The plant responses are explained here according to the theory that auxin controls stem curvature. However, several recent research studies have turned up so many contradictions and exceptions to this theory that it is no longer totally acceptable. As yet, however, there is no new theory that is fully explanatory either.

B. THE EFFECT OF THE HERBICIDE 2,4-D ON STEM GROWTH

Natural auxins are very potent hormones. Extremely small quantities can induce curvature and growth. Plant tissues have means of regulating auxin concentrations, because otherwise normal growth would be disturbed. Certain herbicides have auxin-like activity, but because they are synthetics, the plant doesn't always have the ability to degrade (deactivate) them. Hence, abnormal growth is induced in the plant, and this eventually kills it. This is the basis of such well-known herbicides as 2,4-D (2,4-dichlorophenoxyacetic acid) and related compounds, such as mecoprop, silvex, and 2,4,5-T (formerly used to defoliate trees and shrubs).

ACTIVITY 2 **Observe** plants that have been sprayed with 2,4-D. Most dicot plants are susceptible to 2,4-D. It is absorbed by the leaves and carried through the phloem. **Note** the abnormal stem bending (the condition is called epinasty). Other symptoms include stem splitting, adventitious root formation, swollen shortened roots, and strap-shaped leaves.

ACTIVITY 3 **Observe** radish seedlings that have been grown in four boxes. Two of the boxes have a window made of a glass monochromatic light filter or gelatin paper that allows light of specific wave lengths to pass. One filter allows blue wave lengths of light to pass into the box; another filter allows red wave lengths of light to pass into the box. The third box has clear glass or cellophane, admitting full light (all wave lengths). The fourth box has no window, so the plants are in darkness. A lamp has been placed near the windows of the three boxes. The plants inside these boxes receive light from one direction.

According to the auxin theory, the stem tip is one of the major centers of auxin synthesis. Under natural conditions (in which the plant receives light from many directions), the auxin moves downward and is distributed evenly across the stem. As a result, growth is even and the young stem grows more or less straight. When auxin is not evenly distributed, growth is uneven; the stem elongates more on one side than the other, resulting in a bend or curve.

OBSERVATIONS

1. In which of the boxes is the distribution of auxin apparently even? _____ _____

2. In which of the boxes is the auxin distribution apparently uneven? _____ _____

3. Which wave length of full light has apparently altered the even distribution of auxin? _____ _____

4. Which wave length of full light apparently has no effect on the distribution of auxin? _____ _____

5. In those seedlings showing curvature, which side on the stems has grown more, the shady side or the illuminated side? _____ _____

6. This means, therefore, that there must be a greater concentration of auxin on which side of these stems? _____

7. Remember that auxin is a mobile substance moving *downward* under the influence of gravity. This particular demonstration indicates that auxin also moves _____ under the influence of _____, specifically the _____ wave length.

8. This demonstration shows that the phototropic response of plants is a differential growth rate caused by the _____ (color) wave length of _____.

9. What other effect(s) of light is (are) shown in this demonstration? _____ _____

10. Auxin promotes cell elongation. Which plants show the maximum effect of cell elongation? _____

11. Which plants, therefore, show no inhibition of auxin activity? _____

12. You have seen the effect of light on auxin *distribution*. What can you conclude as to the net effect of light on auxin *activity*? _____ _____

13. What wave length is necessary for the synthesis of chlorophyll? _____

Record the results of this experiment in Figure 18.1. Let the symbol ♀ represent a seedling, and sketch the seedlings in the trays, showing their posture. Record color and stem sturdiness.

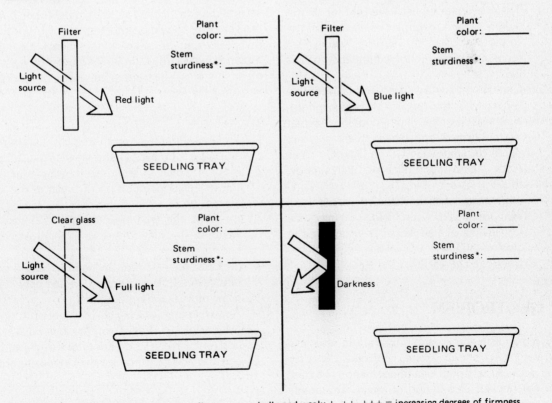

*Use the following symbols for stem sturdiness: o = spindly and weak; +, ++, +++ = increasing degrees of firmness

Figure 18.1   Results of exposure of radish seedlings to unidirectional light of different wavelengths (color) and to darkness.

B. ANOTHER ILLUSTRATION OF THE INFLUENCE OF LIGHT ON AUXIN DISTRIBUTION

ACTIVITY 4   **Observe** two pots of seedlings of oats (*Avena sativa*), barley (*Hordeum vulgare*), or other suitable plants, one set on a rotating platform (clinostat) and the other left stationary. Both plants are illuminated from one direction. Explain the growth response of these two sets of plants in terms of the auxin control theory. _____

_____

_____

_____

C. PHOTOTROPIC RESPONSE OF FUNGI TO LIGHT

ACTIVITY 5   **Observe** the culture of the fungus *Phycomyces blakesleeanus* that has been exposed to light coming from one direction.

1. Does the fungus show a phototropic response?   **Yes   No.**
2. The green plant must have light for photosynthesis, and therefore, its positive phototropic response has survival value. Are fungi green and photosynthetic?   **Yes No.**
3. Can you think of any probable value for this response of the fungus? Although photosynthesis requires light, could it be that normal growth and development also depend to some degree on light?   **Yes No.**   Does the fungal response support this?   **Yes   No.**   Quite so! Many organisms, animals included, depend on light for a variety of biological functions. Organisms of this planet have evolved in close relationship with the spectral energy from our nearest star, the sun.

III. GEOTROPISM

Geotropism is a growth curvature in **response** to gravity. Like phototropism, geotropism was for many decades accepted as a phenomenon resulting from the uneven distribution of auxin. The actual mechanisms are now known to be much more complex, and there is still no generally accepted alternate theory.

A. GEOTROPISM IN GERMINATING SEEDS

ACTIVITY 6   **Observe** corn seeds that have been germinating inside petri dishes and kept moist and stationary by a backing of wet filter paper. The seeds were originally set in the dish with no attempt made to orient them all the same way. The dish has been kept vertical.

OBSERVATIONS

1. In what direction do the roots turn? _____; the shoots? _____.
2. Have any roots or shoots made a complete U turn?   **Yes   No**
3. Of what value to us is this innate geotropic response of the young root and shoot of the seed? _____

B. GEOTROPISM IN THE WHOLE PLANT

ACTIVITY 7   **Observe** a potted specimen of coleus (*Coleus blumei*), geranium (*Pelargonium hortorum*), *Kalanchoe* sp., or other suitable herbaceous plant that has been set on its side for several days or (if class time permits) for an hour or so. The plant should be knocked out of its pot so that the roots can be seen. If the plants have been grown in peat cubes, they make even better specimens for this illustration because the roots are more readily seen.

OBSERVATIONS

1. In what direction are the root tips turned? _____; the shoots and leaves? _____
2. If auxin moves in response to gravity, where is the higher concentration of auxin in the root? _____; in the shoot? _____
3. In order for the root to "bend" downward, which of its sides (upper or lower) would have had to grow more? _____
4. In order for the shoot to "bend" upward, which of its sides (lower or upper) would have had to grow more? _____
5. Approximately the same distribution (concentration) of auxin exists on the lower side of the horizontal roots and shoots of this plant. How can you account then for their opposite growth response? _____

_____

_____

_____

IV. TURGOR MOVEMENTS

Turgor movements result from rapid changes in cell size because of loss of turgor pressure. Turgor pressure drops because of loss of water. When turgor pressure is regained by the cells, the movement of the plant parts affected is reversed.

ACTIVITY 8

1. **Observe** the reaction of the leaves and leaflets of the sensitive-plant (*Mimosa pudica*) to various stimuli and intensity of stimuli, such as touching them with a pencil or holding a lighted match near (but not *too* near).
2. **Observe** specimens (if available) of Venus's flytrap (*Dionaea muscipula*) or sundew (*Drosera rotundifolia*). The reactions of these plants occur in response to the pressure of the weight of insects who light on them. These plants are insectivorous. Changes in leaf position form traps for insects, which are then digested by enzymes secreted by the leaves.
3. **Examine** demonstration microscopes with slides of cross sections of leaves of corn (*Zea mays*) showing the leaf in the folded and open positions. **Locate** the bulliform cells whose collapse due to loss of water pressure causes the corn leaf to fold lengthwise. Under what environmental conditions would the leaf fold? _____ _____. This folding is beneficial because it reduces water loss. How is this reduction accomplished?

_____

_____

EXERCISE 18   **Student Name** _____

QUESTIONS

1. What are two environmental factors to which plants respond by a growth curvature?

    (a) _____; (b) _____
2. Which wave length(s) of full light is (are) responsible for plants bending toward light?_____

    _____

3. Is light necessary for the synthesis of chlorophyll? _____ any specific wave length?

    _____

4. The word auxin is derived from the Greek, *auxein,* to increase. Give the specific action

    exerted on the cell by auxin (which is the basis for its name). _____

    _____

5. The synthesis and development of the 2,4-D family of herbicides grew out of basic botanical

    and chemical studies of _____ in plants.
6. If you were a chemist charged with the research project to develop a synthetic growth
    hormone to be used as an herbicide to interfere with weed seed germination, *what growth*

    response of the seed would you target in on? _____ (There is such
    an herbicide. It is naptalam; it blocks the action of auxin, and the seed loses its normal
    behavior to this growth response.)
7. The same concentration of auxin occurs on the lower side of both stem and root of a
    plant placed horizontally. On this lower side (1) is root cell elongation (a) inhibited or
    (b) promoted? Circle (a) or (b). (2) Is shoot cell elongation (a) inhibited or (b) promoted?
    Circle (a) or (b).
8. From your observations of the root and shoot growth response of the horizontally placed

    plant, it can be concluded that the _____ concentration of _____

    that inhibits the growth of the _____ also promotes the

    growth of the _____.
9. Do auxins control turgor movements?   *Yes   No.*

10. Define plant geotropism. _____

    _____

    _____

PART II

Survey of the Plant Kingdom. Life Cycles: Their Significance

INTRODUCTION TO PART II:
SIGNIFICANCE OF LIFE CYCLES

In the exercises that follow you will study the details of the life cycles of representative plant groups. These details can be confusing and often a colossal bore to many beginning students. But, **basically, all life cycles are the same and they're all involved with survival of the group.**

Any plant group has four basic requisites for survival, which it meets as shown in the table.

These can be restated as *four survival phases:*

1. Multiplication 3. Assimilation
2. Dispersal   4. Genetic variation

The details of every life cycle fall into one of these four phases. Variety is the "spice of life," and every group has its own style of going through these phases. The style may differ but the essence of each phase is the same.

Genetic variation is very important in life cycles and needs a bit more explanation here. Meiosis and fertilization (or syngamy) work together. Just as a card dealer shuffles the cards between each game so that the players can receive different combinations of cards, so meiosis rearranges genes and gene groups so that the resulting haploid nuclei receive new gene combinations. These new gene combinations contribute to genetic variety. Meiosis also segregates two chromosome sets so that each daughter cell gets one set.

Sexual fertilization brings together two haploid nuclei, both bearing somewhat different gene combinations in their otherwise matching chromosome sets.

RELATIONSHIP OF LIFE CYCLE PHASES TO SURVIVAL

REQUISITE FOR SURVIVAL	ACCOMPLISHED BY
1. It must increase its numbers	1. Asexual reproduction (primarily)
2. It must disperse its offspring	2. Motile cells or air- or animal-carried cells or seeds
3. It must obtain and assimilate foods for energy	3. Decomposing and consuming organic matter, if a fungus; photosynthesis, if a green plant
4. It must put some genetic variety into its populations	4. Sexual reproduction: fertilization and meiosis

BACTERIA AND BLUE-GREEN ALGAE

EXERCISE 19

INTRODUCTION

Bacteria make newspaper headlines today. Big firms have made major investments in a new industry called genetic engineering of bacteria. It's all the result of the incredibly successful research of the 1970s in which scientists, using bacterial cells, perfected gene-splicing techniques that allow them to control what they want the cell to produce.

There is enormous potential in this procedure of harnessing microorganisms to produce genetically specified compounds. Microbes are easy and cheap to grow, and the potential exists for using them to synthesize a variety of valuable substances normally made by higher organisms, such as certain hormones (e.g., insulin), enzymes, antibiotics, and so on. Also possible is the bacterial synthesis of numerous commodity chemicals that have traditionally been manufactured by much more expensive and technically elaborate methods.

In this research, in which the goal is to splice genes and recombine them in specific ways (called recombinant DNA), scientists needed a simple cell type by which they could carry out and test these manipulations. The cell of the bacterium (and to a lesser extent, the blue-green alga) was the perfect instrument. Cells of higher organisms are just too complex to serve this purpose.

So here is a unique case in which one of the oldest and simplest of all organisms—the bacterium—is used in one of the most advanced fields of technology, a technology that (as one scientist put it) "suggests a new realm of 'bacterifacture' by which rapid, controlled growth of microorganisms is coupled to the production of specific products." Keep this in mind as you take your first look at these simple organisms.

Although the prospects of genetic engineering of microbes are dazzling, we must emphasize the reality of the ancient role that bacteria have had in nature. It is they and other microbes that are responsible for the recycling of matter between the soil, the air, and living organisms. The quantity of chemical elements on earth is fixed, and any of these elements that have become incorporated into a living organism eventually are released when that organism dies. It is the action of bacteria, fungi, and other microorganisms that releases elements from the complex compounds of the dead tissues or waste products of living things and makes them available for reuse by other organisms. Bacteria, of course, are also the causative organisms in many plant and animal diseases.

SIMILARITY OF BACTERIA AND BLUE-GREEN ALGAE (CYANOBACTERIA)

Bacteria and blue-green algae are so similar that scientists no longer consider the blue-greens to be true algae; hence the name cyanobacteria. Bacteria and blue-green algae stand together, and differ from all other plants because they have a distinctly different cell type that shows a much lower level of organization than that of other plants. This primitive cell is called procaryotic. The cell of all other plant groups is the type studied in other exercises. It is called eucaryotic.

In the procaryotic cell, nuclear material is not enclosed by a membranous envelope but instead spreads throughout the cytoplasm. The chromatin material is an aggregate of fibrils that, during cell division, is seen to form a single, circular DNA molecule sometimes called a ring chromosome.

In addition to this unique "chromosome," the procaryotic cell has a wall structure, a plasmalemma, and other cytoplasmic components that are strikingly different from those of any eucaryotic cell. For these reasons (among others), scientists believe they are warranted in dividing the plant world into two groups, the procaryotes and the eucaryotes (often spelled prokaryote and eukaryote). In fact, procaryotes are considered to be so distinct that they are usually placed in a separate biological subkingdom called the Monera.

Aside from the similarity in cell organization in bacteria and blue-green algae, no clear-cut relationships can be easily demonstrated between them. Because both are ancient and preceded eucaryotic plant life in the earth's history, and because both have a simpler cellular organization than the eucaryotes, most scientists feel that they probably gave rise to eucaryotes, but there is no consensus as to which organisms or what steps were involved in this development.

I. BACTERIA (DIVISION SCHIZOMYCOPHYTA)

In spite of the modest variation in size and shape, there is tremendous variety among bacteria. We find modes of life among them far more physiologically varied than in all other plant types together. There are unicellular and colonylike forms. Many resemble fungi, slime molds, protozoans, or algae. Some are aquatic, some terrestrial. A few are autotrophic. Most are heterotrophic, and of these more than 99 percent are harmless saprophytes.

We have time to study only the most common bacteria, which also happen to be among those with the simplest forms.

A. EXAMPLES OF SOME COMMON BACTERIAL TYPES

Three cell shapes are found in the majority of bacteria. These are (1) coccus—a spherical cell (plural, cocci), (2) bacillus—a rod-shaped cell (plural, bacilli), and (3) spirillum—a rod-shaped cell with one or more curves (plural, spirilli).

Some characteristics of the most common bacteria are:

1. Unicellular and microscopic
2. Nonphotosynthetic
3. Saprophytic
4. Rigid cell walls
5. Reproduction by simple fission

ACTIVITY 1 **Examine** a prepared slide showing the three cell shapes found in different species of bacteria. Do the cells always occur as single isolated cells? *Yes No.* Do you see any chains of cocci? *Yes No.* These chains are considered colonies. They are called STREPTOCOCCI. Certain species of bacteria characteristically grow in chains of cells. Do you see any chains of bacillus forms? *Yes No.* Are all bacilli the same thickness? *Yes No.*

How many cells would form a DIPLOCOCCUS colony? _____ Are there any on your slide? *Yes No.* Cocci that grow in clusters forming flat, irregular plates are called STAPHYLOCOCCI.

B. NITROGEN-FIXING BACTERIA

Nitrogen is unique among the major plant nutrients in that its reservoir is not found in the soil's solid particles or water solution but in the atmosphere, including the air spaces in the soil. The only way that this atmospheric nitrogen can be made available for higher plant nutrition is through certain bacteria. It is the nitrogen-fixing bacteria that convert (fix) this free atmospheric nitrogen into nitrates that can be used by the higher plants. In some circumstances, particularly in very wet or flooded soil, blue-green algae process (fix) the nitrogen that is dissolved in the water into the nitrate form usable by higher plants.

Rhizobium sp. is. the most important soil organism for the fixation of atmospheric nitrogen. However, the bacterium cannot fix the nitrogen unless it is living in the roots of plants, especially the legume plants, for example, alfalfa (*Medicago* sp.), sweet clover (*Melilotus* sp.), clover (*Trifolium* sp.), bush clover (*Lespedeza* sp.), birdsfoot trefoil (*Lotus* sp.), soybean (*Glycine* sp.), pea (*Pisum* sp.), and vetch (*Vicia* sp.).

Species of *Rhizobium* invade the root hairs of these plants (and a few other nonleguminous plants). Their presence stimulates the plant to form small tumorlike growths called **NODULES** around them. Inside the nodules, the bacteria consume food compounds made by the plant. They also fix nitrogen and synthesize organic nitrogen compounds that the plant uses. The association is **MUTUALISTIC**, both plant and bacteria deriving substantial benefits from it.

After harvest or plowing under of these crops, the soil is enriched by the nitrogen compounds released from the dead plant debris and the bacterial cells that eventually die. This is the basis for rotating (alternating) legume crops with other crops. The introduction of clover into customary crop rotations in the 16th century revolutionized agricultural practices throughout the world.

ACTIVITY 2 **Examine** the roots of leguminous plants and note the nodules. If time permits, **crush** open a nodule in a drop of water on a slide. Transfer some of the liquid to another slide and cover with a cover slip and examine for bacterial cells.

II. BLUE-GREEN ALGAE—DIVISION CYANOPHYTA OR CYANOBACTERIA

The Cyanophyta are the blue-green algae, the most primitive of the chlorophyll-bearing plants. Like bacteria, they have the procaryotic type cell, and in this and other ways they are more similar to bacteria than to algae or other plants.

General facts about blue-green algae are

Usual color Blue-green, but many are red, brown, or purple.
Distribution Mostly in ocean and brackish waters.
Habitat and niches Most species grow in tidal marshes and mud flats; many occur in upper 1–3 cm of the soil surface. A few species are found in caves, hot springs, geyser basins, wet cliffs, waterfall ledges, and occur as plankton of ponds, lakes, rivers, and so on.
Motility None. However, in some there is a gliding motion.
Body types Most are filaments; some are loose colonies; some are unicellular.
Value Have both beneficial and detrimental effects. Excess growth in fresh waters creates nuisance and depletes oxygen content.

There isn't much cellular detail visible in a blue-green alga. Nor are there any complicated life cycles. Reproduction is asexual only. It occurs by cell divisions, spores, and fragmentation of the thallus. Let's take a look at a few plants and see what is visible.

Review the section on page 174 on the **significance of life cycles.**

A. GLOEOCAPSA

ACTIVITY 3 **Get** a clean slide and make a wet mount from a culture of Gloeocapsa. **Cover** with a cover slip. **Examine** under the microscope. What body type does Gloeocapsa have? _____ _____. What shape are the CELLS, rectangular or spherical? _____.

Each cell is surrounded by a mucilaginous capsule. All the cells together in a group are also embedded in one common capsule. How many cells do you see in a group? Too many to count? *Yes No.* How many? Circle one: (a) 2 to 8 cells; (b) more than 8; (c) more than 20. Can you see any nuclei? *Yes No.* Any chloroplasts? *Yes No.* Are any cells swimming? *Yes No.* **Put** a drop of methylene blue stain on one edge of the cover glass. By holding a bit of paper toweling at one edge of the cover glass, you can draw water out and pull the stain under the cover glass. The stain will be absorbed by the SHEATH. Can you see individual cell sheaths? *Yes No.* Is the entire COLONY surrounded by a sheath? *Yes No.* **Label** the colony of Gloeocapsa in Figure 19.1, using the terms capitalized in the preceding description.

B. OSCILLATORIA

ACTIVITY 4 **Make** a wet mount of Oscillatoria. What do you see, (a) colonies or (b) filaments? Circle (a) or (b). Is there a SHEATH surrounding the plant? *Yes No.* Which is the bigger dimension, (a) the length of each CELL or (b) the width of each? Circle (a) or (b). How would you describe the cell contents, (a) evenly granular or (b) with various different inclusions? Circle (a) or (b). Do the filaments move back and forth? *Yes No.* What does the name Oscillatoria imply? _____ _____ _____.

Oscillatoria reproduces simply by fragmenting the filament. It happens this way: frequently a cell in the filament dies and the dead cell becomes a weak, breaking point or SEPARATION DISK in the filament. The short chain of cells between any two disks is called a HORMOGONIUM (plural, hormogonia). When a filament breaks, the hormogonia (in some inexplicable way) are able to move through the water, thereby accomplishing dispersal. Cell divisions within the hormogonium lengthen the chain and produce a new filament. (Hormogonium is a name coined from two Greek words meaning roughly a "reproducing necklace.")

Label the filament of Oscillatoria in Figure 19.2, using the terms capitalized in the preceding description.

C. ANABAENA

ACTIVITY 5 **Make** a wet mount of living specimens of Anabaena and cover with a cover slip. Is this (a) a filament or (b) a colony? Circle (a) or (b). Do all the cells in the plant look alike as in Oscillatoria? *Yes No.* Is there a SHEATH around the plant? *Yes No.* Are the cells disk-shaped? *Yes No.* Look for large, oval cells that are clear or transparent. These are **HETEROCYSTS**. They are typical of many blue-green algae. When a filament breaks into smaller fragments, it always breaks at a heterocyst. In some species heterocysts are at the end of the filament (terminal heterocysts). In others they are interspersed (intercalary heterocysts) among the **VEGETATIVE CELLS**. The position varies with different species and is helpful in classifying some of the blue-green algae.

Label the filament of *Anabaena* in Figure 19.3, using the terms capitalized in the preceding description.

D. NOSTOC

Nostoc is a filamentous alga whose filaments grow extensively intertwined with each other, forming larger and larger colonies that are ball-shaped. They have a very thick, firm, gelatinous sheath. Some colonies become large enough to be seen with the naked eye.

ACTIVITY 6 **Make** a wet mount of living *Nostoc* and examine the individual filaments. Do you see any heterocysts? *Yes No.* **Observe** a few macroscopic, jellylike colonies.

EXERCISE 19   **Student Name** _____

QUESTIONS

1. In what fundamental way do bacteria and blue-green algae differ from all other plants?
    _____

2. On the basis of the cell type, name the two subkingdoms of the plant world. _____
    _____

3. What subkingdom consists of the blue-green algae and bacteria?   _____
    _____

4. Is there any evidence that bacteria and blue-green algae are related?   *Yes   No*

5. Name the three basic bacterial shapes.   _____
    _____

6. Is atmospheric nitrogen directly usable by higher plants?   *Yes   No*

7. In what chemical form must nitrogen exist in order for it to be utilized by higher plants?
    _____

8. What kind of bacteria convert atmospheric nitrogen to a form usable by higher plants?
    _____

9. When coccus-type bacteria occur in chain colonies, they are called _____
    _____

10. Do we have any evidence that procaryotes were ancestral to eucaryotes?   *Yes   No*

11. Did procaryotes precede eucaryotes in time in the earth's history?   *Yes   No*

12. Were you able to see any nuclei in the cells of either bacteria or blue-green algae?   *Yes
    No*

13. Are blue-green algae photosynthetic?   *Yes   No*

14. What particular group of plants characteristically harbor nitrogen-fixing bacteria in their
    roots? _____

15. Microbes today are playing a new role in the technology of genetic engineering, but what
    role have they always played in nature? _____
    _____

ONE CELL

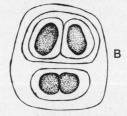

COLONY OF CELLS

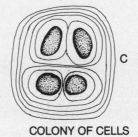

COLONY OF CELLS

Figure 19.1 *Gloeocapsa* sp.

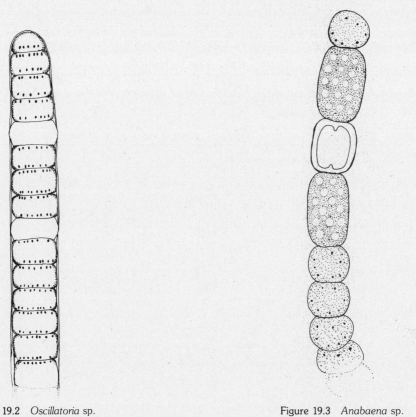

Figure 19.2 *Oscillatoria* sp.

Figure 19.3 *Anabaena* sp.

FUNGI I: Primitive Fungi and Slime Molds

EXERCISE 20

INTRODUCTION

Perhaps today you should visit a cheese factory, a bakery, a brewery, or a pharmaceutical laboratory where antibiotics such as penicillin, streptomycin, and Terramycin are made. If time permits, you might stop at a soft drink or food processing plant and observe the tremendous quantities of citric acid that go into the making of these beverages and foods.

On another day you might go to other laboratories or factories and watch the production of industrial alcohol, glycerol, and a host of organic acids used in paints, plastics, foods, varnishes, adhesives, and pharmaceuticals.

Would this be botany? Would it be Exercise 20, the study of fungi? Indeed it would. You wouldn't absorb the details of fungal structure and function, but you would be flabbergasted by the fact that *all these economically valuable products, consumed or used by millions of people every day, are the products of fungi or of fungal action!*

But fungi earn some bad marks as well as good ones. On farms, in orchards, and in greenhouses. people are forever battling fungi. If it weren't for fungicide sprays, dusts, and soil drenches, many fruits, vegetables, grains, and flower crops would never reach the market.

Parasitic fungi can so devastate certain varieties of food crops that farmers, commercial growers, and home gardeners have given up and grow only disease-resistant varieties that have been developed in research laboratories. Development of new disease-resistant varieties is a continuous research effort and one that occupies the lifetimes of several hundred scientists.

Yet the value of soil fungi cannot be omitted. Practically every kind of crop (like most higher plants) derives considerable benefit from certain soil fungi that grow in or on its roots in a mutualistically beneficial union known as mycorrhiza.

Although fungi are traditionally included for study in elementary botany, they don't look much like plants. In fact, fungi are so different from any other living thing that many people think they constitute a separate kingdom of organisms unrelated to plants, animals, or bacteria.

For one thing, fungi never form tissues, and all parts of the fungus body consist entirely of a system of tubes! Fungi lack chlorophyll. Instead of being synthesizers of food like green plants, they absorb food. No, they're not like animals. Animals ingest food, digest it, and then absorb it. Fungi reverse the sequence. They produce digestive enzymes but secrete them on the substrate, which is then digested; the products of digestion are then absorbed by the fungus.

Of the approximately one-quarter of a million known species of fungi, most are beneficial to the environment and the human economy. It is they who rid the earth of dead organisms, wastes, and other debris by decomposing them. In the process, minerals and organic compounds are returned to the land and water to be recycled by other organisms in nature. These fungi are called the saprophytes.

We know how to manipulate only a few dozen or so of these fungi, as in the making of bread, cheese, beer, wine, antibiotics, organic acids, industrial alcohols, and other compounds.

A minority of fungi live on living material—plants, animals, and humans. These are the parasites. As with all "culprits," the destructive activity of parasites receives far more attention from people than the beneficial action of the saprophytes.

Many more beneficial fungi await discovery. Every scientist, every discoverer, started out like you today, observing and learning the ways of fungi.

I. SOME TERMS USED FOR FUNGI

FUNGUS Term for a single plant (plural, fungi).

MOLD A commonly used term for fungus.

HYPHA A branching, threadlike tubule (plural, hyphae) that, in huge numbers, comprises the mycelium.

MYCELIUM Vegetative or assimilative phase of the thallus. Consists of white masses of microscopic branching filaments called hyphae. A mycelium is perennial. It lives for several years and usually isn't seen because it grows under the soil surface or inside the host material (plural, mycelia).

FRUITING BODY Reproductive part of the thallus that bears the sexually produced spores. In some species it is a microscopic or extremely small single sac. In other species it is a conspicuous, fleshy organ. It may live for a few days or several years.

THALLUS Name for the entire plant body, both vegetative parts and reproductive parts. (This term is also used for algae and bryophytes.)

COENOCYTE A structure, such as the mycelium, with numerous nuclei and no cell walls separating them. The adjective is coenocytic.

SEPTATE With cross walls in the hyphae.

NONSEPTATE Without cross walls; refers specifically to filaments or tubules in the mycelium. Means the same as coenocytic.

PERFECT STAGE The phase of the life cycle in which sexual reproduction occurs. Takes place only once a year in most fungi. Brings about genetic variation in the population.

IMPERFECT STAGE The phase of the life cycle in which asexual reproduction occurs. Takes place many times during the season. It is the chief means for increasing the population size.

II. THE BIG PICTURE OF THE WORLD OF TRUE FUNGI

Before diving into details, let's get a general idea, a "big picture" of the world of the true fungi. The variety of types is a reflection of their evolution. In this evolution we can see that four major trends or lines of advance have occurred:

1. A trend from aquatic life to terrestrial life.
2. A trend from small inconspicuous size to large conspicuous size (especially in the fruiting body).
3. A trend from a simple sac (that merely releases spores) to complex discharge and dispersal mechanisms (that spread spores more efficiently).
4. A trend from a haploid thallus to a diploid thallus.

Figure 20.1 and the accompanying Table 20.1 illustrate the general differences between the classes of true fungi (Eumycophyta). These classes are:

1. Class Oomycetes—Water molds and their relatives. Examples: downy mildews, potato blight fungus, white rusts, pythium.
2. Class Zygomycetes—Primitive terrestrial molds without fruiting bodies. Examples: bread mold, fly fungi.
3. Class Ascomycetes—Terrestrial molds, most of which have fruiting bodies that are spherical, flask-, or cup-shaped. Examples: morel, cup fungi, dutch elm disease fungus, apple scab fungus, ergot.
4. Class Basidiomycetes—Terrestrial molds, most of which have fruiting bodies generally larger than those of ascomycetes and also very diverse in form. Examples: mushrooms, puffballs, rusts, smuts, stinkhorns.

Review the section on the significance of life cycles on page 174.

III. DIVISION EUMYCOPHYTA— CLASS OOMYCETES

A. AQUATIC OOMYCETES

Mycelium—The Vegetative Part of the Thallus

ACTIVITY 1 **Examine** prepared slides or water cultures containing living water molds, such as species of *Saprolegnia* or *Achlya*. The mold will be growing on dead hemp seeds or dead insects in the water. **Examine** the MYCELIUM under the microscope. What is the name for the filaments or tubes that compose this mycelium? _____.
Can you see any cross walls in these tubes?   *Yes*
No.   Give the adjective used to describe this type of mycelium. _____. What survival phase does the mycelium represent? _____. In Table 20.2, **enter** the name mycelium in the box opposite the survival phase it represents.

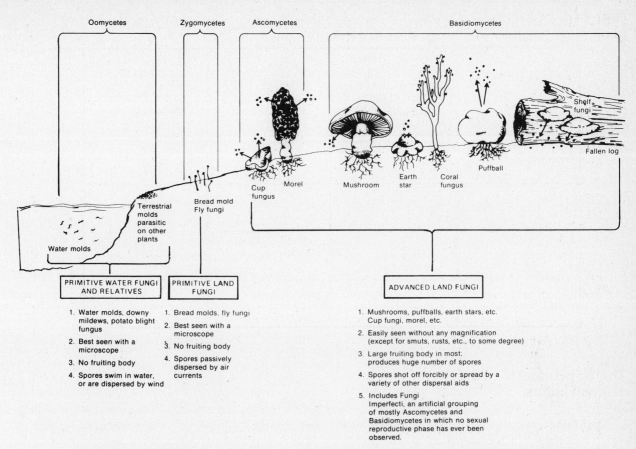

Figure 20.1 The "big picture" of the fungi. A generalized scheme of the habitat, relative size, and evolutionary status of the various groups.

Table 20.1 DIFFERENCES AMONG THE TRUE FUNGI (EUMYCOPHYTA)

	Class	OOMYCETES	ZYGOMYCETES	ASCOMYCETES	FUNGI IMPERFECTI	BASIDIOMYCETES
	Habitat	Some aquatic, some terrestrial	Mostly terrestrial	Mostly terrestrial	Mostly terrestrial	Mostly terrestrial
	Mycelium	Nonseptate, growth not extensive	Nonseptate, growth not extensive	Septate, extensive in growth	Septate, extensive in growth	Septate, extensive in growth
TECHNICAL CHARACTERISTICS	Sexual spore stage	Thick-walled resting stage, called OOSPORE, not in a fruiting body	Thick-walled resting stage, called ZYGOSPORE, not in a fruiting body	ASCOSPORES (usually 8) borne inside a sac (ascus), usually in a fruiting body; many lack the sexual stage	None	BASIDIOSPORES (usually 4) borne on the outside of a club-like cell (basidium), usually in a fruiting body
	Asexual spore stage	Motile spores (ZOOSPORES) produced inside a sporangium	Nonmotile spores (APLANOSPORES) produced inside a sporangium	Nonmotile spores (CONIDIOSPORES) formed on the tips of specialized filaments (CONIDIOPHORES); some lack an asexual stage	Nonmotile spores (CONIDIOSPORES) formed on the tips of specialized filaments (CONIDIOPHORES)	Nonmotile spores (CONIDIOSPORES) formed on the tips of specialized filaments (CONIDIOPHORES); many lack an asexual stage
	Motile cells	Present	None	None	None	None

Table 20.2 LIFE CYCLE CELLS, STAGES, OR EVENTS RELATED TO METHODS OF SPECIES SURVIVAL (REVIEW PAGE 174)

SURVIVAL PHASE	ACCOMPLISHED IN *SAPROLEGNIA* SP. BY WHAT CELL(S), STRUCTURE(S), PHASE(S), OR EVENT(S)	ACCOMPLISHED IN *RHIZOPUS* SP. BY WHAT CELL(S), STRUCTURE(S), PHASE(S), OR EVENT(S)	ACCOMPLISHED IN *PHYSARUM* SP. BY WHAT CELL(S), STRUCTURE(S), PHASE(S), OR EVENT(S)
Multiplication			
Dispersal			
Assimilation			
Genetic variation			

The Reproductive Parts of the Thallus

Examine a prepared slide showing the asexual and sexual organs of *Saprolegnia* sp. or *Achlya* sp. **Look** at Figure 20.2 as a guide.

Asexual Structures   **Look** for oval, cigar-shaped sacs at the tip of a hypha. Each such sac is a SPORANGIUM. It contains many cells that are released and swim about or are carried about by water currents. These motile spores are called ZOO-SPORES. These settle down and encyst for a while. Later the encysted cells germinate into secondary zoospores. Each secondary zoospore can grow into a new thallus. What two survival phases are represented here? (a) _____; (b) _____ . In Table 20.2 **enter** the names of these structures in the boxes opposite the survival phases they represent.

Sexual Structures   **Look** for round bodies attached along the sides of the hyphae. These are the female sex organs, called OOGONIA (singular, OOGONIUM). MEIOSIS occurs within the oogonium and 5–10 haploid EGG NUCLEI are formed. Male sex organs, called ANTHERIDIA (singular, ANTHERIDIUM) look more like an ordinary hypha. MEIOSIS occurs within the antheridia, producing haploid SPERM NUCLEI. When an antheridium comes in contact with an oogonium, it puts out fingerlike extensions, called FERTILIZATION TUBES. These can pierce the oogonium, and each tube enters one egg and delivers a sperm nucleus directly to the egg. If fertilization has occurred, you will see four to five thick-walled, opaque zygotes inside the oogonium. These zygotes, termed OOSPORES, are diploid; they have two sets of genes—one from the sperm and one from the egg. After a period of dormancy (which typically coincides with adverse drought conditions or the cold season), each OOSPORE germinates into a short tube with a terminal sporangium, the GERM SPORANGIUM. Several germ sporangia may be formed. The germ sporangium produces cells that, when released, are motile ZOOSPORES. These are carried about by water and eventually can grow into a new thallus.

These structures studied in this section fall into what two survival phases? (a) _____; (b) _____. **Enter** the names of these structures in the appropriate boxes in Table 20.2. **Complete** the labeling of the life cycle of *Saprolegnia* in Figure 20.2, using the terms capitalized in the preceding description. **Label** the four survival phases in Figure 20.2.

B. TERRESTRIAL OOMYCETES

Certain of these molds cause serious plant diseases. *Phytophthora* causes a large number of diseases, among which is blight in potatoes, which brought on the potato famine in Ireland in the 1840s; *Pythium* causes root rot in many greenhouse and field crops and "damping-off" disease of seedlings in all plants. Downy mildew disease is caused by a variety of molds, one of which almost wiped out the grape-wine industry in France in 1882.

IV. DIVISION EUMYCOPHYTA— CLASS ZYGOMYCETES

Zygomycetes are primitive terrestrial molds without fruiting bodies. Some of these fungi are saprophytes, such as the bread molds, and others are parasitic on insects, such as the fly fungi. The genus *Rhizopus* is very important in the manufacture of organic acids. Lactic acid, which is widely used as flavoring in food and in the manufacture of adhesives and pharmaceuticals, is synthesized by a species of *Rhizopus*. Another *Rhizopus* species is grown for its production of fumaric acid, which is used in the manufacture of plastics and varnishes. *Rhizopus stolonifera,* which you will examine, plays a major role in the decomposition of organic debris.

ACTIVITY 2 **Examine** living specimens of *Rhizopus stolonifera.* **Look** at Figure 20.3 as a guide. **Obtain** an intact portion of the mycelium that has been grown on 1-cm² sections of cellophane paper over agar. **Mount** the cellophane section with the mycelium on a *dry* slide and examine *uncovered* under medium power (100×) on your microscope.

The Mycelium and the Asexual Structures

Note that the MYCELIUM is a white or grayish-white mass of branched filaments not very different from those of *Saprolegnia.* Is this mycelium haploid or diploid? _____ (See Figure 20.3.) What survival phase does this mycelium represent? _____. **Enter** the name mycelium in Table 20.2 in the appropriate box opposite the survival phase it represents.

If the culture is old enough, you will be able to see tiny black spherical bodies, the SPORANGIA (singular, sporangium) attached to erect hyphae, called SPORANGIOPHORES. Sporangia contain haploid cells, SPORANGIOSPORES. These have been formed from simple mitotic division of the haploid nuclei of the thallus. Do they contain the same gene combinations as the parent thallus? *Yes   No.*   When released, they are dispersed by air currents and later each can germinate into a new thallus. Will this new thallus be haploid or diploid? _____. What survival requisite does asexual reproduction accomplish? _____.

Add a drop of water to the mycelium on the slide and cover with a cover glass. **Examine** the sporangia and spores under the microscope. **Note** the cluster of rootlike hyphae at the base of each sporangiophore. These are called RHIZOIDS. They anchor the hyphae. Coming off laterally from the base of the sporangiophore are horizontal-growing hyphae, known as STOLONS. What survival phase(s) are these structures involved in? _____

_____. **Enter** the names of these structures in the appropriate boxes in Table 20.2.

Sexual Structures and the Process of Conjugation

Look at Figure 20.3 as a guide again. **Examine** a prepared slide of *Rhizopus stolonifera* or a living culture of the organism showing sexual reproduction. When two different strains of mycelia (we call them "plus" and "minus" strains) come in contact, all (+) and (−) HYPHAE that touch each other send out a side bulge at the point of contact. This bulge is called a PROGAMETANGIUM. Each progametangium grows larger and pushes against the other. They form a bridge between the hyphae. Soon the nuclear and cytoplasmic contents of each progametangium are walled off from the rest of the hypha. Each walled-off part is now called a GAMETANGIUM. This word simply means a "sac of gametes," and that's what it is—a sac of nuclei that will behave as gametes, the sperm or eggs.

When gametangia are mature, the wall between them breaks down, and the nuclear contents of the two fuse—that is, fertilization, CONJUGATION, takes place. **Look** to see if you can find a dark (dense) sphere in the "bridge" formed by the gametangia. This sphere is the ZYGOSPORANGIUM. It contains the diploid nuclei formed by nuclear fusions inside the gametangia. Of these, only one diploid nucleus survives, giving rise to a single ZYGOSPORE within the zygosporangium. This zygospore remains dormant for 1–3 months. Then when it germinates, it cracks open the zygosporangium and produces a SPORANGIOPHORE that bears a GERM SPORANGIUM at its tip. Meiosis takes place during the process of zygospore germination. Germ sporangia contain either all (+) or all (−) spores or a mixture of the two types. Most of these structures described here represent what survival phase? _____. In Table 20.2 **enter** the names of these structures in the box opposite the appropriate survival phase they represent.

Complete the labeling of the life cycle of *Rhizopus stolonifera* in Figure 20.3, using the terms capitalized in the preceding description.

ACTIVITY 3   If available, **examine** living cultures of plus and minus strains of *Phycomyces blakesleeanus* that have sexually crossed and produced zygospores. **Note** that the zygospores are formed only along the line of contact of the two strains.

V. MYXOMYCOPHYTA—
THE SLIME MOLDS

Slime molds are a small, insignificant group of very little economic importance. Botanists call them fungi because, like fungi, they produce sporangia. Zoologists think they are protozoans because the thallus looks and behaves like a giant amoeba. However, they are usually studied by botanists. Since they are not totally funguslike or animallike, opinions differ about their identity. So they are classified along with the protozoa as a group totally separate from either plants or animals. This group is thus usually placed in the division or kingdom Protista.

ACTIVITY 4   **Examine** a living slime mold such as *Physarum polycephalum* growing on agar or filter paper in a petri dish. **Use** 100 × magnification with the microscope. *Do not* expose the plasmodium to bright light, or within 30 minutes it will go into a sporangial stage. The thallus or assimilative phase is called the **PLASMODIUM**. This is a broad, flat-lying mass of protoplasm covered by a thin plasma membrane and a gelatinous slime sheath. Portions of it fan out wider than others and you can see "veins" of streaming nuclei and particles of cytoplasm.

The plasmodium is a coenocyte. It flows over the surface by protoplasmic streaming and engulfs bacteria and other minute particles in its path. During its reproductive phase, it stops moving and produces erect, saclike structures called **SPORAN-GIA**. Cells inside these divide by meiosis to form **HAPLOID SPORES**. These haploid spores are released, blown around, and lie about for a while going through a **PERIOD** of **DORMANCY**. Later they germinate, and two kinds of cells are formed, flagellated **SWARM CELLS** and amoebalike **MYXA-MOEBA**. These behave like gametes—two swarm cells unite or two myxamoeba unite. Their proto-plasms fuse but not their nuclei. This fusion of protoplasm is known as **PLASMOGAMY**. Later the nuclei fuse. This is known as **KARYOGAMY**. Sub-sequently a **NEW PLASMODIUM** develops.

Enter the names of the structures described here in the appropriate boxes in Table 20.2 aside the corresponding survival phase each represents. *Note:* No asexual reproduction has been described for this organism.

Complete the labeling of the life cycle of *Physarum polycephalum* (Figure 20.4), using the terms capitalized in the preceding description. **Label** also the survival phases in Figure 20.4.

EXERCISE 20   Student Name _____

QUESTIONS

1. What does asexual reproduction achieve for any plant group in terms of its survival?
   _____

2. What is the general term for the entire plant body of a fungus? _____

3. What is the term for the vegetative part of the fungus plant body? _____
   _____

4. What does meiosis achieve for any plant group in terms of its survival? _____
   _____

5. If a fungus has no motile cells, how then is it dispersed? _____
   _____

6. Name two important organic acids produced by species of *Rhizopus*. _____
   _____

7. Which are more primitive, aquatic or terrestrial fungi? _____

8. Which is the more common event, sexual or asexual reproduction? _____
   _____

9. What is the term used to describe a tube that has cross walls? _____

10. What is the term used to describe a multinucleate structure without any cell walls?
    _____

11. What is the term used to describe the gametic union of protoplasm but not the accompany-
    ing union of the nuclei? _____

12. What is the term used for the fusion of two nuclei? _____

13. Every life cycle can be reduced to four survival phases. List them:
    (a) _____; (b) _____
    _____; (c) _____; (d) _____
    _____

14. Which plays the greater role in adding to the population size, sexual reproduction or
    asexual reproduction? _____

15. Why isn't a slime mold considered to be a "100 percent" fungus? _____
    _____
    _____

16. Name four different industries in which fungi play a major role. _____
    _____

17. Do Zygomycetes have large, conspicuous fruiting bodies? _____

18. List the four evolutionary trends seen in the fungi: (a) _____
    _____; (b) _____
    _____
    _____; (c) _____ _____
    _____; (d) _____
    _____

19. Using general terms, name the cells and/or events of a fungus life cycle that are involved in bringing about genetic variation. _____

_____

_____

20. What structure in the life cycle of a fungus is equivalent to the plasmodium of a slime mold? _____

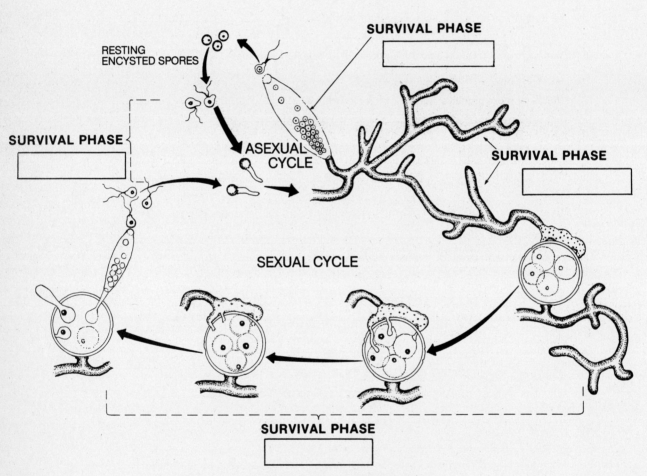

Figure 20.2   Life cycle of *Saprolegnia*.

EXERCISE 20   **Student Name** _____

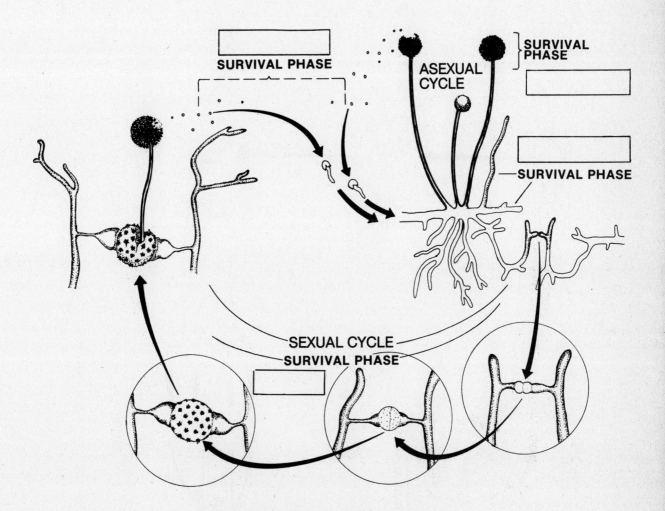

Figure 20.3   Life cycle of *Rhizopus stolonifera*.

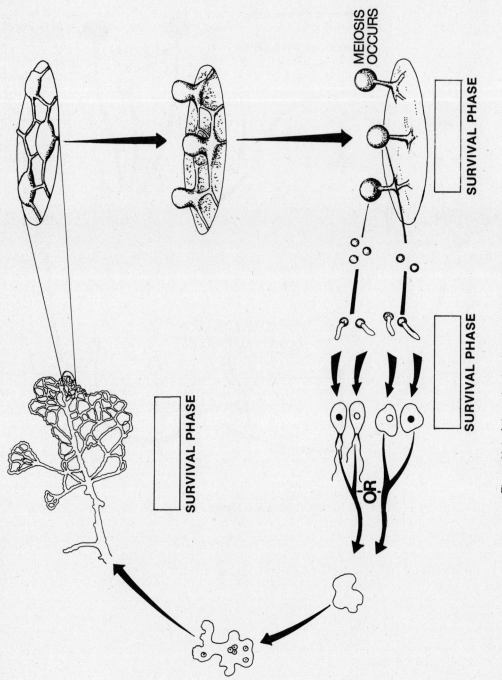

MEIOSIS OCCURS

SURVIVAL PHASE

SURVIVAL PHASE

OR

SURVIVAL PHASE

Figure 20.4   Life cycle of a slime mold, *Physarum polycephalum*.

FUNGI II:
The Higher Fungi

EXERCISE

21

INTRODUCTION

In this exercise you will examine the more complex (that is, the so-called higher) fungi, the Ascomycetes and the Basidiomycetes. The edible mushroom, the shelf fungi that grow on old tree trunks, the famous French truffles sniffed out by trained pigs, and the edible gourmet morel—these are the fungi most familiar to people in general.

This visible part of the fungus is, of course, only the fruiting body. What few people realize is that hidden within the old log or below the soil surface is most of the mass of the organism—the mycelium.

If you look back at Figure 20.1 and Table 20.1 in the previous exercise, you will see that a third group, the Fungi Imperfecti, is included in the group of advanced land fungi. These imperfect fungi are all those that seem to lack sexual reproduction in their life cycle. Either they have lost their ability to reproduce sexually (through evolutionary reduction) or scientists have simply failed to find the sexual stages.

The Fungi Imperfecti is an artificial category created as a classification pigeonhole for those fungi in which only asexual reproduction has been observed. Among the Fungi Imperfecti are some important species, namely, the soil-living species of *Streptomyces* from which we get the antibiotics streptomycin and Aureomycin and also vitamin B_{12}. In this group also are the species *Fusarium* and *Verticillium*, notorious plant pathogens that cause serious wilt diseases in a wide variety of plants.

I. CLASS ASCOMYCETES— THE SAC FUNGI

Ascomycetes are terrestrial molds, most of which have fruiting bodies that are spherical or flask-or cup-shaped, and whose sexually formed spores are borne in a sac, called an **ASCUS**.

Ascomycetes are the largest group of fungi. Economically they are very important, for example, the yeasts (*Saccharomyces* sp.), which are vital in the baking industry and the operations of breweries and wineries. From yeasts also we derive our commercial source of B vitamins as well as glycerol, a compound with many different industrial uses. Some ascomycetes, such as truffles (*Tuber* sp.) and morels (*Morchella esculenta*), are prized as food delicacies.

As discussed in the Introduction, the sexual stages of many fungi aren't known. For some genera, it is assumed that they would be ascomycetes if the sexual stages *were* known. *Aspergillus* is an economically important fungus in this group. From it we produce citric acid in vast quantities for the soft drink industry as well for use in foods. *Aspergillus* is also our source of gluconic acid, used in medicines and foods, and it is a source of valuable enzymes in certain industries. Another probable ascomycete is *Penicillium,* from which we get the antibiotic penicillin. Another very large number of ascomycetes cause tremendous damage to field and orchard crops, timber, and ornamental plants.

Their life cycle has a **SEXUAL** or **ASCUS STAGE** and an **ASEXUAL** or **CONIDIAL STAGE.** Many ascomycetes don't seem to have a sexual stage in their life cycle. They have either lost this stage during their evolution, or they do really have it but we've just not ever seen it. Species without a known sexual stage are very numerous and cause many plant diseases. As noted earlier, we classify them all in one group known as the Fungi Imperfecti (because they lack the so-called "perfect" or sexual stage). Comparative features of ascomycetes with other major fungal groups are given in Table 20.1 in the preceding exercise.

191

A. THE ASCOCARP— THE SEXUAL FRUITING BODY OF THE ASCOMYCETES

In many, but not all, of the ascomycetes, the sexually formed fruiting body has become a large, fleshy organ called an **ASCOCARP**. Many ascocarps forcibly "shoot" the spores out with a puffing action. This is repeated many times until all the spores are ejected. The puffing action is brought on by changes in moisture conditions in the ascocarp. Not all ascocarps are built exactly the same. Let's look at a few.

Open-Type Ascocarp—Apothecium

ACTIVITY 1 **Examine** dried or preserved specimens of the fungus *Peziza*, which has a simple, open ascocarp called an **APOTHECIUM** (Figure 21.1). The apothecium is cup-shaped and wide open across the top. Any fungus having this type of ascocarp is called a "cup fungus." Below the **CUP** part is a slender, supporting **STALK**. When the fungus is alive, the ascocarp is often brightly colored (and not a dull, ugly black or brown color like the preserved ones). It also may be fleshy, leathery, or gelatinous. Within the shallow cup are tiny sacs, each called an **ASCUS**. These contain the spores, called **ASCO-SPORES**.

Examine a prepared slide of a cross section of the Ascocarp of *Peziza* sp. under the microscope. **Find** the ASCI (singular, ascus). These are vertically elongate, tubelike sacs. Each contains eight haploid spores, called **ASCOSPORES**. When mature, the ascus ruptures and releases the spores. **Note** the hyphae, called **STERILE HYPHAE**, between the asci. These are thought to flex and move and so help in releasing the spores from the surface of the ascocarp. Asci and sterile hyphae together form the fertile layer, called the **HYMENIUM**. A hymenium is present in all ascocarp types. **Label** the *Peziza* apothecium in Figure 21.1, using the terms capitalized in the preceding descriptions.

ACTIVITY 2 **Examine** morels (*Morchella esculenta*). These are fungi that superficially resemble mushrooms, though they are more slender and columnar. The surface has ridges, so the morel appears spongy. The depressions between the ridges are lined with the **HYMENIUM**. Each depression is the same basic **OPEN-TYPE ASCO-CARP** or **APOTHECIUM** as in *Peziza*.

Closed-Type Ascocarp— Cleistothecium

ACTIVITY 3 **Examine** a lilac (*Syringa* sp.) leaf infected with powdery mildew—a disease caused by a variety of ascomycete fungi. **Find** the minute, pepper-grain-size black bodies. These black bodies are cleistothecia, closed-type ascocarps. They are ball-like objects that are hollow inside. This hollow cavity is lined by the hymenium, consisting of the asci and sterile hyphae. **Note** that the cleistothecia have **RIGID APPENDAGES** with branched tips. These aid in dispersal. **Put** some cleistothecia on a slide with water and a cover glass. **Press down** on the cover glass in order to rupture the CLEISTOTHECIUM. ASCI will squeeze out of each cleistothecium. Powdery mildew fungi attack many ornamental plants such as roses, chrysanthemums, snapdragons, African violets, and turfgrasses, including the bluegrasses. Whitish-gray patches of mycelia on leaves, flowers, or stems are evidence of powdery mildew fungi. If sections of your lawn are heavily shaded by trees, they are very likely to develop powdery mildew infection.

Label the powdery mildew cleistothecium in Figure 21.2, using the terms capitalized in the preceding description.

Partially Open-Type Ascocarp—Perithecium

ACTIVITY 4 **Examine** living cultures of *Sordaria fimicola* growing on an agar plate. These are ascomycetes whose ascocarp is a **PERITHECIUM**, a closed, sphere-shaped organ with a single pore at the top. **Mount** a portion of the fungus on a slide, add some water, cover, and examine. **Press down** on the cover slip, and you should be able to split some of the perithecia. ASCI and STERILE HYPHAE are inside the perithecium. **Use** prepared slides if living material is not available.

Label the perithecium of Figure 21.3, using the terms capitalized in the preceding description.

B. LIFE CYCLE OF AN ASCOMYCETE

Life cycles of ascomycetes are very complicated, and many are not well understood. A generalized cycle is shown in Figure 21.4. **Look** at this figure as you read the description that follows.

Sexual Phase

Evidence indicates that the site of both fertilization and meiosis is the **YOUNG ASCUS** (Figure

21.4A). When sexual reproduction occurs (about once a year), the mature mycelium forms an **ASCO-CARP**. Each young ascus within the ascocarp contains two **HAPLOID NUCLEI**. These fuse to form a **DIPLOID ZYGOTE NUCLEUS** (Figure 21.4B). The zygote nucleus divides by meiosis. Eventually, each of the eight **HAPLOID NUCLEI** in the ascus (Figure 21.4E) becomes invested by protoplasm and a cell wall, and thus eight **ASCOSPORES** are formed in the **ASCUS** (Figure 21.4F).

When mature, the ascus opens at its tip, and each ascospore is released (and often forcibly ejected by the ascocarp). It is carried away by air currents or other dispersal agents. Thus ascospores, like conidiospores (see below), play a major role in plant dispersal. Eventually the **GERMINATION** of the **ASCOSPORE** occurs, and a **YOUNG MYCE-LIUM** takes form.

The growing mycelium digests and assimilates food materials from the substratum upon which it lives. It becomes an extensive, cottonlike mass of white hyphae permeating the substratum. When mycelia are mature and food reserves have accumulated, they reproduce, either sexually (as just described) or asexually (as more often happens).

Asexual Phase

The **MATURE MYCELIUM** produces erect hyphae, called **CONIDIOPHORES**. These form **CONIDIOSPORES** (by mitotic cell divisions) at their tips. Before splitting off, conidiospores commonly remain together in chains. The conidiophores often form branching patterns characteristic of the species (see Figure 21.6). Conidiospores are dispersed by air currents. The **GERMINATION** of a **CONIDIOSPORE** results in a young mycelium, and the cycle cranks on again many times more.

ACTIVITY 5 **Complete** the labeling of the ascomycete life cycle in Figure 21.4, using the terms capitalized in the preceding description. **Label** the survival phases in Figure 21.4. **Enter** in Table 21.1 the names of the structures involved in each of the four survival phases of an ascomycete.

C. YEAST

Yeasts are ascomycetes in which the entire thallus is just one cell. The unicellular condition of yeasts appears to have developed in connection with life in sugary solutions. Certain other fungi that are not ascomycetes will assume a "yeast" condition if grown in sugary solutions.

ACTIVITY 6 **Examine** *Saccharomyces cerevisiae*, the common brewer's yeast, by placing a small drop of yeast culture on a slide. **Cover** with a cover glass and examine under high power.

Asexual reproduction takes place by budding. A NEW CELL occurs as a BUD, which later breaks off of the PARENT CELL. Commonly, though, new cells themselves start to bud before they break off the parent cell.

You can't see the nucleus in this preparation, but a large central vacuole is usually visible.

We will not consider sexual reproduction of yeast.

Label Figure 21.5, using the terms capitalized in the preceding description.

II. CLASS FUNGI IMPERFECTI (DEUTEROMYCETES)

Most of the members of this group are probably ascomycetes. A minority are basidiomycetes. They reproduce only by **CONIDIA**, as far as scientists have been able to determine. It is theorized that either they have lost their sexual stages during their evolution or we've just never found these stages.

Some of the worst fungal pests are in this group. To mention a few, we have *Phomopsis*, which causes serious dieback in conifers and canker in gardenias. *Fusarium* causes a wilt of asters and carnations (both of which are important commercial greenhouse crops); *Fusarium* also causes a serious blight in turfgrass, flax, cabbage, and tomato. *Botrytis* infects the leaves of chrysanthemums, begonias, carnations, and other commercially valuable greenhouse crops. *Helminthosporum* causes blight and leaf spot diseases of turfgrasses. *Rhizoctonia* is a common cause of root rot disease in greenhouse plants and field crops.

In contrast, a *Penicillium* sp., the blue-green mold commonly seen on decaying citrus fruit, is the source of penicillin (the first antibiotic). Some species are used in the making of Roquefort, Camembert, and blue cheese. *Aspergillus* is another extremely common and valuable fungus. It will grow on almost anything, provided there's a little moisture.

ACTIVITY 7 **Use** prepared slides or **obtain** from an agar plate culture a strip of cellophane paper containing the mycelium of *Penicillium* or *Aspergillus*. **Mount** it on a dry slide and examine under medium magnification. **Note** the CONIDIO-PHORES and the CHAINS of CONIDIA. Are the conidiophores of *Aspergillus* and of *Penicillium* arranged the same way? **Yes No.**

Make a wet mount of the mycelium by adding a drop of 70 percent ethanol and then a drop of 5 percent potassium hydroxide (KOH). **Cover** with a cover glass and examine under the microscope. *Caution:* DO NOT allow any KOH to come in contact with the lens of the objective on the microscope. **Note** the branching (if any) of the conidiophore and the arrangement of the conidia.

Label the conidiophores and conidia of *Penicillium* and *Aspergillus* in Figure 21.6.

III. CLASS BASIDIOMYCETES

Basidiomycetes are the second largest group of fungi. Most of the large and conspicuous fungi seen in fields and woods are basidiomycetes. These include mushrooms, stinkhorns, puffballs, jelly fungi, and shelf fungi. In this class there are also numerous microfungi, the most important of which are the smuts and rusts. These latter are extremely virulent plant pathogens that cause extensive damage to many crops and ornamental plants.

Distinguishing characteristics of the basidiomycetes are outlined in Table 20.1 of the preceding exercise.

A. THE BASIDIOCARP— THE SEXUAL FRUITING BODY OF THE BASIDIOMYCETES

The fruiting body or **BASIDIOCARP** is the part of the fungus that people see and recognize— not as a basidiocarp, but as a mushroom, toadstool, puffball, stinkhorn, jelly fungus, or shelf fungus on the side of an old tree stump, a dead log, or ground.

ACTIVITY 8 **Examine** a mushroom specimen— a BASIDIOCARP. The basidiocarp can be compared to an umbrella. The "handle" is the STIPE and the caplike portion is the PILEUS. On the underside of the pileus you will see things that look like, and are called, GILLS. The gills are platelike. On their surface is the HYMENIUM, the spore-forming layer. You can't see anything smaller than the gills without preparing sections on slides and using the microscope.

Label all the parts of the mushroom shown in Figure 21.7, using the terms capitalized in the preceding description.

ACTIVITY 9 If available, **examine** spore prints that the instructor may provide or that you could make yourself. A spore print shows the arrangement of the spore-bearing surfaces, of which there is a great variety among the basidiomycetes. The spore print of a mushroom can be made in the following way.

1. Set the pileus of a ripe mushroom (with its gill side down) on a piece of white paper.
2. Cover it with a glass or bowl so that air currents won't blow away the spores as they drop from the gills.
3. Let the fungus remain for an hour or perhaps longer on the paper.
4. Remove the cover and examine the spore pattern (print).

B. LIFE CYCLE OF A BASIDIOMYCETE

Sexual Phase

Look at Figure 21.7 as a guide in this description. The vegetative or assimilative phase is (as with other fungi) the **MYCELIUM**. The **PRIMARY MYCELIUM** is the first stage, a mass of **HYPHAE** formed by the germination of **HAPLOID SPORES**. Commonly there are different strains of mycelia that for convenience we will call "plus" and "minus" strains. Hyphae of the primary mycelium are septate and uninucleate.

When hyphae of primary mycelia of different strains come in contact, they merge. The result of this is the formation of a **SECONDARY MYCELIUM.**

Cells of the hyphae of the secondary mycelium have two nuclei, one from each of the two primary mycelia. The secondary mycelium is very long-lived compared with the primary. Eventually, the growth of several hyphae becomes compactly organized into the **BASIDIOCARP**. Spores are released from the basidiocarp. They germinate into primary mycelia, and the cycle starts over.

Spore Formation

Spore formation takes place on the surface of the **GILLS**. **Look** at Figure 21.7 again as a guide in this description. A terminal cell of a hypha starts to become a **YOUNG BASIDIUM** (plural, basidia). Its two nuclei fuse (KARYOGAMY) and fertilization takes place. This finally brings together the two sets of genes from the original two spores that produced the primary mycelia.

The diploid nucleus then divides by **MEIOSIS.** Four **HAPLOID NUCLEI** are formed.

The **YOUNG BASIDIUM** cell enlarges and produces four knoblike protuberances. As the knobs enlarge, the haploid nuclei move up into them, one nucleus into each. These four enlarged structures, each with one nucleus, are the **BASIDIO-SPORES**. The number of spores is characteristically four. Each spore is attached to the *outside* of the basidium by a slender stalk. This differs from ascomycetes, which produce spores *inside* the ascus.

When a spore is mature, it is forcibly ejected from the surface of the gill. (Remember, this spore formation has been taking place on the gills on the underside of the basidiocarp.) The catapulting of each spore throws it away from the gill surface so that it is more likely to be picked up by air currents. A little droplet of water forms at the base of the spore. Changes in surface tension of this droplet of water is the triggering mechanism that catapults the spore off the basidium and the gill.

ACTIVITY 10 **Using** the microscope, examine a prepared slide showing a section through the cap of a gill-type mushroom such as *Coprinus* sp. Under low magnification, **note** the central STIPE surrounded by the radiating GILLS. **Change** to high magnification and examine the HYMENIUM with the BASIDIA and BASIDIOSPORES. The mature spores are much larger than the basidia. They will also be scattered about and not necessarily all attached to the basidia.

Label the enlarged gill sections shown in Figure 21.7, using the terms capitalized. **Complete** the labeling of the life cycle of a basidiomycete in Figure 21.7, making sure to use all the terms capitalized in the preceding description. In Table 21.1 **enter** the names of the structures involved in each of the four survival phases of a basidiomycete. **Label** the survival phases in Figure 21.7.

Asexual Reproduction

Basidiomycetes reproduce asexually by a variety of methods. Many seem to have no asexual cycle. One method is by the production of conidia as found in the ascomycetes. We will not go into the details of this.

C. SMUTS AND RUSTS

These are basidiomycetes that are microfungi. They all parasitize plants and cause tremendous damage, particularly to fruits and cereal grains. Their fruiting bodies are microscopic but occur in such large masses that their presence is revealed by rust-colored lesions or black, sooty-looking malformed tissues of the infected plants.

ACTIVITY 11 **Examine** plants infected with rusts or smuts.

EXERCISE 21

Student Name _____

QUESTIONS

1. Name two economically important ascomycetes. _____ _____

2. Which class of fungi is the largest? _____

3. Nuclear fusion is termed _____

4. Which group exhibits the greater variety of fruiting bodies, (a) the ascomycetes or (b) the basidiomycetes? Circle (a) or (b).

5. Basidiomycetes and ascomycetes are all terrestrial fungi. Circle one: (a) true; (b) false.

6. In ascomycetes, which is the more common method of reproduction, by conidia or by ascospores? _____

7. Contrast karyogamy and plasmogamy. _____ _____ _____

8. Name the assimilative phase of fungi. _____

9. Name a cell in the dispersal phase of an ascomycete. _____

10. Name a cell in the dispersal phase of basidiomycete. _____

11. Conidia are the only known cells for dispersal in the Fungi Imperfecti. Circle one: (a) true; (b) false.

12. What is the typical number of ascospores formed in an ascus? _____

13. What is the typical number of basidiospores formed on a basidium? _____

14. Name two economically important members of Fungi Imperfecti. _____ _____

15. Basidiospores are "shot" off by surface tension changes in a drop of _____ located at the base of the _____

16. Smuts and rusts belong to what class of fungi? _____

17. Fungi Imperfecti consist mostly of species belonging to which class of fungi? _____

18. Conidia and conidiospores are found only in ascomycetes. Circle one: (a) true; (b) false.

19. Name two common examples of basidiomycetes: (a) _____ ; (b) _____

20. Name the *fungus class* for each of the following.
    a. Spores of the sexual life cycle are commonly found in groups of eight and borne *inside* the spore-forming structure: _____
    b. Spores of the sexual life cycle are formed in groups of four and borne *outside* the spore-forming structure: _____

Table 21.1 LIFE CYCLE CELLS, STAGES, OR EVENTS RELATED TO METHODS OF
SPECIES SURVIVAL

SURVIVAL PHASE	ACCOMPLISHED IN AN ASCOMYCETE BY WHAT CELL(S), STRUCTURE(S), PHASE(S), OR EVENT(S)	ACCOMPLISHED IN A BASIDIOMYCETE BY WHAT CELL(S), STRUCTURE(S), PHASE(S), OR EVENT(S)
Multiplication		
Dispersal		
Assimilation		
Genetic variation		

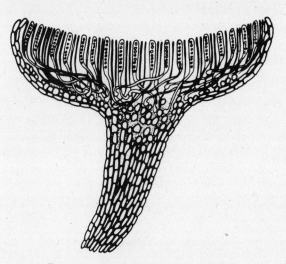

Figure 21.1 Apothecium type of ascocarp.

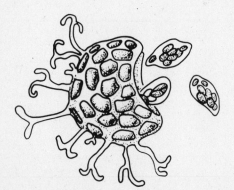

Figure 21.2 Cleistothecium type of ascocarp.

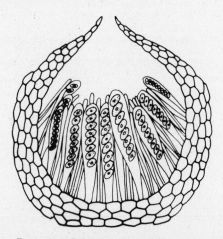

Figure 21.3 Perithecium type of ascocarp.

EXERCISE 21   **Student Name** _____

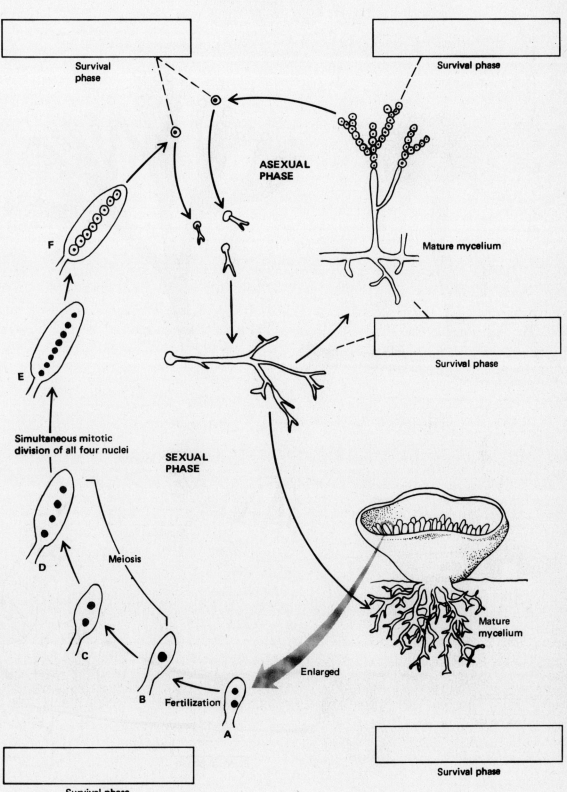

Figure 21.4 Generalized life cycle of an ascomycete.

Figure 21.5 Asexual reproduction of brewer's yeast (*Saccharomyces cerevisiae*).

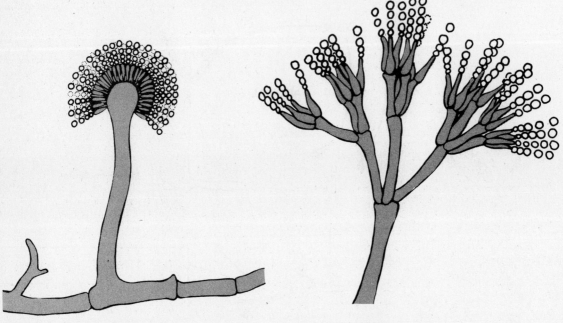

A. *Aspergillus* sp.

B. *Penicillium* sp.

Figure 21.6 Two members of the Fungi Imperfecti showing two types of conidiophore branching patterns.

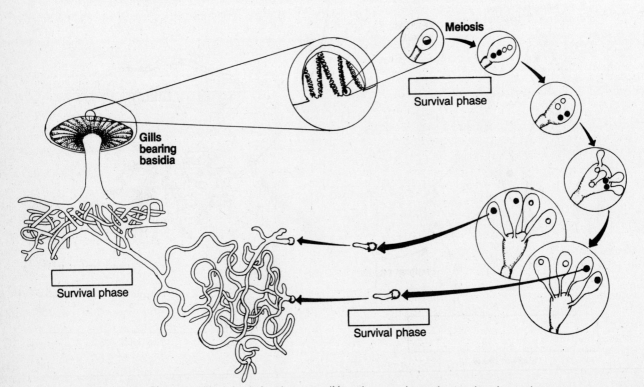

Figure 21.7 Generalized life cycle of a basidiomycete. (Note: An asexual reproductive phase has not been observed in most mushrooms.)

GREEN ALGAE, EUGLENOIDS, AND LICHENS

INTRODUCTION TO ALGAE AS A GROUP

Algae are chiefly aquatic plants, found in both fresh waters and the sea. Many live as free-floating organisms in the plankton on or near the water's surface. Others live in the deeper zones as benthic plants attached to some rock or other object. Still, there are many whose habitat is terrestrial. You can find them in the soil, on tree bark, in hot springs, in or under the snow or ice, and inside the bodies of other plants or animals; a few live in a communal association with fungi, forming "plants" called lichens.

Algae, by their sheer numbers, are responsible for more photosynthesis than any other group of plants. Vast quantities of oxygen are also produced as a by-product of algal photosynthesis. Thus, algae play a major role in the production of energy and oxygen supplies for all forms of life on earth.

Higher (land) plants are remarkably uniform in the kinds of photosynthetic pigments they contain, but the algae are not. The different groups of algae have characteristic colors that form the basis for their common names. Thus, we have the blue-green algae, the green algae, the reds, the browns, and the golden-brown algae.

These varied colors are due chiefly to the presence of various auxiliary pigments. These are light-capturing pigments that are auxiliary to chlorophyll a, the primary photosynthetic pigment. These auxiliary pigments have the ability to absorb different wave lengths of light under conditions of reduced light, as found at various depths in the waters or soil. Thus they facilitate the process of photosynthesis by capturing supplemental light energy and transferring it to the photosynthetically active pigment, chlorophyll a.

On the negative side, algae can create problems by their excessive growth in water supplies, filtering systems, swimming pools, ponds, and so on. Occasionally they have also been the cause of extensive fish kills through toxins produced by population explosions of certain species.

Algae are a huge, diverse assemblage of fairly simple, photosynthetic plants whose origins date back over 4 billion years. Scientists recognize several distinct groups (divisions). Differences among the groups are based on size, habitat, structural complexities, reproduction, physiology, and biochemistry of the plants.

Some terms first encountered in descriptions of algae, many of which are used in descriptions of other plant groups, follow.

ANTHERIDIUM In algae, a single cell that contains sperm (plural, antheridia).

CILIUM Short, hairlike process on a cell whose vibrating or lashing movements serve to propel the cell through water (plural, cilia).

COLONY A thallus type composed of a group of closely associated but independently functioning cells or unicellular organisms among which there are no marked structural differences and little or no division of function.

EYESPOT A small pigmented structure thought to be sensitive to light.

FILAMENT A thallus type in which the cells occur in a threadlike row.

FLAGELLUM A very long whiplike process on a cell whose vibrating or lashing movements serve to propel the cell through water (plural, flagella).

GAMETANGIUM   A general term applied to any cell within which gametes are formed.

HETEROGAMY   A type of sexual reproduction in which male and female gametes differ in size, structure, and behavior.

ISOGAMY   A type of sexual reproduction in which all gametes are alike in size and structure and often in behavior; considered primitive (adjective, isogamous).

OOGAMY   A type of sexual reproduction involving a larger, nonmotile egg and a smaller, motile sperm (adjective, oogamous).

OOGONIUM   A single cell that contains one or several eggs (plural, oogonia).

PYRENOID   A center of starch formation in certain chloroplasts.

THALLUS   The name for the entire plant body.

ZOOSPORE   A flagellated motile cell.

ZYGOSPORE   A diploid spore formed by the fusion of isogametes (or isogametangia); often goes into a resting or dormant stage.

Review the section on the significance of life cycles on page 174.

I. DIVISION CHLOROPHYTA— GREEN ALGAE

Next to the diatoms, the green algae are the second largest group of algae. They are the most diverse of all algae. Some approach the structural complexity of simple land plants. Individual plants are microscopic, but colonies of them can form masses visible to the eye.

The function, structure, and chemical makeup of their cells are similar, and in many cases identical, to those of higher plants. Taking all these features into account, it is generally agreed that the green algae are probably ancestral to all land plants.

General facts about green algae are

Usual color	Green.
Distribution	Mostly in fresh waters and on land.
Habitat and niches	Most species are floaters in the plankton of rivers, lakes, reservoirs, and creeks. Also, in soil, snow, on rocks, tree bark, flower pots, and fish tanks. A few species are floaters in plankton of the neritic zone (continental shelves) and littoral (coastal) zones.
Motility	By flagella.
Body types	Single cells, filaments, and colonies.
Value	Important source of oxygen and food for aquatic organisms; some are eaten by humans.

A. LINES OF EVOLUTION IN THE GREEN ALGAE

Two general trends are recognizable in the evolution of green algae. These are:

1. Increase in plant size and form from a single-cell type to a filamentous or colonial type.
2. Change in sexual reproduction from isogamy to oogamy.

The increase in size and form is seen in the range of growth forms found today. These growth forms or body types illustrate some of the probable stages of evolution. Each is a successful way of life that, within certain limits, has survived to the present.

Unicellular organisms are the most primitive. Filamentous and colonial forms probably arose from them by cell divisions occurring in different planes as indicated in Figure 22.1.

Study the generalized life cycle shown in Figure 22.2 and **review** it as you examine the various plants included in this exercise.

B. REPRESENTATIVE GREEN ALGAE

Chlamydomonas or *Carteria*—Single-Cell-Type Isogamous Green Alga

Chlamydomonas (as well as *Carteria*) is a single-celled plant. Thus the thallus is one cell. Its sexual reproduction is isogamous.

ACTIVITY 1   **Get** a clean slide and add to it a drop of methyl cellulose and a drop of water from the *Chlamydomonas* or *Carteria* culture. **Stir** the two drops together gently, using the edge of a cover glass. The methyl cellulose is viscous and will slow down the motility of the organisms. **Cover** with the cover glass. **Use** low power on the microscope and locate the cells. Then **examine** them under higher magnification.

After observing the movement of the cells, **use** some iodine to kill the cells and stain the parts. **Put** a drop of iodine on one edge of the cover glass. **Hold** a piece of rough-torn paper toweling on the opposite edge of the cover glass. The iodine will be drawn under the cover glass.

Because the cell is small, you may not be able to see all its parts. Probably you can see at least some of the following: the CELL WALL, the NUCLEUS, the cup-shaped CHLOROPLAST, the EYESPOT (an

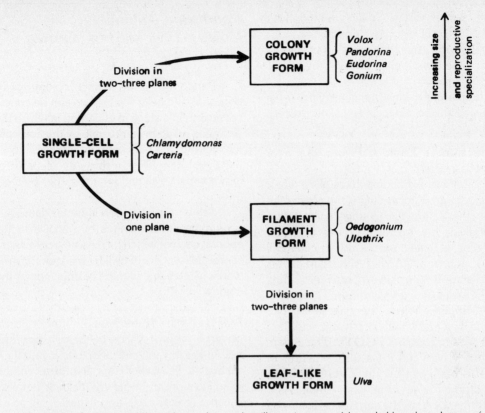

Figure 22.1 Some of the different forms of green algae illustrating some of the probable paths and stages of evolution.

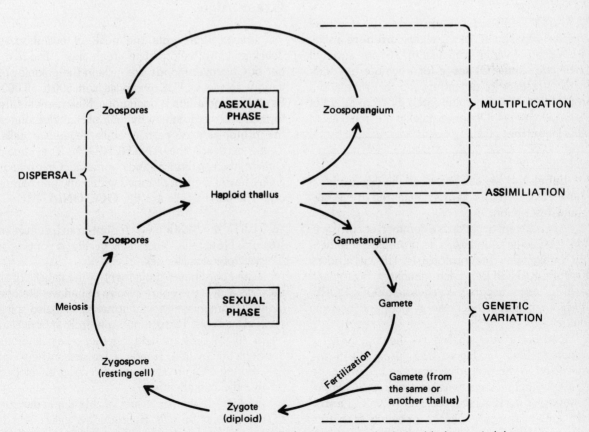

Figure 22.2 Generalized life cycle of green algae. Each part is concerned with one of the four survival phases of the population—dispersal, multiplication, assimilation, and genetic variation.

orange or red spot), the PYRENOID (starch body in the lower part of the chloroplast), and two anterior FLAGELLA (four in *Carteria*). If you cannot see the flagella, reduce the light coming through the microscope and look again.

Label the mature thallus (the central figure) in Figure 22.3, using the terms capitalized in the preceding description.

Reproduction   Look at Figure 22.3 as a guide for this description.

The sexual phase commonly takes place when adverse conditions arise, for example, the pond dries up or cold weather sets in. In this stage, a vegetative cell will subdivide its protoplast into 8 to 32 protoplasts. These remain within the original cell wall. This cell is now a **GAMETANGIUM**. Later the protoplasts are released as separate cells. They develop two flagella (four in *Carteria*). These cells are **ISOGAMETES**. When the flagella of two gametes come in contact, they intertwine, and then the two cells fuse to form a **ZYGOTE**. The zygote develops a thick wall and goes into a period of dormancy. This resting cell is the **ZYGOSPORE**. It can resist drought or other adverse conditions.

When dormancy ends, the zygospore divides by meiosis and forms four **ZOOSPORES**. Zoospores swim about, increase in size, and become typical adult cells.

ACTIVITY 2   In order to see the action of sexual reproduction, if material is available, **prepare** a slide with a drop of water from a culture of one strain of *Chlamydomonas*. **Observe** for a minute to check for the presence of organisms. To this slide add a drop of water from a culture of a different strain of *Chlamydomonas*. **Observe** quickly the behavior of cells. **Describe:** _____
_____
_____

Label the sexual phase of the life cycle in Figure 22.3, using the terms capitalized in the preceding description.

Asexual reproduction is a common occurrence. A vegetative cell subdivides its protoplast into two, four, or eight daughter protoplasts. This cell is now a ZOOSPORANGIUM. Each daughter protoplast, when released, becomes a flagellated ZOOSPORE. These swim about, increase in size, and become adult cells.

Complete the labeling of the life cycle of *Chlamydomonas* in Figure 22.3, using the terms capitalized in the preceding description. **Enter** the terms designating the survival phases in the *Chlamydomonas* life cycle in Figure 22.3. In Table 22.1 **enter** the names of the cells, phases, or events of this life cycle that are involved in each of the four survival phases.

Pandorina—Simple-Colony-Type Heterogamous Green Alga

Pandorina is a simple colony of *Chlamydomonas*-like cells. *Pandorina* can be considered as representing a stage in evolution in which the thallus has been made bigger simply by the aggregation of a few cells with a fairly simple degree of coordination between them. Four or thirty-two cells are arranged in a compact ball and kept together by a gelatinous envelope.

Sexual reproduction is heterogamous, and the zygote produced divides by meiosis to form zoospores, each of which forms a new colony. In asexual reproduction, each cell in the colony produces a whole new colony within the envelope of the parent colony.

ACTIVITY 3   **Make** a wet mount of *Pandorina* or obtain a prepared slide of the organism. **Examine** first under low magnification and locate the colonies that may be swimming. **Examine** under higher magnification and note the cellular details such as eyespot, chloroplast, and flagella.

Volvox—Complex-Colony-Type Oogamous Green Alga

Volvox represents the peak of evolutionary development of colony-type green algae. The number of *Chlamydomonas*-like cells in the colonies of various species of *Volvox* ranges from 500 to 50,000. Sexual reproduction is oogamous. Male gametes are formed in large numbers by multiple subdivisions of the protoplasts of certain adult vegetative cells. These cells are the **ANTHERIDIA**. The female gamete, the egg, is large and nonmotile. Only one or a very few of them are formed within any one parent cell. The parent cells are the **OOGONIA**.

ACTIVITY 4   **Make** a wet mount from a culture of *Volvox*. **Note** that each cell of the colony is a *Chlamydomonas*-like cell.

The coordinated movement of the flagella of all the cells allows the entire colony to behave like one individual organism, moving forward as it also spins on a vertical axis. The colony varies from spherical to ovoid. All the cells are held together by a gelatinous envelope and by thin strands of protoplasm between neighboring cells. In addition, each cell is enveloped by an individual sheath.

Volvox shows some specialization and division of labor among the cells. Reproductive cells occur at the base or hind end of the colony, while the forward or leading end has only vegetative cells. Eyespots in the vegetative cells are larger than those of the reproductive cells. Small daughter colonies formed as a result of either sexual or asexual reproduction are usually seen floating in the center of the larger parent colony. When the parent colony disintegrates, the daughter colonies are released and become mature colonies.

Ulothrix—Simple-Filament-Type Isogamous Green Alga

Thallus   The thallus of *Ulothrix* is a simple, unbranched (unbranched is considered primitive) filament with only the basal cell, the holdfast cell, differentiated from the rest of the filament. It attaches the filament to some rock, pebble, or other object.

ACTIVITY 5   **Examine** a prepared slide or living specimens of *Ulothrix*. **Note** that the cells are all alike. The CHLOROPLAST in each cell is C-shaped and may have one or more PYRENOIDS. No eyespot is present. Why not?＿＿＿＿＿＿＿＿＿

＿＿＿＿＿＿＿＿＿＿＿＿＿＿＿＿＿＿＿ The NUCLEUS may or may not be visible.

Label the thallus of *Ulothrix* in Figure 22.4, using the terms defined in the preceding description.

Reproduction   Any cell except the holdfast cell in the filament is capable of reproducing the plant, either asexually or sexually. **Look** at Figure 22.4 as a guide in this description. In the asexual phase the protoplast of a cell divides by mitosis to form four or eight daughter protoplasts. These are released and become flagellated ZOO-SPORES. The parent cell that produced these zoospores is called a ZOOSPORANGIUM. After swimming about, each dispersed zoospore settles down, loses its flagella, and undergoes mitotic divisions to form a new FILAMENT.

In the sexual phase the protoplast of a cell divides many times to form 32 to 64 daughter protoplasts. When freed from the parent cell (the GAMETANGIUM), these daughter protoplasts function as GAMETES. They are ISOGAMETES. Each has two flagella, not four as in zoospores. They are also smaller than zoospores.

Eventually two isogametes from different filaments fuse to form a quadriflagellate ZYGOTE. The zygote goes through a dormant period as a ZYGO-SPORE. Later it divides by meiosis to form four haploid ZOOSPORES. Each of these will grow into a new filament.

Is there any basic difference in the cyclic events of this life cycle and of *Chlamydomonas*?   **Yes**

No.   Is there any real difference in individual cell behavior?   **Yes   No.**   What, then, is the only difference between the *Chlamydomonas* plant and the plant *Ulothrix?*＿＿＿＿＿＿＿＿＿＿

＿＿＿＿＿＿＿＿＿＿＿＿＿＿＿＿＿＿＿

Complete the labeling of the life cycle of *Ulothrix* in Figure 22.4, using the terms capitalized in the preceding description. **Enter** the terms designating the four survival phases in the *Ulothrix* life cycle in Figure 22.4. In Table 22.1 **enter** the names of the cells, phases, or events of this life cycle that are involved in each of the four survival phases.

Oedogonium—Simple-Filament-Type Oogamous Green Alga

Thallus

ACTIVITY 6   **Examine** a prepared slide or living specimens of *Oedogonium*. *Oedogonium* is, like *Ulothrix*, a simple, unbranched filament. But it reproduces by oogamy, an advanced method compared to the isogamy of *Ulothrix*. Cells in the filament are mostly elongate. The CHLOROPLAST is reticulate, that is, in a delicate meshwork. PYRENOIDS are located at the intersections of the reticulum, and you see them as scattered dark bodies in the cell. The NUCLEUS is usually centrally placed. A HOLDFAST cell is present in some species.

Reproduction   **Look** at Figure 22.5 as a guide in this description.

ACTIVITY 7   To study the sexual phase, **examine** prepared slides with filaments of *Oedogonium* showing antheridia and oogonia. Here and there among the elongate vegetative cells in the filament are groups of much smaller disklike cells. Each of the short disklike cells is an ANTHERIDIUM. Each forms two PROTOPLASTS. After release, these protoplasts become SPERM. They are small, ovoid cells with a ring of flagella at one end.

Locate an oogonium. This will be a single cell whose entire protoplast has become a large, globose, opaque EGG. The oogonium has a FERTILIZA-TION PORE through which a sperm cell can enter to fertilize the egg.

After fertilization, the ZYGOTE that is formed remains within the oogonium. It develops a thick wall, goes dormant, and is known as an OOSPORE. As the filament decays, the oospore is released but remains dormant for several months. Later the oospore divides by meiosis to form four haploid ZOOSPORES. These have a ring of flagella at one end. They resemble sperm but are larger cells. They swim about, and later each grows into a new filament.

The asexual phase can occur in two ways (1) by fragmentation of the filament, each fragment growing into a new filament; or (2) by the transformation of the protoplast of a vegetative cell into a ZOOSPORE as shown in Figure 22.5. After release from the old cell wall, the zoospore swims about and then grows into a new filament.

Complete the labeling of the life cycle of *Oedogonium* in Figure 22.5, using the terms capitalized in the preceding description. **Enter** the terms designating the four survival phases in the *Oedogonium* life cycle in Figure 22.5. In Table 22.1 **enter** the names of the cells, phases, or events of this life cycle that are involved in each of the four survival phases.

Ulva—Leafy Type Green Alga

Ulva (sea lettuce) is leafy. It exemplifies another evolutionary line in green algae. The thallus is a thin flat sheet of cells instead of a chain or a cluster. It grows attached to other objects and has a holdfast with rhizoid- or fingerlike extensions.

ACTIVITY 8 If available, **examine** preserved specimens or dried herbarium specimens of *Ulva*.

II. DIVISION EUGLENOPHYTA— EUGLENOIDS

Euglenoids are unicellular organisms whose classification as plants or animals can be debated equally either way. They exhibit both plant and animal characteristics. They are not related to any algae, but because they occupy the same kinds of habitats as algae and are of comparable simplicity, they are included in this study of algae. Zoologists include them in studies of single-celled animals known as protozoa.

General facts about euglenoids are

Usual color	Green; a few are colorless.
Habitat	Fresh water, predominantly.
Body types	Single cell
Cell wall	Variable; often a flexible pellicle.
Motility	By one long whiplike flagellum.
Value	A food source for fish and other aquatic animals.

ACTIVITY 9 **Put** a drop of methyl cellulose on a clean slide and add to it a drop of water from a culture of *Euglena*. **Using** the edge of the cover glass, stir gently. **Cover** with a cover glass and under low magnification, locate the organisms.

Note the swimming motion. **Reduce** the light and you may be able to see a long flagellum at one end of the cell. Near the base of the flagellum you may be able to see a GULLET. Some euglenoids are thought to ingest food into the gullet.

Note that the cell is elongate with a pointed end and a blunt end. The blunt end is the leading end of the cell that is propelled through the water by the flagellum. A light-sensitive, pigmented body, called the EYESPOT, is near the blunt end. Can you see it?   *Yes   No*.   Do you see CHLORO-PLASTS?   *Yes   No*.   Is a NUCLEUS present?   *Yes   No*.   Does the organism change shape?   *Yes   No*.   Does it have a rigid wall, then?   *Yes   No*.   The changes in cell shape are possible because the protoplast is bounded by a flexible PELLICLE.

III. LICHENS

A lichen is a plant made up of a fungus and an alga living together as one unit. The association is mutualistic, each organism deriving substantial benefits from it. The alga synthesizes carbohydrates and other food materials. The fungus absorbs and retains moisture. Most of the lichen is composed of fungal hyphae. The algal cells are nested among the hyphae and are permeated by rootlike growths from them. The fungus is usually an ascomycete, and the alga is usually one of the single-cell-type blue-green (Cyanophyta) or green (Chlorophyta) algae.

There are three growth forms or body types of lichens:

1. CRUSTOSE   The lichen forms just a crust on the surface on which it is growing.
2. FOLIOSE   The lichen is more or less leafy in appearance.
3. FRUTICOSE   The lichen is shrublike with multibranching and intertwined fibrouslike parts.

General facts about lichens are

Usual color	Variable. Can be gray-green, white, orange, yellow, brown, or black. Many bright colors, especially in the reproductive parts.
Distribution	Terrestrial, worldwide.
Habitat and niches	Wide variety of habitats and climatic conditions—tree trunks, rocks, any unpainted, weather-exposed wood, and certain types of poor soils.
Body types	1. Crustlike—crustose 2. Leafy—foliose 3. Shrubby—fruticose

Value (biological and economic) — Pasturage for reindeer and caribou in the Arctic; human consumption in the Orient; variety of medicinal and industrial uses; used for dying cloth by primitive peoples; useful in detection of certain air pollutants.

ACTIVITY 10 **Examine** species of lichens which show the crustose, foliose, and fruticose growth forms. **Note** the brightly colored, usually cup-shaped, fruiting bodies. These fruiting bodies are ASCOCARPS (look back at Exercise 21 if you've forgotten what ascocarps are). They are functionless for the lichen but useful to us in identifying the plant.

If fresh specimens are available, **examine** the lichens with a dissecting microscope.

ACTIVITY 11 **Examine** a prepared slide of a section of the foliose-type lichen *Physcia* (or other available material) and note the internal organization. **Look** at Figure 22.6 as a guide in this description while you examine the slide.

The Thallus or Body

The body or thallus of the lichen has four layers:

1. **UPPER CORTEX** The upper surface layer, composed of fungal hyphae so highly twisted and compact that the cortex appears cellular.
2. **ALGAL LAYER** Just below the upper cortex, composed of clumps of algal cells lying among loosely interwoven fungal hyphae.
3. **MEDULLA** A thick layer of loosely packed fungal hyphae.
4. **LOWER CORTEX** Similar to the upper cortex but with RHIZOIDS (root-like hyphal growths).

The Ascocarps

Examine the slide again and this time look for saucerlike structures growing up from the top of the thallus. These structures (you may see only one) are the fruiting bodies or ASCOCARPS. In the depression on the top side of the ascocarp is the HYMENIUM, a layer of ASCI and STERILE HYPHAE. Although ascocarps are present, they play no role in reproducing the lichen. Instead, lichens reproduce by fragments breaking off the plant or by special, regenerative clumps of hyphae and algae called soredia. Fragments and soredia get blown about and grow into new lichens.

Label the diagram of the lichen *Physcia* in Figure 22.6, using the terms capitalized in the preceding description.

EXERCISE 22   Student Name _____

QUESTIONS

1. Do the life cycles of green algae contain the same four survival phases as fungi? _____

2. List these phases. Use just one word or phrase to identify each. (a) _____
   _____; (b) _____; (c) _____
   _____; (d) _____

3. What is the term for the most primitive level of gamete specialization? _____
   _____

4. What are the three body types or growth forms found among green algae? (a) _____
   _____; (b) _____; (c) _____

5. Which growth form or body type is considered the most primitive? _____
   _____

6. Give at least two reasons why green algae are considered ancestral to land plants. (a)
   _____
   _____; (b) _____
   _____

7. Are green algae ordinarily visible to the naked eye? _____

8. Are the sperm and eggs of green algae produced by (a) multicellular organs or (b) single
   cells? Circle (a) or (b).

9. What component of the cell has a distinct and recognizable shape for each genus of the
   green algae studied? _____

10. In which stage is there a rearrangement of chromosomes and gene combinations, in (a)
    mitosis or in (b) meiosis? Circle (a) or (b).

11. Which cells, therefore, carry new genetic "blueprints" for making the new filament, the
    zoospores of the (a) sexual phase or (b) asexual phase? Circle (a) or (b).

12. Name a colonial green alga with oogamous reproduction. _____

13. Why wouldn't you expect to find eyespots in cells in filamentous colonies? _____
    _____

14. Name the two cell types in the green alga life cycle that function in dispersal of the popula-
    tion. (a) _____; (b) _____
    _____

15. Briefly, what are the two broad, general evolutionary trends recognizable in green algae?
    (a) _____; (b) _____
    _____

16. Would you expect to find many species of green algae in the ocean? _____

17. Are euglenoids algae? _____

18. Which phase of the life cycle of the green alga is mainly responsible for maintaining and
    increasing the number of individuals in the population? _____

19. What are the two events in the life cycle that introduce genetic variation into the population?
    (a) _____; (b) _____

20. Lichens are easily recognized by certain brightly colored parts. What is the name of these
    colored parts? _____

21. Lichens typically have ascocarps. They therefore reproduce by cells called ascospores.
    Circle one: (a) true; (b) false.

22. Which layer of a lichen thallus is photosynthetic? _____

23. The fungus portion of most lichens belongs to which class of fungi? _____

    _____

24. Lichens are very sensitive to air pollution. If you live in the city or suburb, think for a minute.

    Do you see many trees with their barks encrusted with lichens? _____

25. The bulk of a lichen plant is formed by which of its component members (a) the alga or (b) the fungus? Circle (a) or (b).

Table 22.1 LIFE CYCLE CELLS, STAGES, OR EVENTS RELATED TO METHODS OF SPECIES SURVIVAL

SURVIVAL PHASE	ACCOMPLISHED IN *CHLAMYDOMONAS* SP. BY WHAT CELL(S), STRUCTURE(S), PHASE(S), OR EVENT(S)	ACCOMPLISHED IN *ULOTHRIX* SP. BY WHAT CELL(S), STRUCTURE(S), PHASE(S), OR EVENT(S)	ACCOMPLISHED IN *OEDOGONIUM* SP. BY WHAT CELL(S), STRUCTURE(S), PHASE(S), OR EVENT(S)
Multiplication			
Dispersal			
Assimilation			
Genetic variation			

EXERCISE 22

Student Name _____

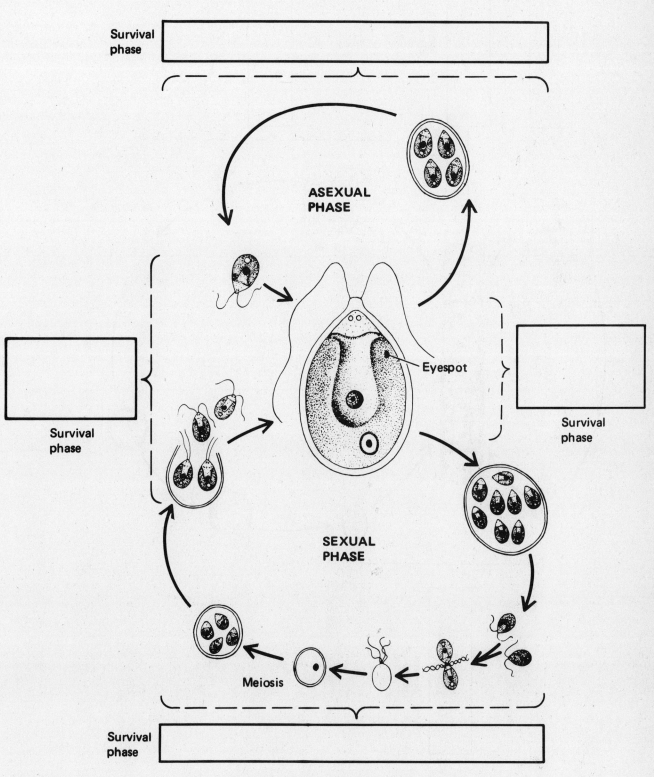

Figure 22.3 Generalized life cycle of *Chlamydomonas.*

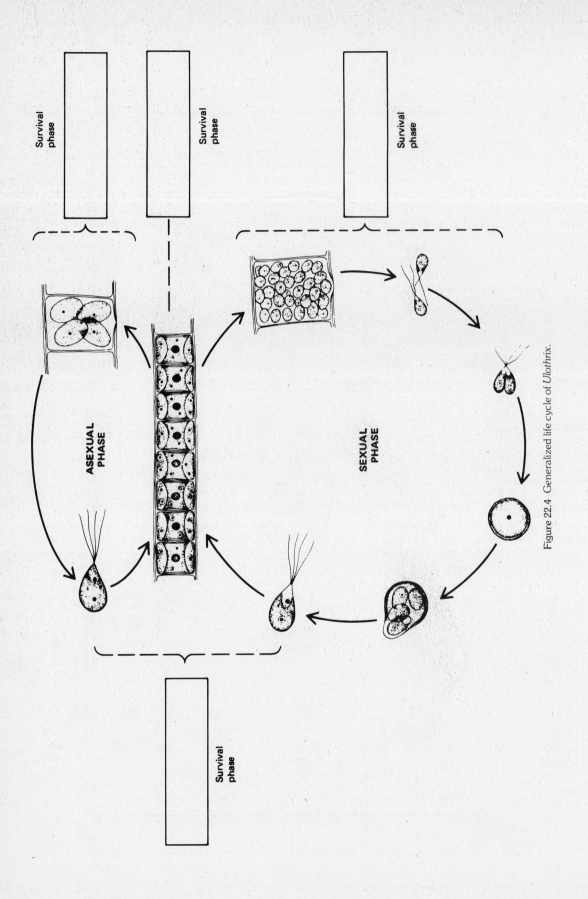

Survival phase

Survival phase

Survival phase

Survival phase

ASEXUAL PHASE

SEXUAL PHASE

Figure 22.4 Generalized life cycle of *Ulothrix*.

EXERCISE 22   **Student Name** _____

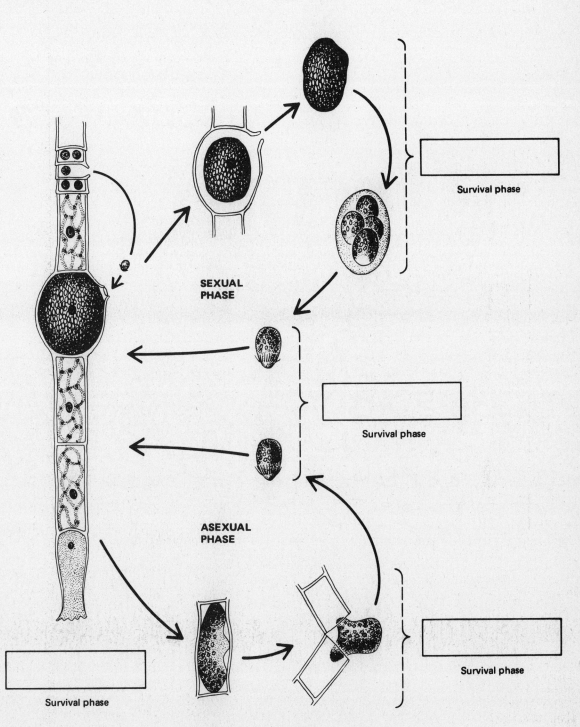

Figure 22.5 *Generalized life cycle of* Oedogonium.

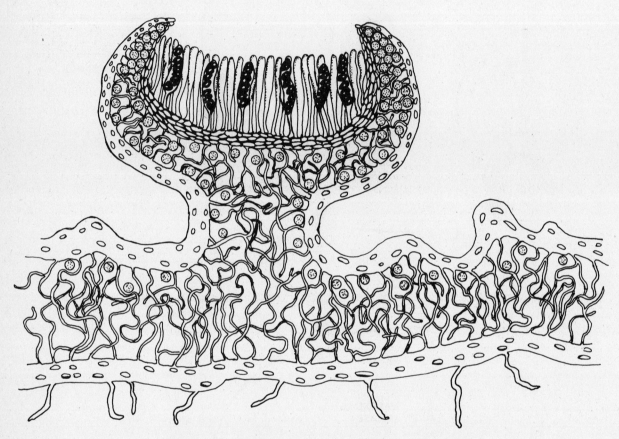

Figure 22.6 Cross section of a lichen with cup-fungus type fruiting body.

IDENTIFICATION OF SOME COMMON FRESHWATER ALGAE

EXERCISE

23

This exercise consists of a key designed to enable you to make an approximate identification of a variety of the most common freshwater and terrestrial algae. Several hundred genera have been deliberately left out to simplify the key. Except in a few cases, the majority of algae you'll find in any freshwater or soil sample will belong to the genera in this key.

You may discover in your sample numerous "little round green things" (LRGTs) that even under the highest magnification of your microscope, are still LRGTs. They defy identification by visual appearance alone. Even experts on algae (phycologists) often need to culture LRGTs for months and study their life cycles before the genus and species can be identified.

Section IV shows examples of desmids, diatoms, and other algae. These are so variable that the key would lose its simplicity if we were to include them. If you have trouble going through the key, then look at Section IV, and you may discover that your unknown is a diatom or a desmid. Remember that there are thousands of species of desmids and diatoms, and Section IV illustrates just generalized representatives.

I. KEY TO SOME COMMON FRESHWATER ALGAE[1]

Directions for Using Key: This key consists of pairs or couplets of choices, 1a–1b, 2a–2b, and so on. Compare your plant specimen with the descriptions given in couplet 1a–1b. Choose whichever statement (1a or 1b) is more applicable to your specimen and go on to the next statement as indicated by the number at the extreme right. Continue through the key, each time choosing between a pair of choices until you terminate at the name of a plant.

1a.   Cells blue-green, sometimes olive green or purplish; pigmentation evenly spread throughout the cell; no nucleus or other small bodies visible within the protoplast; cell wall thin and usually with a mucilaginous sheath .... **Blue-green algae**
(Section II)

1b.   Cells green to yellow green; pigments contained within definite plastids; nucleus and plastids visible; cell wall distinct and typically without a sheath .... **Green algae**
(Section III)

---

[1]Courtesy of Dr. Richard L. Smith, Eastern Illinois University, Charleston, IL.

II. BLUE-GREEN ALGAE (or CYANOBACTERIA or CYANOPHYTA; included here for completeness of coverage, though they are not truly algae)

1a.	Unicellular or colonial	2
1b.	Filamentous	6
2a.	Cells ellipsoidal or cylindrical	**Gloeothece**
2b.	Cells spherical or hemispherical	3
3a.	Cells arranged in rows, forming a more or less rectangular sheet	**Merismopedia**
3b.	Cells solitary or united in round to irregular colonies	4
4a.	Numerous small cells in a colony; sheath watery, often indistinct	**Polycystis**
4b.	Few (less than 50) cells in a colony; sheath thick, often in layers around cells and/or colony	5
5a.	Sheath colored	**Gloeocapsa**
5b.	Sheath colorless	**Chroococcus**
6a.	Without heterocysts	7
6b.	With heterocysts	10
7a.	Without an evident mucilaginous sheath	8
7b.	With an evident mucilaginous sheath	9
8a.	Filaments straight, with cross walls	**Oscillatoria**
8b.	Filaments spiral, cross walls indistinct	**Spirulina**
9a.	One filament within a sheath	**Lyngbya**
9b.	Many filaments in a sheath	**Microcoleus**
10a.	Heterocysts terminal	11
10b.	Heterocysts intercalary	12
11a.	Filaments tapering	**Rivularia**
11b.	Filaments not tapering	**Cylindrospermum**
12a.	Filaments showing apparent branching at the heterocysts	**Tolypothrix**
12b.	Filaments unbranched	13
13a.	Heterocysts round; filaments within globular mucilaginous mass	**Nostoc**
13b.	Heterocysts oval; filaments not in a globular mucilaginous mass	**Anabaena**

III. GREEN ALGAE AND EUGLENOIDS

1a.	Plants unicellular or colonial	2
1b.	Plant filamentous	14
2a.	Motile by flagella	3
2b.	Nonmotile	8
3a.	Unicellular	4
3b.	Colonial	5
4a.	Cells long and tapering when swimming, round when resting; one flagellum and a red eyespot	**Euglena**
4b.	Cells oval; two flagella; chloroplast cup-shaped	**Chlamydomonas**
5a.	Colony a flat plate, 4–8–16 cells	**Gonium**
5b.	Colony spherical	6
6a.	Cells close together, usually 16	**Pandorina**
6b.	Cells remote from each other	7
7a.	32 (usually) to 128 cells	**Eudorina**
7b.	Hundreds of cells	**Volvox**
8a.	Unicellular	9
8b.	Colonial	10
9a.	Cells in two symmetrical halves connected by a narrow isthmus (see Section IV)	**Desmids**
9b.	Cells spherical to oval, sometimes in irregular masses; on wood or moist soil	**Protococcus**

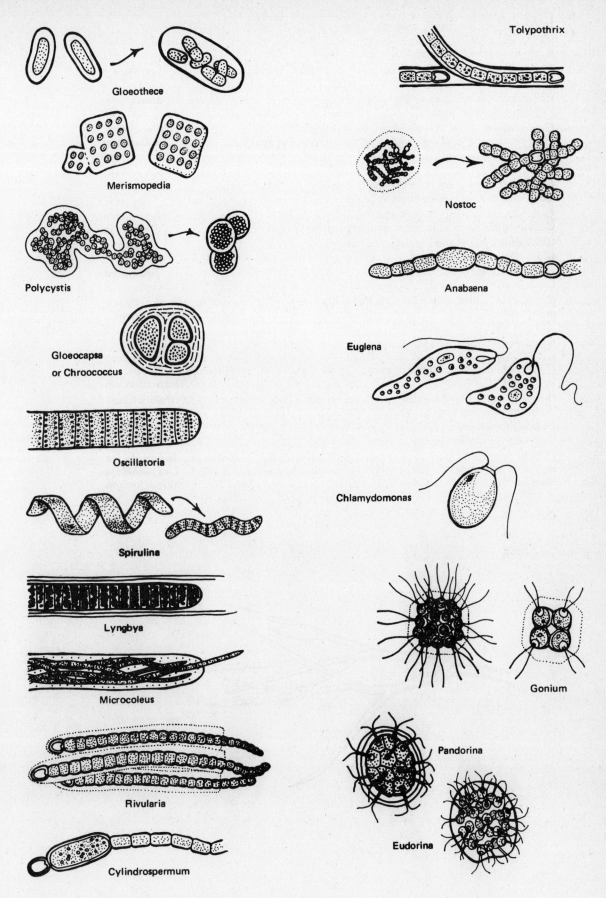

Gloeothece

Tolypothrix

Merismopedia

Nostoc

Polycystis

Anabaena

Gloeocapsa
or Chroococcus

Euglena

Oscillatoria

Chlamydomonas

Spirulina

Lyngbya

Microcoleus

Gonium

Rivularia

Pandorina

Eudorina

Cylindrospermum

10a.	Colonies with four cells (sometimes eight) in a row; spines often on end cells . . . . **Scenedesmus**
10b.	Colonies with more than four cells 11
11a.	Colonies mucilaginous; cells in groups of four within mucilage. . . . . **Tetraspora**
11b.	Colonies not mucilaginous. 12
12a.	Cells forming a net often visible to the unaided eye . . . . **Hydrodictyon**
12b.	Cells forming a flat plate 13
13a.	Plate irregular; some cells with long, sheathed bristles. . . . . **Coleochaete**
13b.	Plate regular; marginal cells with lobes, horns, or short spines . . . .**Pediastrum**
14a.	Filaments unbranched 15
14b.	Filaments branched. 21
15a.	Cells short, usually not twice as long as broad 16
15b.	Cells long, at least twice as long as broad . . . .:17
16a.	Cells with thick walls; chloroplast diffuse . . . . **Microspora**
16b.	Cells with thin walls; chloroplast in a ring around interior of cell. . . . . **Ulothrix**
17a.	Chloroplasts star-shaped, usually two per cell. . . . . **Zygnema**
17b.	Chloroplasts not star-shaped 18
18a.	Chloroplasts spiral . . . . **Spirogyra**
18b.	Chloroplasts not spiral 19
19a.	Cell wall thin; chloroplast a flat plate that is broad in surface view and appears as a thin line in side view. . . . . **Mougeotia**
19b.	Cell wall thick 20
20a.	Chloroplast more or less uniform; some cells with apical caps; swollen oogonia . . . . **Oedogonium**
20b.	Chloroplast dense and granular; cells large, with a very few short rhizoidal branches. . . . . **Rhizocolonium**
21a.	Branches short, with a bulblike base tapering into a long spine . . . . **Bulbochaete**
21b.	Branches relatively long 22
22a.	Cells with thick walls . . . . **Cladophora**
22b.	Cells with thin walls 23
23a.	Plant body showing marked differentiation between a single row of large cells forming the main axis and numerous tufts of short lateral branches with small cells. . . **Draparnaldia**
23b.	Plant body not so differentiated . . . . **Stigeoclonium**

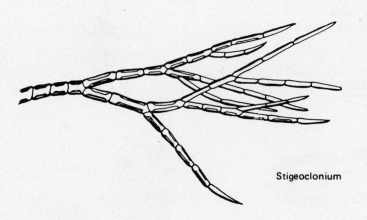

Stigeoclonium

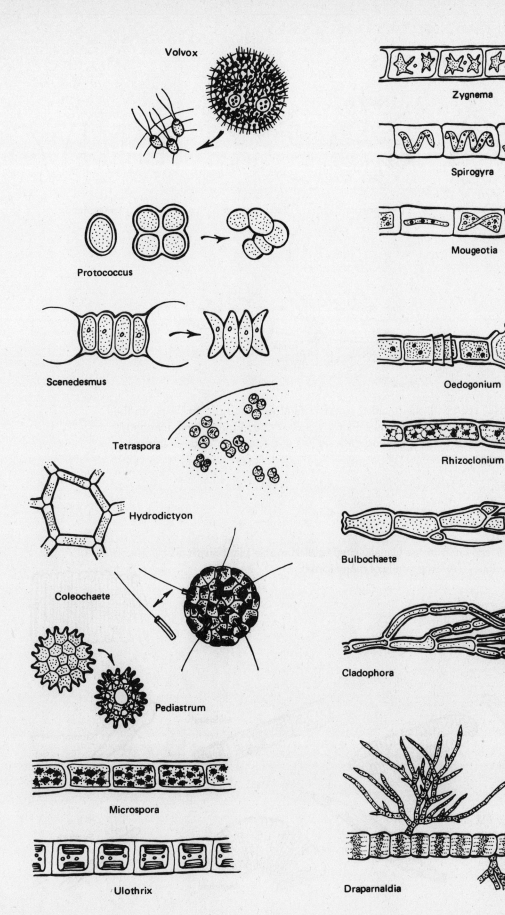

Volvox

Zygnema

Spirogyra

Protococcus

Mougeotia

Scenedesmus

Oedogonium

Tetraspora

Rhizoclonium

Hydrodictyon

Bulbochaete

Coleochaete

Cladophora

Pediastrum

Microspora

Ulothrix

Draparnaldia

IV. OTHER ALGAE NOT INCLUDED IN KEY

A. GREEN ALGAE—DESMIDS

Closterium

Cosmarium

Micrasterias

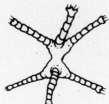

Staurastrium

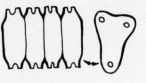

Desmidium

B. CHARA

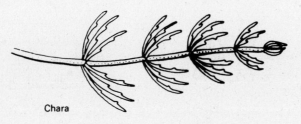

Chara

A large alga easily seen with the unaided eye. Plants attached to the soil beneath the water and often forming extensive meadows. Plants may be 0.3 to 1 meter long. Plant body consists of a cylindrical axis bearing a whorl of branches at its several "nodes." Internode lengths up to 5 cm long.

C. DIATOMS

Often yellow or brown in mass. Usually single-celled motile plants with glassy silicon walls showing striae (lines). A large family of many forms.

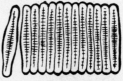

Fragilaria

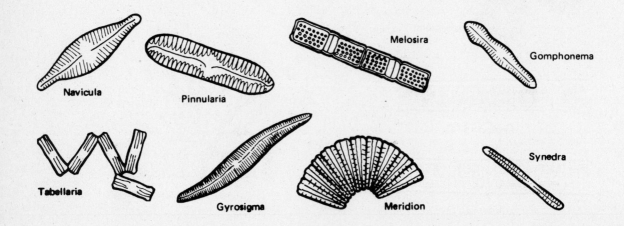

Navicula

Pinnularia

Melosira

Gomphonema

Tabellaria

Gyrosigma

Meridion

Synedra

D. OTHER ALGAE

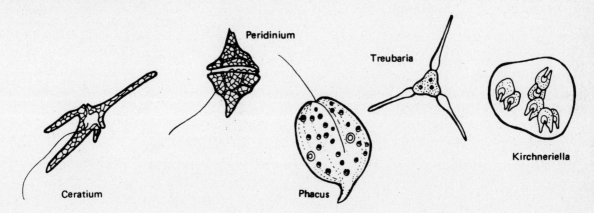

Ceratium

Peridinium

Phacus

Treubaria

Kirchneriella

BROWN ALGAE, RED ALGAE, AND GOLDEN-BROWN ALGAE

EXERCISE 24

INTRODUCTION

Those of you who live along coastal areas will be familiar with some of the brown algae of today's exercise. They are what most people recognize as seaweeds. Red algae, because of their smaller size and deep water habitat, are not so commonly known.

Of the golden-brown algae, only one group, the diatoms, is familiar to most people. Even so, the majority of people are only vaguely aware of the organisms themselves. To most people, diatoms are known better by their skeletal remains called diatomaceous earth. This is a white or yellow, chalklike silica material composed of the remains of the shells of these one-celled plants. The "earth" occurs as vast beds of marine sediments laid down over millions of years. Diatomaceous earth is used in making dynamite, insulating material, and a variety of other products.

I. DIVISION PHAEOPHYTA— BROWN ALGAE

The Phaeophyta are the brown seaweeds. They range from microscopic filamentous species to big shrublike plants. Some of the large ones have an internal anatomy similar in some ways to land plants, though no one suggests they are related.

General facts about brown algae are

Usual color	Brown
Distribution	Marine, predominantly; mostly in the cooler to very cold oceanic waters or their cold currents in warmer latitudes.
Habitat and niches	Majority are in rough waters of the tidal zone. Attached to rocks of the shoreline or to the ocean bottom at 15- to 20-meter depths. Some are microscopic in the plankton; some are large floating masses.
Motility	Flagellated cells in reproduction stages.
Body types	Range from microscopic filaments to large shrublike plants with broad, leathery blades.
Value	Of some value. Are the source of alginates, which have wide industrial uses; also used as supplementary feed for livestock and as soil fertilizer.

Review the section on significance of life cycles on page 174.

A. THE ROCKWEEDS

Rockweeds live in the rough waters of the tidal zone. Their tough, leathery thallus is well adapted to survive the battering waves and the periodic drying experienced at low tide.

ACTIVITY 1 **Examine** specimens of *Fucus*, a common rockweed. **Note** that the THALLUS has a HOLDFAST, a rootlike dish that firmly attaches the thallus to rocks. Otherwise the plants would be torn away by the rough water. Above the holdfast is a short stemlike part, the STIPE. The remainder of the plant consists of a much branched BLADE. The branching is a simple, equal-forking pattern. **Note** the AIR BLADDERS along the blade. These keep the plant afloat at high tide. At the tips of the branches are swollen, pointed sacs, the RECEPTACLES. These are covered with small pores that lead into cavities that contain GAMETANGIA. **Feel** how slippery the plants are. This is due to mucilage that protects the plant from desiccation when it is exposed to air at low tide.

Label the *Fucus* thallus in Figure 24.1, using the terms capitalized in the preceding description.

B. THE KELPS

ACTIVITY 2 **Examine** specimens of *Nereocystis* or *Laminaria* if available. Kelps are much larger plants than rockweeds. Many of the same basic plant parts are identifiable in the kelps as you saw in the rockweeds. You should be able to recognize the HOLDFAST, STIPE, BLADE, and AIR BLADDER. Again, **note** the tough, leathery, slimy feel to the plants.

Label Figures 24.2 and 24.3, using the terms capitalized in the preceding description.

C. SARGASSUM

ACTIVITY 3 If available, **examine** *Sargassum,* a floating brown alga in tropical waters. **Note** its stemlike STIPE, leaflike BLADES, and berrylike AIR BLADDERS. These plants float in large numbers, often extending for hundreds and thousands of hectares over the sea's surface. The ships of Christopher Columbus sailed right into one such area, much to the crew's dismay. They feared their ships would be dragged under by it. The Sargasso Sea is named after these plants.

Label Figure 24.4, using the terms capitalized in the preceding description.

II. DIVISION RHODOPHYTA— RED ALGAE

Red algae are small, fragile, delicate plants, rarely more than 0.7 meter long. They look like feathery, or lacy tufts of plants. Most of the species are filamentous. They have complex life cycles that we will not consider.

General facts about red algae are

Usual color	Bright red or pink, but many are also bright green or brown
Distribution	Marine; the oceans of the temperate zone, subtropics, and tropics.
Habitats and niches	In calm waters; attached to rocks or other plants in the calmer, deeper waters beyond the tidal zone; a few at shallower depths.
Motility	None.
Body types	Filaments or leaflike parts that form delicate, branched, tufted, membranous or somewhat fleshy plants.
Value	Of some value. Source of carrageenin widely used commercially for its gelling properties; source of agar used in a variety of manufactured goods and processed foods.

ACTIVITY 4 **Examine** the various specimens of red algae available, including *Stenogramma* (note the method of branching); *Gelidium* (a source of agar); and *Rhodymenia* (one of the larger species).

III. DIVISION CHRYSOPHYTA— GOLDEN-BROWN ALGAE

The Chrysophyta include three different but related algal groups. Time permits study of only one group, namely, the diatoms.

General facts about diatoms are

Usual color	Golden brown or greenish brown.
Distribution	Most are aquatic; freshwater and marine species approximately equal in numbers. Some are terrestrial.
Habitat and niches	As plankton of the seas and fresh waters. They are one of the two most abundant algae of the plankton of midocean; also found on soil surface (as the most abundant of all algae), as well as in hot springs, wet rock, and so on.
Motility	A gliding motion in some forms.
Body types	Mainly unicellular, a few are weakly colonial. Cell walls composed of two overlapping valves that fit together in a way similar to the parts of a candy box. Cells may be circular, triangular, or rectangular in shape.
Value	Extremely valuable. They are the principal primary food producers of the sea, the basic link in all marine food chains, the major single world source of photosynthetically produced food and oxygen. Their skeletal fossil remains are mined as diatomaceous earth, which has many industrial uses.

Diatoms are single-celled plants with extremely novel walls. The living protoplast of the cell is enclosed by two quite glassy walls that fit one over the other like the two halves of a candy box. These two wall halves are called **VALVES**. Valves typically have markings in a wide variety of patterns. After diatoms die, their glassy valves remain intact, and over the past millennia have accumulated in vast rock deposits known as diatomaceous earth.

Looking down at a scattering of diatoms, you often see two basic shapes, round ones and rectangular ones. Triangular-shaped ones are also present but are basically round in symmetry. The rectangular-shaped ones are termed **PENNATE**, and the round and triangular-shaped ones are called **CENTRIC**.

ACTIVITY 5 **Make** a wet mount from a culture of living soil diatoms. (A good place to obtain diatoms is from the surface soil of a potted plant.) **Examine** the slide under low, and then high magnification. Do you see any pennate diatoms? *Yes No.* Are there any centric forms? *Yes No.* Look at the illustrations of Exercise 23, Algae Key. Do any

of your specimens resemble the illustrated ones? *Yes No.* Can you make any tentative identification of your specimens? (a) _____ _____; (b) _____; (c) _____; (d) _____ _____

Pennate forms move by a gliding motion. **Observe** a few cells for a few minutes and you will be able to see this movement in some of them.

Make a wet mount, this time using some diatomaceous earth mixed in a little water. Do you see any pennate forms? *Yes No.* Are there any centric forms? *Yes No.* If you see centric forms, then your sample of diatomaceous earth is of marine origin, because centric forms occur almost exclusively in the sea. From your observations of these two samples, what would you say about the distribution of pennate forms? Are they organisms of (a) the sea only, (b) fresh water only, or (c) both seas and fresh waters? Circle (a), (b), or (c).

Label the pennate and centric type diatoms in Figure 24.5.

EXERCISE 24   **Student Name** _____

QUESTIONS

1. Rough tidal water is a common habitat for the kelps and rockweeds. Give two adaptations (structural and otherwise) found in these plants which help them survive in this habitat.

    (a) _____

    _____; (b) _____

    _____

2. Red algae characteristically live in quiet, calm water depths below wave action. What effect

    has this had on their growth forms? _____

    _____

3. Kelps are pretty big, weighty plants. What keeps them afloat? _____

    _____

4. There are no delicate, filamentous brown algae. Circle one: (a) true, (b) false.
5. Red algae are predominantly marine. Circle one: (a) true; (b) false.
6. Suppose you're hired as the nature counselor at a summer camp near a big lake.

    a. Should you get out your old botany notes and "bone up" on *all* the algae? _____

    b. Would your fishing line get snagged in rockweeds? _____

    c. Could you find any diatoms to show the campers? _____. You'd expect to find (a) pennate types; (b) centric types; (c) both types; (d) neither. Circle (a), (b), (c), or (d).

    d. As you lead the way under a waterfall, what group of algae will you know enough to point

        out growing on the wet ledges? (division name) _____

    e. The camp has a microscope and you've brought your algae key from this manual. You

        collect some lake water. Would *most* of the key be of use to you? _____

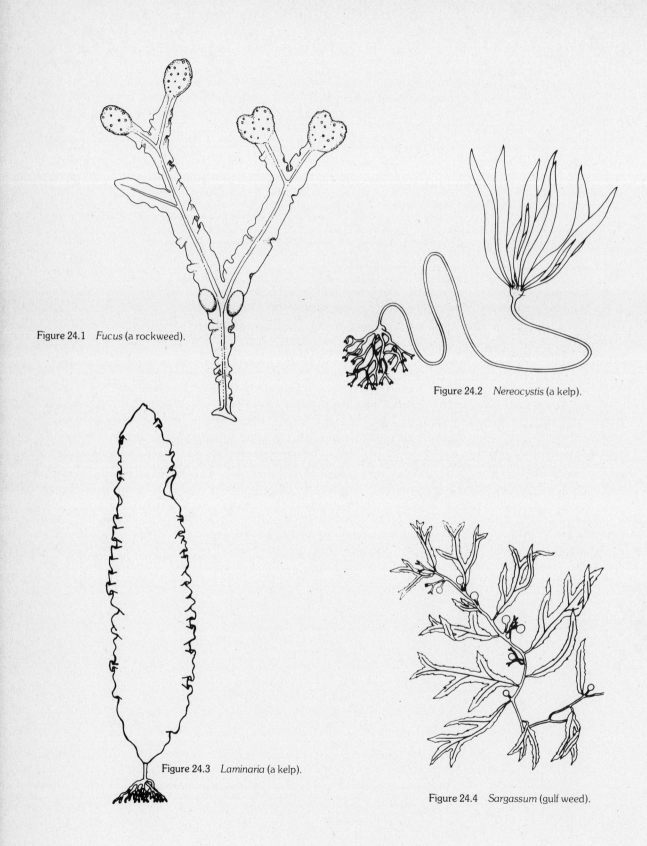

Figure 24.1   *Fucus* (a rockweed).

Figure 24.2   *Nereocystis* (a kelp).

Figure 24.3   *Laminaria* (a kelp).

Figure 24.4   *Sargassum* (gulf weed).

Student Name _____

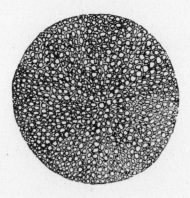

Figure 24.5   Two examples of diatoms.

MOSSES AND LIVERWORTS, AND AN INTRODUCTION TO LAND PLANTS

EXERCISE

25

INTRODUCTION TO LAND PLANTS

Conditions on land are extremely variable. Temperature, moisture, wind, exposure—these vary from place to place and from season to season. This is in sharp contrast to the aquatic world, in which conditions vary much less and habitats are much more uniform. But the multiplicity of land habitats with their varying life-support requirements means that, for life on land, plants must have certain structural and functional features that allow them to survive and thrive in a particular habitat.

This multiplicity of land habitats has evoked a multiplicity of plant types. How have these structural and functional features of the land plant arisen? Land plants have developed a life cycle "style" that allows for substantial increase in genetic variation in the populations. Any population endowed with greater genetic variation has preadaptive potential for survival in new or changing habitats on land.

This increase in genetic variation is due largely to the increase in the number of meiotic events in the life cycle. The more meiotic events (that is, the more cells dividing by meiosis), the more the chromosomes and genes are shuffled into new gene combinations. New gene combinations and gene mutations in individuals lead to genetic variation in a population. Thus, ultimately, very different organisms with different genetic makeup have come to occupy and survive in the varied land habitats.

Compare the generalized life cycles in Figures 25.1 and 25.2. In the land plant (Figure 25.1) you can see that for each zygote formed, there are many meiotic events generated, not in the zygote itself, but because each zygote produces a multicellular plant (the sporophyte) that in turn produces thousands of spore mother cells, each of which divides by meiosis. Compare this with the one meiotic event generated for each zygote formed in the life cycle of the basic alga or fungus (Figure 25.2).

The thallus (or plant body) that produces the spores in the land plant is the *sporophyte* or *sporophyte generation*. The thallus (or plant body) that produces the gametes (eggs and sperm) is the *gametophyte* or *gametophyte generation*. So, in the life cycle of the land plant we find an *alternation of two generations:* a sporophyte and a gametophyte.

Review the section on significance of life cycles on page 174.

DIVISION BRYOPHYTA (BRYOPHYTES)—MOSSES AND LIVERWORTS

Mosses and liverworts make up most of the bryophytes, which are tiny plants averaging less than 10 cm in height. Their small size is largely due to their lack of tissues (xylem and phloem) for support and for internal conduction. The absence of these tissues also limits them to relatively few types of habitats. Most grow in moist or wet shady places. Some mosses, though, are adapted to dry, sunny exposed sites.

In bryophytes, the gametophyte thallus is the chief photosynthetic body in the life cycle. But it is also the thallus that produces the gametes. Its small, ground-hugging growth form favors fertilization because a surface film of water is necessary for sperm transport to the egg. But its small size automatically limits the photosynthetic surface, which in turn puts a limit on the potential for growth.

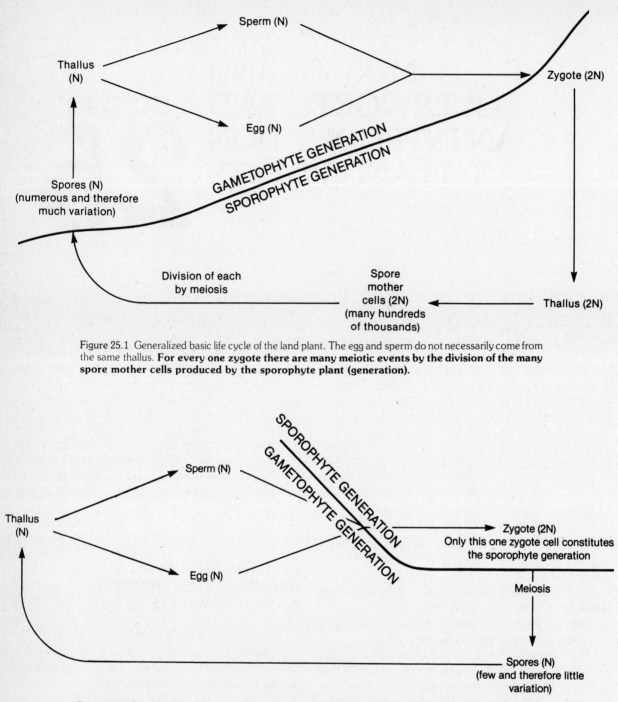

Figure 25.1 Generalized basic life cycle of the land plant. The egg and sperm do not necessarily come from the same thallus. **For every one zygote there are many meiotic events by the division of the many spore mother cells produced by the sporophyte plant (generation).**

Figure 25.2 Simplified life cycle approximating the basic life cycle of an alga and fungus. The egg and sperm do not necessarily come from the same thallus (many algae and fungi have more complex cycles, and some of these are comparable to the cycle in the land plant shown in Figure 25.1). **For every one zygote there is only one meiotic event.**

Also, the general lack of a cuticle makes it liable to dehydration. See the summary in Table 25.1.

The sporophyte thallus is totally supported and partially nourished by the tiny gametophyte thallus. Thus the sporophyte can never grow any larger than the gametophyte. This combination of features has led to a blind alley as regards evolution in bryophytes (see Table 25.1). The first bryophytes may have evolved from certain highly advanced green algae. In

time, many became rather elaborate little plants that thrived long before the dinosaurs. However, they were probably always suited to only a limited range of terrestrial habitats, and as time went on they have declined into the very simple, uncomplicated types we see today.

In this exercise you will study representatives of two classes of bryophytes, the mosses (class Musci) and the liverworts (class Hepaticae).

Table 25.1 SUMMARY OF THE PROGRESSIVE AND PRIMITIVE FEATURES OF THE BRYOPHYTES AS THEY RELATE TO SUCCESSFUL LIFE ON LAND

PROGRESSIVE FEATURES OVER THE ALGAE	PRIMITIVE FEATURES
1. Eggs and sperm enclosed by protective organs, the archegonia and antheridia. These prevent dehydration and death of the gametes 2. Embryo protected within the archegonium and nourished by the female parent plant 3. Introduction of a sporophyte generation into the life cycle leads to more genetic variation through meiotic divisions of numerous spore mother cells	1. The combination of photosynthesis and sexual reproduction in one plant (the gametophyte) is not a practical union of functions. The reason is that the necessity for water for sexual reproduction keeps the gametophyte a low, ground-hugging plant to facilitate sperm transfer by rain and water droplets. This in turn limits the potential for assimilative functions and growth and restricts the range of habitats 2. No vascular or supporting tissue, therefore limited in growth 3. Little or no cuticle, therefore susceptible to dehydration 4. Water necessary for fertilization, therefore restricted plant distribution

GENERALIZED LIFE CYCLE OF BRYOPHYTES

Study Figure 25.3, which is a generalized life cycle of bryophytes, as you proceed through the exercise. Also **study** Table 25.2, which shows how the life cycle stages, structures, and cells are involved in species survival.

Table 25.2 LIFE CYCLE STAGES, STRUCTURES, AND CELLS RELATED TO THE FOUR SURVIVAL PHASES IN SPECIES OF BRYOPHYTES

SURVIVAL PHASE	ACCOMPLISHED BY
Multiplication	1. Asexual reproduction (gemmae, shoot fragments, and new branches from the gametophyte) 2. Spore production by the sporophyte
Dispersal	1. Spores from the sporophyte 2. Gemmae and shoot fragments from the gametophyte (to some extent)
Assimilation	Gametophyte thallus
Genetic variation	1. Meiosis by spore mother cells in the sporangium of the sporophyte 2. Sexual fusion in the gametophyte generation

I. THE MOSSES—CLASS MUSCI

A. *THE GAMETOPHYTE THALLUS*

The gametophyte is multifunctional. It serves in asexual and sexual reproduction; it photosynthesizes, manufacturing foods for its own growth and that of the embryo and sporophyte; and it supports the sporophyte. As we have seen, though, such a combination of multiple functions is not very practical and limits the plant in several ways. **Look** at Figure 25.4 as a guide as you examine the moss plant and its parts.

Vegetative Parts

ACTIVITY 1 **Examine** the leafy gametophyte plants of species of *Mnium, Polytrichum,* or *Funaria.* If live plants are provided, you can see that numerous tiny gametophytes form a dense, cushiony, carpetlike mat. The mat has been collected from a forest floor or other such shady ground.

With forceps, **pull up** one of the gametophytes. *Do not* pull up any plant that has a sporophyte on it (that is, any plant having a thin, leafless filament and capsule growing at the tip). Pick only those plants that are leafy all the way to the top.

The plant typically has a slender, erect, stemlike axis crowded with leaflike appendages. The rootlike, multicellular filaments at the base are RHIZOIDS. (You may not see any rhizoids because they usually break off when the plant is pulled up). **Remove** a "LEAF" and mount it on a slide for microscopic examination. How many cells thick is the leaf? _____. Are there stomates? *Yes No.* Is there a midrib? *Yes No.* Can you see xylem cells? *Yes No.* Is there a vein system? *Yes No.* Is this a true leaf? *Yes No.* Does it carry on photosynthesis? *Yes No.* **Look** at a rhizoid on a slide under your microscope. Is the rhizoid made up of one or of several cells. (Circle one.) Is there any xylem? *Yes No.* Any root hairs? *Yes No.* Is the rhizoid a true root? *Yes No.*

Label A and B of the moss life cycle in Figure 25.4, using the terms capitalized in the preceding descriptions.

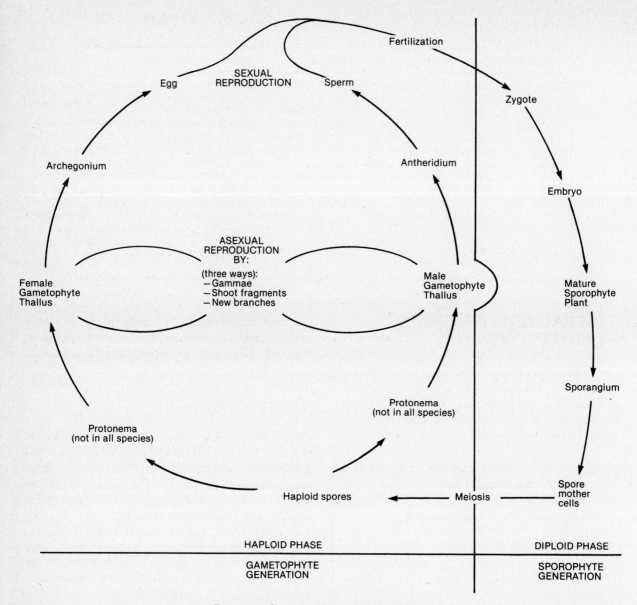

Figure 25.3 Generalized life cycle of bryophytes.

Reproductive Parts

Examine the tip of a leafy plant. If the tip has a flat, "flowerlike" cluster of leaves, the plant is a **MALE GAMETOPHYTE**, and the male reproductive organs are located in this flat circlet of leaves. If the leaf arrangement at the tip is no different from the rest of the stem, the plant is a **FEMALE GAMETOPHYTE**, and the female reproductive organs occur hidden among the leaves at the tip.

Rain drops splattering over the plants carry sperm from male plants to the female. As a droplet of rain sits in the cuplike circlet of leaves on the male head, sperm are released into it. Subsequent ricocheting or splattering of these sperm-laden

droplets results in some of them landing on female plants. The motile sperm then swim through the water toward the female organs. **Label** the male plant and the female plant in Figure 25.4, using the terms capitalized in the preceding description.

Reproductive Organs
Antheridia

ACTIVITY 2 **Examine** prepared slides of the tip of a male gametophyte showing the **ANTHERIDIA** (plural), the male reproductive organs that produce the sperm. An **ANTHERIDIUM** (singular) is broadly or narrowly oval with a one-layered outer coat, the

JACKET LAYER. Virtually all the antheridia that are intact and recognizable on your slide contain tightly packed, angular cells that are in some stage of forming sperm. These are the SPERM-FORMING CELLS. Intermingled among the antheridia are "leaves" and special slender filaments known as paraphyses (singular, paraphysis) or sterile hairs. These are erect and hairlike with expanded, club-like tips. What function do you suppose the "leaves" and paraphyses serve? _____

_____

What is the function of the jacket layer? _____

_____.

Label the antheridium and its parts in Figure 25.4, using the terms capitalized in the preceding description.

Sperm

ACTIVITY 3 **Obtain** a male gametophyte plant of *Mnium* or *Polytrichum* or other moss and note the characteristic cup-shaped circlet of leaves at the tip. This tip is designed to catch and hold a droplet of rain or moisture. The collected water will eventually cause the antheridia to split open and release sperm. When mature, sperm develop two flagella by which they maneuver through the watery film or droplets covering the plants. **Place** the plant on a slide and, using a needle, gently press the base of the head. Squeeze out a few antheridia and examine them under the microscope. After a while the antheridia will probably rupture and discharge sperm. **Keep** the slide and check it periodically to see if sperm are released.

Archegonia

ACTIVITY 4 **Examine** a prepared slide of a moss gametophyte showing ARCHEGONIA (plural). Find an ARCHEGONIUM (singular). This is shaped something like a bowling pin or a long necked vase. It consists of (1) an oval, or globe-shaped base, the VENTER, within which is one EGG, and (2) a long, slender chimney of cells, called the NECK. The archegonium exudes a chemical that attracts sperm cells, which swim down the neck, which in turn channels them toward the egg. One SPERM fertilizes the egg and a ZYGOTE is formed. The zygote develops by many cell divisions into the EMBRYO. The embryo continues growth, developing into the sporophyte. The early stages of growth occur within the archegonium. Because of the minute size of the archegonium, your slide section may not have archegonia that are all complete, and you may see necks without venters attached and vice versa.

Label the archegonium and its parts, both before and after fertilization, in Figure 25.4, using the terms capitalized in the preceding description.

B. THE SPOROPHYTE THALLUS

The sporophyte produces and discharges the spores. These are then carried by air currents (in some cases, water currents) away from the parent plant. A benefit to this dispersal is that the new gametophyte plants developing from spores are thus less likely to crowd and compete with the old parent population for light, nutrients, and space.

ACTIVITY 5 **Examine** plants bearing sporophytes. The SPOROPHYTE plant is designed well for its function of spore discharge. There is a little CAPSULE or SPORANGIUM borne at the top of an upright stalk, the SETA. The base of the seta is called the FOOT. The foot is not usually visible to you. It is embedded in an archegonium at the tip of a gametophyte. The elevated capsule is a definite advantage for spore dispersal by air. The capsule serves three functions: the basal part is photosynthetic; the central core produces spore mother cells that divide by meiosis to form the spores; and the tip is designed to discharge spores.

Dissect a capsule. With forceps, gently pinch the very tip of the capsule and pull off the loosely fitting hood, the CALYPTRA. The calyptra is the remnant of the archegonium that was carried upward by the growth of the sporophyte. Next you can see a trim little, tight-fitting lid, the OPERCULUM. **Remove** the operculum with forceps. Transfer the capsule to a small blob of Vaseline on a slide and, using the low power of the microscope, orient the capsule so that you are looking down into the mouth of the capsule. Or cut off the upper end of the capsule with a sharp razor blade and mount it in water on a slide. Look at the toothlike projections encircling the rim of the mouth. These projections are the PERISTOME. In many mosses, like *Mnium*, the teeth forming the peristome are hygroscopic (sensitive to changes in air moisture), and they bend in and out in response to these changes. This bending in and out of the teeth flips the spores out of the capsule. In other mosses, like *Polytrichum*, the teeth do not bend in and out. Instead, they are all joined to a membrane that stretches across the mouth, and the spores sift out through tiny holes in the membrane between the teeth tips. In this case, the teeth serve to slow down the escape of spores, thus prolonging the period of dispersal. What is the advantage of this? _____

_____

_____

_____

Test the hygroscopic mechanism yourself. Take two capsules from sporophytes. Moisten one on a wet paper towel; allow the other one to dry in the warmth of a lamp (or just let it air dry). Don't lose it! After a few minutes, look at the curvature of the peristome teeth in each capsule. **Describe** what you see. _____
_____ .

Would high- or low-moisture conditions favor release of the spores? (Circle one.) **Label** the sporophyte and the parts of the sporangium in Figure 25.4, using the terms capitalized in the preceding description.

C. SPORES AND PROTONEMA

Spores

ACTIVITY 6   **Crush** a capsule of a moss with a needle and force out some spores into a drop of water on a slide. **Cover** with a cover glass and **examine** under the microscope. Are the spores green? *Yes   No.*   Are they all alike? *Yes   No.*   SPORES are the haploid daughter cells of SPORE MOTHER CELLS that divided by MEIOSIS within the capsule. The reshuffling of chromosome pairs and rearranging of genes that occurs during the meiotic division of spore mother cells produces haploid spores with many different genetic combinations. This serves to produce a degree of genetic variation in spores that will ensure adaptive variability among the gametophyte plants that result from spore germination.

   Complete the labeling of the sporangium and its contents in Figure 25.4, using the terms capitalized in the preceding description.

The Protonema

ACTIVITY 7   **Examine** living spores that are germinating on a nutrient medium. Or if these are not available, examine prepared slides of moss protonema. A protonema is formed by SPORE GERMINATION. The spore divides, and the succeeding cells keep on dividing to form a cylindrical, much-branched, green algal-looking stage of the moss called the PROTONEMA. The protonema lives for a long time. As it matures, it produces small leafy BUDS. Growth of each bud produces an erect leafy gametophyte thallus.

   Label the protonema and its parts in Figure 25.4, using the terms capitalized in the preceding description.

D. PEAT MOSS— SPHAGNUM

Mosses are biologically and ecologically important. However, in terms of direct-use economic value, the only one of importance is the peat moss, *Sphagnum*. This is because it is one of the most absorbent of all materials. When a *Sphagnum* plant is completely dry, it can absorb and hold up to 200 times its own weight in water. This property makes it useful for many purposes. The plant also contains compounds that inhibit bacterial growth. In the past, some American Indians used it for diapering infants; up to the American Civil War doctors used it for dressing wounds. Today it is most commonly used for a variety of purposes in the ornamental plant industry. Commercial growers pack the roots of shrubs and trees in wet peat moss so they won't dry out during shipment. Soils containing high quantities of *Sphagnum* are known as peat. Packaged peat is widely sold for use in commercial greenhouses and home gardening.

The Sphagnum Plant

ACTIVITY 8   **Examine** living, preserved, or dried *Sphagnum* plants. Note the "stems," "leaves," and "branches."

   Determine the water-holding capacity of *Sphagnum*. Weigh a small plastic drinking glass or beaker and record the weight to the nearest 0.1 gram. Fill the glass with dry *Sphagnum* and reweigh. Record the combined weight of glass and *Sphagnum*. Fill the glass with water and allow the *Sphagnum* to soak for 10–15 minutes. Invert and allow the excess water to drain off for 5 minutes. Reweigh the beaker and wet *Sphagnum*. From the increased weight, you can find the amount of water absorbed by the *Sphagnum*. Comparing this to the dry weight of the *Sphagnum,* you can determine the water-holding capacity of the plant.

Weight of beaker + dry *Sphagnum*: _____ gm
Weight of empty beaker: _____ gm
Weight of dry *Sphagnum*: _____ gm
Weight of beaker + wet *Sphagnum*: _____ gm
Amount of water held by *Sphagnum*: _____ gm
Water-holding capacity of *Sphagnum*: _____ gm

Internal Structure of Sphagnum "Leaves"

ACTIVITY 9   The *Sphagnum* leaf has *special* large, water-holding cells. **Mount** a leaf of living or preserved *Sphagnum* in water and examine it under the microscope. **Note** that there are two cell types. Thin, tubelike cells connect with each other in an open meshwork. These cells contain numerous small bodies that you should recognize as the _____. The larger, colorless cells between the tube-shaped cells have small pores in their walls. The walls also show spiral glassy-looking thickenings. Which of the two cell types

forms the greater portion of the leaf? _____
_____. What is the obvious
function of these cells? _____
_____.

II. THE LIVERWORTS— CLASS HEPATICAE

Less commonly encountered, and even less commonly recognized, are the liverworts. In many ways they closely parallel the mosses, but there are essential differences between them. Most liverwort species (more than 80 percent) have leafy foliose gametophytes, and this makes it difficult to readily distinguish them from mosses. A minority of liverworts have what is termed a thallose gametophyte. Thallose gametophytes form broad, branching ribbons that are conspicuously lobed or ruffled. They are generally much larger than leafy foliose gametophytes. Both thallose and foliose gametophytes are primarily prostrate or creeping and not erect as are most moss gametophytes. **Look** at Figure 25.5 as a guide as you examine specimens of liverworts.

A. *THALLOSE-TYPE GAMETOPHYTE*

ACTIVITY 10 **Examine** the gametophytes of species of *Marchantia* or *Conocephalum*. Note that the plant has a dark green, RIBBONLIKE THALLUS and branches in an equal forking pattern. The upper surface has a central groove running its length and it also has diamond-shaped markings that outline the underlying internal air chambers. Photosynthetic cells occupy the air chambers. Each area has a stomatelike pore for air passage. **Examine** the underside of the gametophyte and look at the dense growth of fine, hairlike RHIZOIDS.

Reproductive Parts

Gametophytes of *Marchantia* have three kinds of special structures on the top side of the plants. One kind consists of little cups, GEMMAE CUPS. These serve for asexual reproduction. The other two kinds look like miniature plants growing upright on the surface of the ribbon. These are the male

ANTHERIDIOPHORES and the female ARCHEGONIOPHORES. They occur on separate gametophyte plants.

Antheridiophore **Examine** an ANTHERIDIOPHORE. This consists of a small, flat plate or disk, the ANTHERIDIAL RECEPTACLE, borne at the tip of a short, slender stalk. ANTHERIDIA are embedded in this disk and release sperm through pores at the top.

Archegoniophore **Examine** an ARCHEGONIOPHORE. This is a tiny, umbrellalike structure with several free-hanging, fingerlike lobes, called RAYS. On the underside of each ray, ARCHEGONIA hang suspended, neck down. Sperm are carried from antheridiophores by surface moisture to the archegoniophores, and there they swim up the necks of the archegonia. Fertilization and the development of the sporophyte takes place inside the archegonium as in mosses.

Gemmae Cups Gemmae cups are small, round, bowl-shaped structures. If you look down directly into the open gemmae cup and use a hand-lens or low power of the microscope, you can see the many little disk-shaped GEMMAE. Gemmae are washed or flicked out of the cup by rain and carried away from the parent plant. Under favorable conditions, a gemma will develop RHIZOIDS, and a YOUNG NEW GAMETOPHYTE is produced from it.

Label all the parts of the gametophyte thallus of the liverwort in Figure 25.5, using the terms capitalized in the preceding descriptions.

B. THE SPOROPHYTE

In most cases, the mature sporophyte is simply a small sack, the **CAPSULE** or **SPORANGIUM**, suspended by a short stalk and hanging out of an archegonium on the underside of a ray of an archegoniophore. When the capsule matures, it splits open. Special spindle-shaped cells, known as elaters, occur among the **SPORES**. Their spiral banded walls are hygroscopic, and the elaters perform squirming movements that help to loosen the spore mass and throw spores out of the capsule.

Label all the parts of the sporophyte of the liverwort in Figure 25.5, using the terms capitalized in the preceding description.

EXERCISE 25   **Student Name** _____

QUESTIONS

1. What two completely different functions are performed by the gametophyte? (a) _____
   _____ ; (b) _____
   _____

2. If moss gametophytes were much taller plants, what problems would they probably have
   in regard to fertilization? _____
   _____

3. In the absence of xylem, how does the gametophyte take up soil moisture? _____
   _____

4. Give at least two reasons why the gametophyte plant is not the best designed structure for
   efficiency in photosynthesis. (a) _____ ;
   (b) _____

5. What prevents the sporophyte from developing into a taller plant to achieve better spore
   trajectory? _____

6. What is an advantage of dispersing spores? _____
   _____

7. Describe the habitat best suited to the majority of mosses. _____
   _____

8. Does the spore develop into the embryo? _____

9. Is the embryo haploid or diploid? _____

10. During what event of the life cycle of moss are several new gene combinations created by a
    reshuffling of chromosome pairs and rearrangement of genes? _____

11. Which generation develops from the growth of the embryo? _____

12. Does meiosis lead to the formation of gametes or of spores? _____

13. Which of the following is (are) a haploid: (a) spore(s); (b) egg; (c) sperm. Circle your
    choice(s).

14. The zygote divides to form many cells that become organized into the _____
    _____ which then matures into the _____
    generation.

15. Which group of algae are thought to be ancestral to bryophytes? _____

16. Cite at least three different reasons why the structural design of the bryophytes has made
    them inadequately adapted for successful land life. (a) _____
    _____ ; (b) _____
    _____

17. Increase in genetic variation in land plants is essentially due to _____
    _____
    _____

18. For each zygote formed in the life cycle of most algae and fungi, how many chances are
    there for the creation of new gene combinations? (a) one; (b) a few; (c) numerous. Circle (a),
    (b), or (c).

19. In most algae the cell that divides by meiosis is the _____, but in land plants,
    the cell that divides by meiosis is the _____

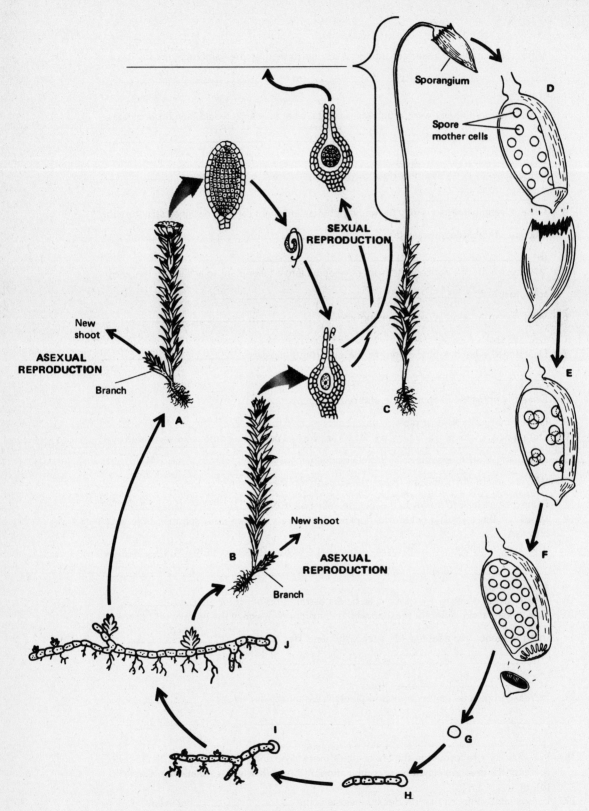

Figure 25.4 Life cycle of a moss (*Polytrichum* sp.) showing sexual and asexual methods of reproduction.

Sporangium

Spore
mother cells

D

SEXUAL
REPRODUCTION

ASEXUAL
REPRODUCTION

New
shoot

Branch

A.

C

E

B

New shoot

ASEXUAL
REPRODUCTION

Branch

F

J

G

I

H

EXERCISE 25   **Student Name** _____

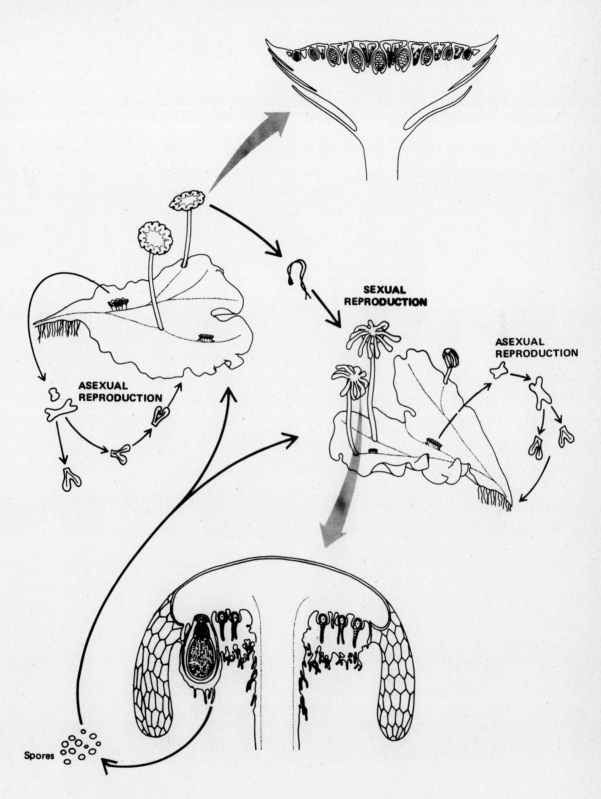

Figure 25.5 Life cycle of a liverwort (*Marchantia polymorpha*), showing sexual and asexual methods of reproduction.

LOWER VASCULAR PLANTS: FERNS, CLUB MOSSES, HORSETAILS, AND *PSILOTUM*

INTRODUCTION

If people valued ancient plants in the same way they appreciate old works of art, then the ferns and other lower vascular plants would all be on display in museums around the world!

Indeed, fossils of some of these ancient groups are kept in several museums throughout the world, the British Museum in London, the Smithsonian in Washington, and the museum of the University of Moscow, to cite a few.

Lower vascular plants are old groups. Not as old as the bacteria, fungi, and algae, but they do date back 400 million years to a geological period called the Carboniferous. In that time, these plants formed great forests over much of the earth. Small herbaceous types existed side by side with the tree forms. They lived in extensive swamps that covered the continents.

Time gradually brought a drastic deterioration of the warm, humid climate of the Carboniferous swamps, and conditions grew colder and drier. Competition with other plants, which were better able to cope with the changes, resulted in the elimination of practically all of these swamp-living lower vascular plants. Today we have only a few living relict types. All, except for the tree ferns, are small plants. All, with the exception of the ferns, play a limited role in the vegetation of the earth.

Old silver dollars are valuable because the U.S. mint no longer makes them. Old cars are still prized for their technologically extinct design. People find them worth looking at. Unlike products of human technology, nature can go on (almost forever) making a few samples of ancient and evolutionarily defunct designs of plants (and animals too). Right now there are only about 208 genera of ferns and other lower vascular plants left. They manage to survive in a world very different from the heyday of their ancestors. Just as old model cars help us to appreciate the development of the modern car, so these old plants help us to understand and better appreciate the modern flowering plants that appeared much later in time and that have sustained the human race throughout its history.

The name lower vascular plant is based on the fact that these are plants that have vascular (conducting) tissues of xylem and phloem, but they are at a lower level of evolution in their methods of reproduction than are the more complex, higher vascular plants, the seed plants.

Lower vascular plants show many adaptations to land life but still retain some primitive features in their life history that are a holdover from their aquatic algal ancestry, the most notable of which is the necessity of water for fertilization.

Read the information given in Table 26.1. It lists the primitive features that were to handicap these plants when the world environment changed, and it also gives a summary of the progressive features exhibited by lower vascular plants beyond those of the bryophytes that allow them to exist on land while reaching a much larger size than the bryophytes.

In lower vascular plants (and higher plants) the sporophyte stage in the life cycle is the dominant phase, not the gametophyte stage as in bryophytes.

By developing the sporophyte stage into the chief photosynthesizing (or energy-making) phase, plants found a winning combination of functions—the functions of photosynthesis and of spore production. Both functions are sustained by the same environmental conditions and structural design—namely, a sturdy, erect plant making use of sun and air for photosynthesis and spore dispersal. This switch to the dominance of the sporophyte stage was a great "innovation" in the evolution of land plants.

Table 26.1   SUMMARY OF THE PROGRESSIVE AND PRIMITIVE FEATURES OF THE LIVING CLUB MOSSES, FERNS, AND HORSETAILS AS THEY RELATE TO SUCCESSFUL LIFE ON LAND

PROGRESSIVE FEATURES OVER THE BYROPHYTES	PRIMITIVE FEATURES
1. The incorporation of photosynthesis and spore dispersal in one thallus (the sporophyte) is a good combination of functions. Both have some of the same environmental needs (sun and air). The sturdy erect plant, the sporophyte, suits both functions equally well 2. The sporophyte is well adapted to life on land: has cuticle, lignin, vascular tissue, and numerous stomates 3. Vigorous rhizome growth produces new shoots. This quickly renews the sporophyte generation and maintains the population in an area. It bypasses the slow cycling through the gametophyte generation	1. Leaf in club mosses and horsetails is small and has only one vein. This may limit the efficiency of translocation of materials and in effect may limit growth of these same two groups 2. Water is still required for fertilization 3. There is alternation of two independent-living generations (gametophyte and sporophyte) having different growth needs and somewhat different habitats 4. Gametophyte is restricted to wet, shady places; lacks cuticle, lignin, and (in most cases) vascular tissue 5. Gametophyte generation is slow-growing, requiring many months to many years before producing a zygote 6. Spore is slow to germinate 7. All are free-sporing (with one known living exception and certain fossil forms)—that is, the spore is released free of the parent plant, and hence its survival and germination depend on chance favorable conditions in the environment.

Review the section on significance of life cycles on page 174.

I. GENERALIZED LIFE CYCLE

Study the diagram of Figure 26.1, which illustrates the general life cycle of the lower vascular plants. Also **study** Table 26.2. Remember that the gametophyte phase (thallus) consists of a very small plant, and most of your work will concern only the sporophyte.

II. *PSILOTUM*

ACTIVITY 1   **Examine** living or preserved specimens or color transparencies of *Psilotum nudum*. *Psilotum* is the most primitive living vascular plant. It grows wild in Florida and Hawaii and other parts of the subtropics and tropics. It is also a fairly popular ornamental plant in greenhouses and homes farther north. For many years it has generally been considered a living relic of an extinct group known as the psilophytes. Some authorities now consider it a primitive fern, but not everyone is in agreement.

The simple plant you see is the sporophyte. The aerial (above ground) STEM is a slender axis that has an equal forking type of branching known as DICHOTOMOUS BRANCHING. This branching pattern is found in many primitive plants. Are leaves present?   *Yes*   *No.*   Are there small appendages that resemble SCALES?   *Yes*   *No.* If you have a living specimen, what color is the stem?

_____. If you don't have a living specimen, what color would you expect the stem to be when you can see that there are no leaves? _____ The aerial shoot originates from an underground stem, the RHIZOME. You probably won't see the rhizome.

Look more closely at the branching axes and **find** the tiny, trilobed, globelike sporangia. Each SPORANGIUM occurs in the axil of a SCALE. MEIOSPORES are released from these sporangia, and each may germinate into a GAMETOPHYTE THALLUS, which produces archegonia and antheridia. A SPERM from an ANTHERIDIUM fertilizes an EGG in an ARCHEGONIUM to form a ZYGOTE that develops into a YOUNG SPOROPHYTE, which remains attached to the gametophyte during its early stages of growth. Later it becomes an independent-living MATURE SPOROPHYTE PLANT.

Complete the labeling of Figure 26.2 using the terms capitalized in the preceding description.

III. DIVISION LYCOPODOPHYTA— CLUB MOSSES

Lycopods are called club mosses because of the mosslike appearance of the leafy stems and the club-shaped cones, although not all species have cones. *Lycopodium* and *Selaginella* are two representatives of the group.

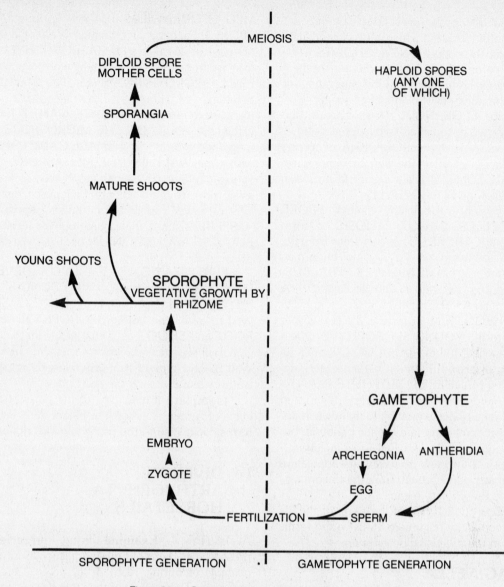

Figure 26.1   Generalized life cycle of the lower vascular plants.

Table 26.2   LIFE CYCLE STAGES, STRUCTURES, AND CELLS RELATED TO METHODS OF SPECIES SURVIVAL IN LOWER VASCULAR PLANTS

SURVIVAL PHASE	ACCOMPLISHED BY
Multiplication	1. New shoots from the rhizome (asexual reproduction) 2. Spore production by sporophyte
Dispersal	1. Spores 2. Rhizome (to some extent) 3. Bulbils (in some species)
Assimilation	Shoots, leaves, and rhizome of the sporophyte thallus
Genetic variation	1. Meiosis by spore mother cells in the sporangium of the sporophyte 2. Sexual fusion in the gametophyte generation

A. *LYCOPODIUM*

ACTIVITY 2 **Examine** living, preserved, or pressed specimens of *Lycopodium,* such as *Lycopodium clavatum, L. lucidulum,* or *L. complanatum.* The leafy plant you see is the sporophyte. **Note** the way the **STEM** branches. Is it an equal forking pattern?   *Yes   No.*   What is this pattern called? _____

_____. Is the stem axis naked as in *Psilotum*? *Yes   No.*   Is the stem densely covered with leaves?   *Yes   No.*   In those species that have cones, the parts of the axes that bear the cones seem almost naked because they have few leaves.

The leafy aerial (above-ground) shoots arise from a horizontal, leafy stem, the **RHIZOME.** The rhizome may grow below the soil's surface or just under the leaf litter of the forest floor. The rhizome forms a dichotomously branching network, and at its growing ends it produces **NEW SHOOTS**. The rhizome is perennial, and its production of new shoots by asexual reproduction is the chief means of increasing the numbers of plants in an area.

Look for **SPORANGIA**. Depending on the species, these will be found either clustered together in **CONES** or as solitary **SPORANGIA** occurring separately any place along the stem. In either case, the **SPORANGIUM** itself is always borne in the axil of a leaf, called the **SPOROPHYLL**.

Within each sporangium diploid **SPORE MOTHER CELLS** divide by **MEIOSIS** to form **HAPLOID MEIOSPORES**. These are freed from the sporangium and are dispersed by wind. Each may germinate into a **GAMETOPHYTE THALLUS**, which produces antheridia and archegonia. A **SPERM** from an **ANTHERIDIUM** fertilizes an **EGG** in an **ARCHEGONIUM** to form a **ZYGOTE** that develops into the **YOUNG SPOROPHYTE**, which remains attached to the gametophyte during its early stages of growth. Later it becomes an independent-living **MATURE SPOROPHYTE PLANT**.

The gametophyte stage of the life cycle can, and often does, require many months to many years for completion. It does introduce genetic variation, but because it requires such a long time for completion, it is not the primary means by which the population increases in numbers. Growth from the rhizome, as described, is the primary means.

Complete the labeling of the parts of the life cycle of *Lycopodium* (Figure 26.3), using the terms capitalized in the preceding description.

B. SELAGINELLA

Selaginella is a genus related to *Lycopodium*, but it has certain more advanced structural and reproductive features, and most species are smaller and more delicate than *Lycopodium*. They are creeping or prostrate in growth habit.

ACTIVITY 3 **Examine** specimens of *Selaginella*, such as *Selaginella uncinata* (rainbow fern), *S. kraussiana brownii* (cushion moss), or *S. lepidophylla* (resurrection plant). **Note** the tiny **LEAVES** in four longitudinal rows. Do the branches show **DICHOTOMOUS BRANCHING?** *Yes No.* **Look** to see if your specimen has any long unbranched, leafless axes coming off the leafy stem and hanging toward the ground. If so, this is a **RHIZOPHORE**. When it touches the ground, the tip will branch and develop underground rootlets.

Note the small **CONES** at the tips of the leafy stems. The cone of *Selaginella* has two kinds of sporangia; one type, a **MICROSPORANGIUM** that forms many **MICROSPORES**, and another type, a **MEGASPORANGIUM** that forms four **MEGASPORES**. Each megasporangium is borne in the axil of a cone scale called a **MEGASPOROPHYLL**; each microsporangium is borne in the axil of a cone scale called a **MICROSPOROPHYLL**. The production of two kinds of spores is called *heterospory*. Each microspore produces a **MALE GAMETOPHYTE** within the **WALL OF THE MICROSPORE**. Each megaspore produces a **FEMALE GAMETOPHYTE** within the **WALL OF THE MEGASPORE**. Microspores containing male gametophytes are blown by air currents to the vicinity of the female megaspore, which is never expelled from the megasporangium. A **SPERM** from a male gametophyte fertilizes an **EGG** in an **ARCHEGONIUM** of the female gametophyte.

FERTILIZATION, EMBRYO DEVELOPMENT, and early growth of the **YOUNG SPOROPHYTE** all take place within the **MEGASPORE WALL**. Later the young **SPOROPHYTE** forms a **ROOT** and **SHOOT** as it emerges from the female gametophyte tissue within the **MEGASPORE WALL**. This type of reproduction closely parallels features found among the seed plants.

Complete the labeling of the parts of the life cycle of *Selaginella* shown in Figure 26.4, using the terms capitalized in the preceding description.

IV. DIVISION ARTHROPHYTA— HORSETAILS

ACTIVITY 4 **Examine** living, preserved, or pressed specimens of *Equisetum,* such as *Equisetum hyemale* (scouring rush) or *E. arvense* (common horsetail). The surface of the stem has conspicuous longitudinal **RIDGES** or **RIBS**. Do the stems have dichotomous branching? *Yes No.* **LEAVES** are reduced to small scales that lie flat against the stem (**APPRESSED LEAVES**). These leaves occur as simple extensions of the ribs. They appear as a rather loose light-colored sheath found at the **NODES** (or joints) along the stem. Arthrophyte means "jointed stem." In branching species such as *Equisetum arvense*, **BRANCHES** as well as **LEAVES** occur at these nodes. A horizontal **RHIZOME** produces new shoots each year and is the primary means by which the population increases in size.

SPORANGIA are always clustered together on special stalked sporophylls in **CONES** located at the tips of shoots. When sporangia are young, the **YOUNG CONE** has a tight, smooth surface. When sporangia and spores are mature, the cone surface becomes more loose and open, exposing the sporangia to the air.

Shedding and dispersal of the spores are aided by the hygroscopic movement of tiny appendages called **ELATERS** that are attached to one end of each **SPORE** and that coil and uncoil in response to changes in atmospheric humidity. Each **SPORE** eventually gives rise to a **GAMETOPHYTE THALLUS** that produces long-necked archegonia, each containing one egg, and antheridia, each containing several sperm. A sperm swims to an archegonium, where fertilization takes place. A **YOUNG SPOROPHYTE** eventually grows from the embryo. In time the **MATURE SPOROPHYTE** is formed. **Complete** the labeling of the life cycle of *Equisetum arvense* shown in Figure 26.5, using the terms capitalized in the preceding description. *Note:* In *Equisetum hyemale* the cones are borne on vegetative shoots and not on special reproductive shoots as in *E. arvense*, which is shown in Figure 26.5.

V. DIVISION PTEROPHYTA— FERNS

Ferns are the most highly evolved of the lower vascular plants. In terms of variety and breadth of distribution, they are the most successful of these plants. Ferns grew side by side with the lycopods and arthrophytes in the world flora of past geological time. Many ferns were trees. As noted earlier, practically all of these plants are now extinct. Of the Arthrophyta, only one genus, *Equisetum* (represented by 25 species), survives. Of the Lycopodophyta, only four genera (of about 800 species) exist. Ferns, however, have persisted with greater success.

We know of about 200 genera (10,000 species), some very ancient, some of more recent origin. It may be that because ferns have large, multiveined leaves, comparable to those of the seed plants, they have been more successful in "holding their own" in competition with the more advanced seed plants.

Like other lower vascular plants, living ferns have the same relatively primitive life cycle with two independent-living sporophyte and gametophyte generations. They differ from other lower vascular plants, as mentioned above, by their large (usually pinnate) leaves. Most ferns are herbaceous. A few tropical genera are treelike, relict types reminiscent of many of their Carboniferous ancestors.

MATURE SPOROPHYTE OF THE FERN

ACTIVITY 5 **Review** the generalized life cycle of the lower vascular plants in Figure 26.1 and **use** the pictorial cycle of fern in Figure 26.6 as a guide in your study.

Examine living or pressed specimens of ferns such as Boston fern (*Nephrolepis exaltata* v. *Bostoniensis*), Christmas fern (*Polystichum acrosti-*

choides), species of maidenhair fern (*Adiantum* sp.), and *Polypodium* sp.

A. THE LEAF

Most of the body of the fern is made up of **LEAVES**. The leaves arise directly from the horizontal **RHIZOME**. There are no aerial (aboveground) stems. Each leaf is called a **FROND**. It consists of (1) a **BLADE** that is pinnately subdivided into smaller parts, called **PINNAE** (singular, pinna); and (2) a **PETIOLE**, a stalk that attaches the blade to the rhizome. Adventitious roots grow from the rhizome. **Label** the frond and its parts as shown in the life cycle of Figure 26.6, using the terms capitalized in the preceding description.

B. THE RHIZOME

Examine living specimens of ferns and note the horizontal stem, the **RHIZOME**, which may be at or just below the ground surface. The rhizome produces new leaves at its growing tip. Young leaves first appear in a tightly coiled mass, and as growth goes on, the coil seemingly "unrolls." This coiled mass is called a "**FIDDLE HEAD**." Do you see any fiddle heads on your plants? *Yes No.* Your chances of seeing them, of course, depend on the season or growth stage of the part.

Examine portions of rhizomes that have been dug up and find the **ADVENTITIOUS ROOTS**. As with all the other lower vascular plants, the primary means of increasing the population size is by growth of new shoots from the rhizome.

Label the rhizome with its adventitious roots and a fiddle head as shown in the life cycle of Figure 26.6.

C. THE SORI AND SPORANGIA

Examine with a hand-lens the undersurface of the pinnae and find brown or orange colored "spots." These are the **SORI** (singular, sorus). Sori are clusters of **SPORANGIA**. They are not always present at all times of the year. They also change in color and surface texture as they mature. Many persons mistake them for bugs.

In many genera (but not in all) the sori are protected by an epidermal flap of tissue, called the **INDUSIUM**. The shape of this flap of tissue differs in different genera. Does your specimen have sori? *Yes No.* Is there an indusium with each sorus? *Yes No.* What is the genus of fern that you are inspecting? _____

_____.

If there is a specimen of *Adiantum* available, **examine** the underside of its pinnae. Here you will see that the edges of each pinna is folded under a little and forms a false indusium over the sori.

Examine leaves of *Polypodium* and determine whether or not the sori are covered by an indusium. Are they?   Yes   No.

D. THE ANNULUS—THE DISPERSAL MECHANISM OF THE SPORANGIUM

ACTIVITY 6   **Using** a dissecting needle, **scrape** the surface of one or two sori so that black specks (the sporangia) fall into a drop of water on a slide. Cover with a cover glass and examine under the microscope. Prepared slides may be substituted here if your ferns have no sori.

Look for golden-brown colored spheres. These are the SPORANGIA. The cells composing the walls of the sporangium are large and transparent, so you should be able to see the SPORES within. These are formed by the meiotic division of diploid spore mother cells.

Release of the spores is an important function. It is accomplished by a special band of cells, called the ANNULUS. The annulus partly or completely encircles the sporangium. It forcibly shoots the spores for some distance. **Find** the annulus. You can recognize it by the thick inner walls and side walls of its cells. **Note** that the outer cell walls are thin. Near the stalk of the sporangium there are a few cells of the annulus with thin walls all around. These are the LIP CELLS. The annulus works as described in the following paragraph.

When the spores are mature, the sporangium loses water and dries. As the annulus cells lose water, the thin outer walls shrink and the thick side walls are pulled toward each other. Water tensions are set up that, acting like a drawstring, straighten out the annulus. It breaks at its weakest point—the lip cells—and then flips backward, ripping open the wall of the sporangium. Almost immediately, the cohesive force of the water molecules that shrank the outer walls is overcome. As a result, the shrunken walls balloon outward, causing the annulus to snap back to its original circular configuration and catapulting the spores ahead and into the air.

ACTIVITY 7   **Moisten** some sporangia and then let them dry out slowly. **Observe** the rupture of the annulus and the discharge of the spores.

Label the parts of the dispersal mechanism of the sporangium of the fern life cycle in Figure 26.6, using the terms capitalized in the preceding description.

E. FERN GAMETOPHYTE

Spores germinate into the gametophyte plant.

This plant is a minute, green disk (a few millimeters in diameter), called the **PROTHALLUS.**

ACTIVITY 8   **Examine** prepared slides of the prothallus. The prothallus is heart shaped, with one pointed end and one lobed end divided by a deep apical notch. It lies flat, and on its underside are RHIZOIDS that form a dense, hairy mat. Scattered among the rhizoids are globular ANTHERIDIA and vase-shaped ARCHEGONIA. Antheridia occur more commonly near the margin of the prothallus, whereas the archegonia are usually near the apical notch. An ARCHEGONIUM consists of a venter and a neck. An EGG occupies the venter. Usually only the neck is visible because the venter is embedded in the tissue of the prothallus. SPERM are visible in antheridia.

FERTILIZATION requires water. The prostrate growth habit of the gametophyte with the sex organs on its lower wet side facilitates the swimming of the sperm to the vicinity of the archegonial necks. Fertilization occurs in the venter.

ACTIVITY 9   **Examine** living fern prothalli. Usually they can be grown quite easily on the surface of clay pots kept moist, on agar plates, or on moistened peat cubes. But their growth requires 1–3 months. **Use** a dissecting microscope and look for antheridia and archegonia. **Label** the parts of the prothallus as shown in the life cycle of Figure 26.6, using the terms capitalized in the preceding description.

F. YOUNG SPOROPHYTE OF THE FERN

ACTIVITY 10   **Examine** prepared slides or living specimens of YOUNG SPOROPHYTES. These will still be embedded in the gametophyte thallus. While inside the archeogonium, the zygote develops into an EMBRYO. The embryo produces the rhizome, ROOTS, and leaves. The first YOUNG LEAF grows upward through the apical notch. With continued growth, the rhizome produces more leaves. **Label** the young sporophyte growing from the gametophyte as shown in the life cycle of Figure 26.6. **Complete** the labeling of the life cycle, using the terms capitalized in the preceding description.

VI. FOSSIL FORMS

ACTIVITY 11   **Examine** compressions or impressions of fossil stems and leaves of Lycopodophyta and Arthrophyta (if these are available).

Examine slides or color transparencies or acetate "peels" showing structural and anatomical detail of fossil vascular plants (if these are available).

EXERCISE 26

Student Name _____

QUESTIONS

1. List three tissues or special features of the sporophyte of the lower vascular plants that help sustain their life on land. (a) _____; (b) _____; (c) _____

2. What is the chief means of maintaining the populations of any of these plants studied? _____ _____

3. Which division of plants is characterized by jointed stems? _____

4. If you found a plant with an equal-forking branching pattern, would you consider it to be a highly evolved plant? _____ What scientific term could you use to describe equal forking? _____

5. List all the haploid structures and cells in the generalized life cycle of Figure 26.1. _____ _____ _____ _____

6. Name the cell in the life cycle that develops into the embryo. _____

7. Name the thallus on which the sex organs are located. _____

8. Which thallus of the life cycle is the chief photosynthetic, assimilative generation? _____ _____

9. Are spore mother cells retained in the sporangium (of most species) or are they expelled? _____

10. Does meiosis occur within the sporangium? _____

11. Name two stages in the life cycle during which genes or groups of genes are either combined or arranged. (a) _____; (b) _____ _____

12. Since populations are chiefly maintained by asexual reproduction through the growth of the rhizome, why then is the sexual phase retained? _____ _____ _____

13. Name one stage in the life cycle during which the chromosome number is (a) doubled _____; (b) halved _____

14. What structure provides the physical and nutritive support for the embryo? _____ _____

15. Name the structure(s) and/or cells that constitute the *dispersal* stage of the life cycle. _____

16. Name the stages or events in the life cycle in which genetic variation is generated in the population. _____

17. Is water necessary (a) for spore dispersal? _____; (b) for fertilization? _____

18. In their life cycle, ferns are similar to what two other divisions of plants? (a) _____ _____; (b) _____ _____

19. In their leaf structure, ferns are similar to what plants? _____

20. Is the sporophyte stage of lower vascular plants suited to life on land? _____

21. Barring some catastrophe, each year stands of ferns in the temperate zone reappear or exhibit renewed life and growth. What structure produces these new fronds? _____

_____

22. A cluster of sporangia on a fern leaf is called a _____

23. Is water necessary for fertilization in ferns? _____

24. Name the structure that functions as the spore dispersal mechanism of the fern. _____

_____

25. What are the two compatible functions that are carried out by the sporophyte plant?

(a) _____; (b) _____

26. Does it make sense to have the gametophyte plant be the primary, photosynthesizing stage of the life cycle? _____ Explain. _____

_____

_____

27. Is the gametophyte prothallus of the fern dependent upon the sporophyte plant for its nourishment? _____

28. Ferns typically inhabit shady, sheltered areas. Is there any correlation between their habitat and their life cycle? _____ Explain. _____

_____

_____

29. Do ferns have vascular tissue? _____

30. Are sori on the upper or lower surface of frond pinnae? _____ What is the advantage of this? _____

EXERCISE 26   **Student Name** _____

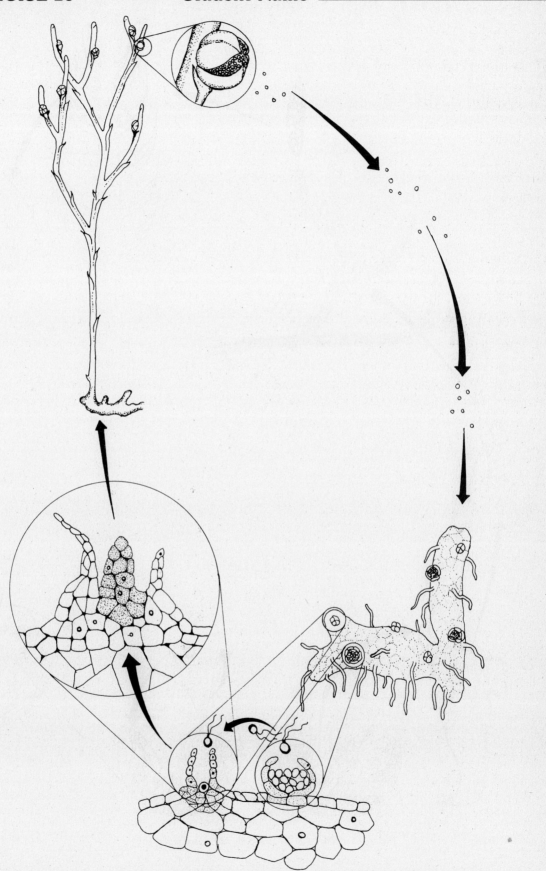

Figure 26.2 Life cycle of *Psilotum*.

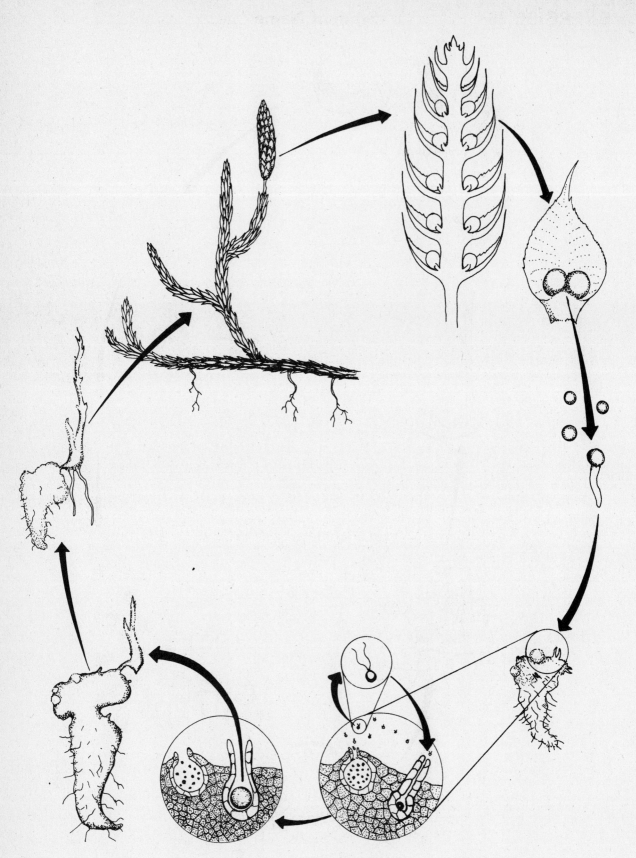

Figure 26.3 Life cycle of *Lycopodium*.

EXERCISE 26   **Student Name** _____

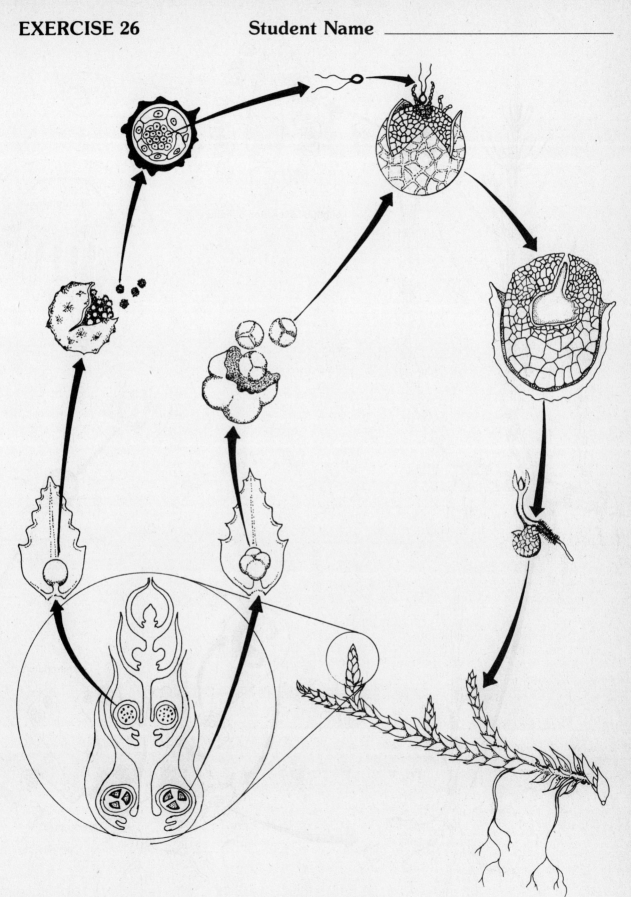

Figure 26.4 Life cycle of *Selaginella*.

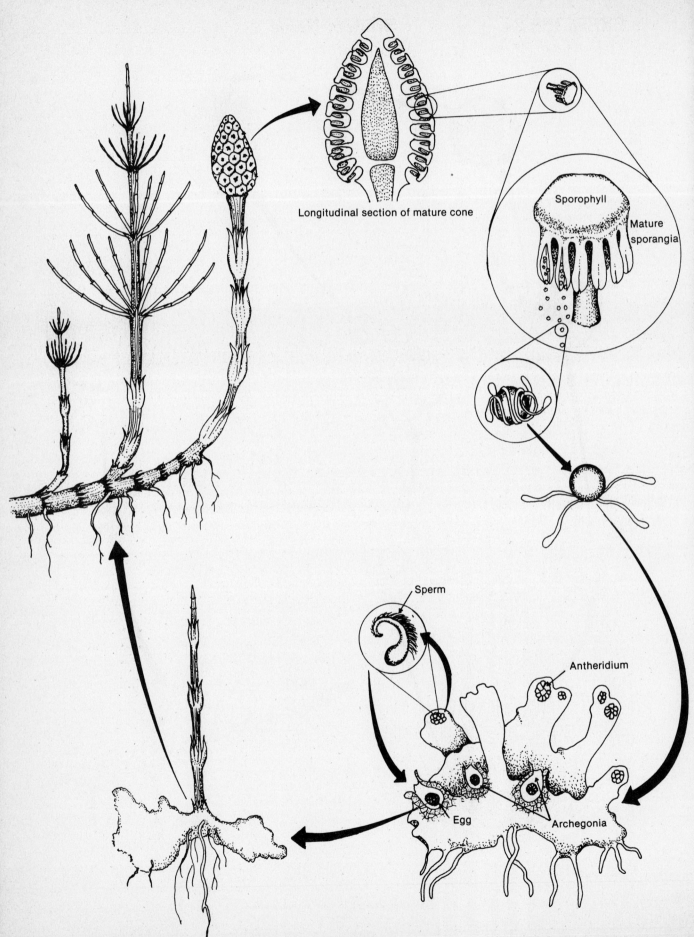

Longitudinal section of mature cone

Sporophyll

Mature sporangia

Sperm

Antheridium

Egg

Archegonia

Figure 26.5   Life cycle of *Equisetum arvense*.

EXERCISE 26   **Student Name** _____

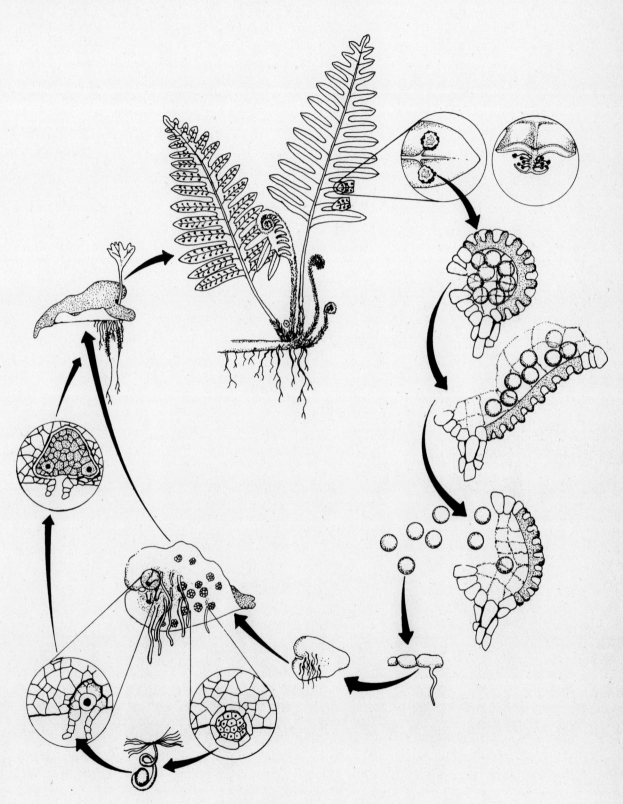

Figure 26.6 Life cycle of a fern.

CONE-BEARING SEED PLANTS

EXERCISE 27

INTRODUCTION

The evolution of the seed was a significant achievement for land plants. The seed contains an embryonic sporophyte supplied with nutrients for quick, vigorous growth. In the life cycle of higher vascular plants, the seed replaces the spore as the dispersal unit. It can survive adverse conditions better than a spore, and it grows and develops under conditions suitable for the mature sporophyte.

Many ancient land plants (now extinct) formed seeds. Living groups that produce seeds are the cone-bearing seed plants (gymnosperms) and the flowering seed plants (angiosperms). Gymnosperm seeds are described as naked because they are not enclosed in fruit as are those of the flowering plants (*gymno,* naked; *sperma,* seed). Some gymnosperms, such as yew (*Taxus*) and *Ginkgo,* do not have cones but instead bear fleshy, fruitlike bodies.

Conifers (division Coniferophyta) are the most numerous gymnosperms today. Examples familiar to you are pine, spruce, hemlock, fir, juniper, and cedar. Other divisions of gymnosperms, which include, for example, *Ginkgo* (Japanese fan tree) and cycads, will not be considered in this exercise.

Sporophytes of the Coniferophyta are trees or shrubs. The large, woody conifer sporophyte cannot, however, be considered a significant evolutionary advance over the lower vascular plants. Many extinct lower vascular plants were of tree size. It is the seed that is the great difference. The origin of the seed has come about by the extreme reduction in the size of the female gametophyte. Correlated with this was the reduction of the male gametophyte.

The female gametophyte is a miniature kernel of cells. The male gametophyte is what we call the pollen grain. Both gametophytes are now totally dependent on but at the same time protected by the

Table 27.1   SUMMARY OF PROGRESSIVE FEATURES OF GYMNOSPERMS GIVING THEM SELECTIVE ADVANTAGE OVER LOWER VASCULAR PLANTS

PROGRESSIVE FEATURES	SELECTIVE ADVANTAGE
1. The seed	1. As a dispersal unit, it has more survival assets than a spore; it protects and nourishes the young sporophyte in its embryonic and seedling stages
2. Germination of spores inside the sporangia so that gametophytes are enclosed in sporangia	2. Change in world climate to drier conditions of Mesozoic Era favored selection of plants that retained their delicate spores and gametophytes inside the sporangia
3. Reduction of the gametophyte plants to miniature units that are tenants of the sporophyte plant	3. Eliminates a vulnerable (free-living gametophyte) stage in life cycle
4. Sperm-carrying pollen tube	4. No longer dependent on water for fertilization; greater assurance of fertilization; less risk to sperm survival
5. Wind-borne pollen	5. Wider range of cross fertilization and greater potential for genetic variation
6. Tough, narrow, water-conserving leaves	6. Survival advantage in dry climate

sporophyte and are borne in the cones of the sporophyte.

The miniature female gametophyte thallus develops into the seed after fertilization. The light, airborne pollen grain, which carries the sperm to the female, frees the plants from dependency on water for fertilization.

You have already studied the stems and leaves of the conifer sporophyte. The emphasis of this exercise, therefore, is on the gametophyte and the cone structures of the sporophyte in which it is found.

Review the section on the significance of life cycles on page 174.

Note the generalized life cycle of pine (*Pinus* sp.) in Figure 27.1 and the pictorial life cycle in Figure 27.2 and use them as guides while you work through the exercise.

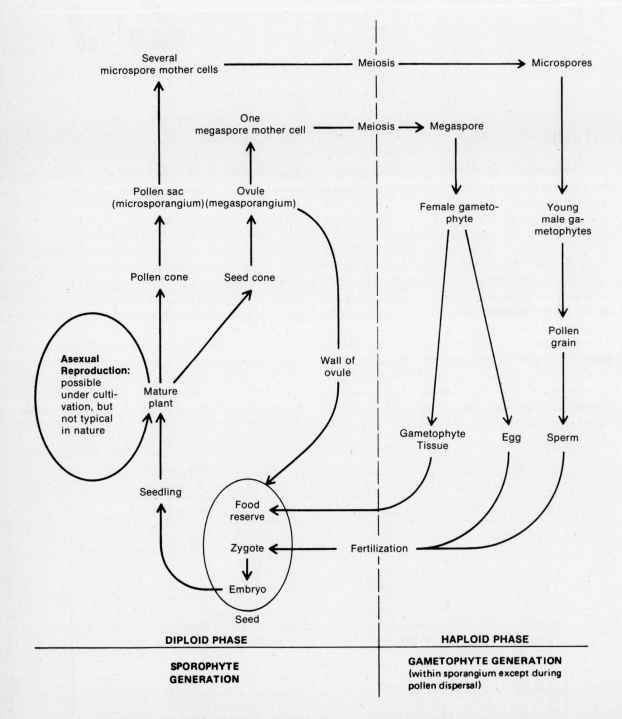

Figure 27.1 Generalized life cycle of pine (*Pinus* sp.), a representative gymnosperm.

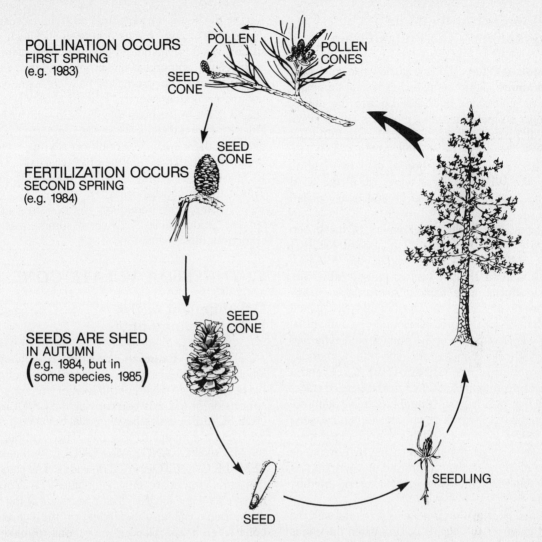

POLLINATION OCCURS
FIRST SPRING
(e.g. 1983)

POLLEN

POLLEN
CONES

SEED
CONE

SEED
CONE

FERTILIZATION OCCURS
SECOND SPRING
(e.g. 1984)

SEEDS ARE SHED
IN AUTUMN
(e.g. 1984, but in
some species, 1985)

SEED
CONE

SEED

SEEDLING

Figure 27.2   Generalized pictorial life cycle of pine (*Pinus* sp.).

I. SPOROPHYTE AND GAMETOPHYTE

A. THE WOODY PLANT— THE SPOROPHYTE

ACTIVITY 1   Weather permitting, the instructor will conduct you on a brief tour of the campus to look at some examples of gymnosperms. The trees and shrubs that you will see represent the mature sporophyte plant. Herbarium specimens or cuttings from different gymnosperm plants can also be examined.

B. THE POLLEN OR STAMINATE CONE

ACTIVITY 2   **Examine** POLLEN CONES of pine or some other conifer. If you have a portion of a twig, you will see that the cones occur in clusters. **Examine** a single cone. It is quite small. **Measure** or **estimate** its length. How long is it? _____ mm.

Examine a prepared slide showing a longitudinal section of a staminate pine cone, *or* **remove** a single cone from the cluster and examine it with a hand-lens and dissecting needle. It is composed of individual segments. (**Look** at Figure 27.3 as a guide while you work.) **Tease** these apart with the needle. Each of these segments is a tiny modified leaf, called a MICROSPOROPHYLL. If your cone is mature, the microsporophyll will be papery. **Place** a few of these microsporophylls on a slide in a drop of water. Cover with a cover slip and examine under the dissecting microscope, or just examine them dry, using a hand-lens. **Find** two elongate sacs embedded on the lower surface of the microsporophyll. These two sacs are the MICROSPORANGIA or pollen sacs. They produce MICROSPORES, which are not shed and dispersed as in lower vascular plants. Instead, they remain within the sporangium, where they germinate into male gametophytes. **Label** Figure 27.3 using the terms capitalized in the preceding description.

Development of the Male Gametophyte—The Pollen Grain

Look at Figure 27.3 as a guide while you work.

Examine either prepared slides of staminate cones or specimens of cones. **Locate** the microsporangia on the undersurface of each microsporophyll. Microsporangia are filled with many POLLEN GRAINS. These are the YOUNG MALE GAMETOPHYTES. They are oval bodies with two lateral, bladderlike WINGS. These wings aid in the wind dispersal of this relatively large and heavy pollen grain.

The stages in the development of the pollen grain are shown in Figure 27.3. Diploid MICROSPORE MOTHER CELLS within the MICROSPORANGIUM (pollen sac) divide by MEIOSIS. Each produces four haploid microspores. Each microspore produces two prothallial cells, which are remnants of the prothallus, and one antheridial cell, the sole remnant of the antheridium. The two prothallial cells degenerate, and the antheridial cell divides to form a GENERATIVE CELL and a TUBE CELL (Figure 27.3C). In this two-celled stage, the young pollen grain (immature male gametophyte) is shed. It drifts down onto the young seed cone, where it remains relatively unchanged for a year. Then it germinates into its mature stage (Figure 27.3D), which is composed of a germination tube within which are two sperm nuclei and the tube nucleus.

Pollen germination is closely correlated with the development of the egg. It occurs when the egg is formed and ready for fertilization.

In the lower vascular plants, the microspore germinates and produces a little multicellular prothallus called the gametophyte plant. This plant has multicellular antheridia that produce sperm. But in gymnosperms the prothallus has (through evolutionary reduction over millions of years) become scaled down to the point where there is no longer a little plant per se but merely token remnant cells. For example, there are only a couple of cells representing the prothallus. These are the prothallial cells. There is only one cell representing the many antheridia. This is the antheridial cell.

A comparable reduction has occurred in the female. Study Table 27.2, which summarizes and compares these conditions.

C. THE SEED (OVULATE) CONE

Development of the Female Gametophyte

ACTIVITY 3 **Examine** prepared slides showing longitudinal sections of ovulate (seed) cones of pine. **Note** the rather thick central axis of the cone. On either side of the axis you can see the OVULIFEROUS SCALES, which have an oval bulge near their point of attachment to the axis. This oval bulge is the MEGASPORANGIUM or OVULE. Within it the FEMALE GAMETOPHYTE develops. The stages in this development are shown in Figure 27.4. You will not see all these stages. The appearance of the cells in the ovule will vary depending on the age of the cone when it was sliced into sections to make the slide. You may be able to see the MEGASPORE

Table 27.2 COMPARISON OF THE DEVELOPMENT OF THE GAMETOPHYTE OF THE LOWER VASCULAR PLANTS AND THE GYMNOSPERMS

PLANT GROUP	PLACE OF GERMINATION OF THE SPORE	PRODUCT OF GERMINATION OF THE SPORE	SIZE OF GAMETOPHYTE	LOCATION OF GAMETOPHYTE	APPEARANCE OF GAMETOPHYTE
LOWER VASCULAR PLANTS* (ferns, lycopods, horsetails)	Outside the sporangium†	Multicellular prothallus plant with multicellular antheridia and/or archegonia	Visible to the naked eye or with aid of a hand-lens	Free-living on the soil	See Figures 26.2, 26.3, and 26.5
GYMNOSPERMS	Microspore: inside the microsporangium (or pollen sac)	Two prothallial cells that are relics of the prothallus; one tube cell; one generative cell from which two sperm nuclei are ultimately derived	Visible with a microscope	Protected inside the pollen sac (microsporangium)	See Figure 27.3D
	Megaspore: inside the megasporangium (or ovule)	Small kernel of cells containing two highly reduced archegonia	Visible with a microscope	Protected inside the ovule (megasporangium)	See Figure 27.4B

*Certain ancient (now extinct) forms produced megaspores and microspores.
†In *Selaginella* (and certain extinct forms) germination occurs inside the sporangium.

MOTHER CELL, a large, light-colored, centrally located cell. You may also be able to see POLLEN GRAINS lodged nearby, or growing into, the ovule.

Young Seed Cones in the First Spring of Growth

Examine preserved young (first-spring) cones or, if in season, pine twigs bearing first-spring cones. These cones are quite small (7–8 mm long). When alive, they are soft and green or reddish. The cone consists of modified branches, called OVULIFER-OUS SCALES. Each scale bears on its upper surface two spore cases, each called a MEGASPORAN-GIUM. In this young stage, the scales are slightly separated and pollen grains sift down between them, falling on the megasporangia. This is pollination. It takes place in the first spring's growth of the ovulate or seed cone (Figure 27.4A), using the terms capitalized in the preceding description.

Young Seed Cones in the Second Spring of Growth

ACTIVITY 4 **Examine** cones that are one year old. Are these cones as soft as the cones of the first spring? *Yes No.* Are the scales (a) loose or (b) close together? Circle (a) or (b). Sometime during the second spring the FEMALE GAMETO-PHYTE completes development. The pollen completes development also by forming a POLLEN TUBE that delivers the SPERM to the EGG in the ARCHEGONIUM. Fertilization takes place in the second spring of growth of the ovulate cone.

Label the young one-year-old seed cone in Figure 27.4B, using the terms capitalized in the preceding description.

Mature Seed (Ovulate) Cones

ACTIVITY 5 **Examine** mature SEED CONES that have shed or are shedding seeds. SEEDS are shed in the autumn that follows fertilization. In some species they are not shed for another year. Are the ovuliferous scales soft or woody? _____ _____ Is the cone closed or open? _____

_____. Do you find any seeds on the surface of the scales? *Yes No.* How many on each scale? _____.

If you don't find the seeds, what can you see? _____

Wouldn't it be better to have the seeds located on the lower side of each scale so they could drop out easily? *Yes No.* What is the advantage of having them on the upper side of the scale? _____

II. SEED

ACTIVITY 6 **Dissect** a seed of piñon pine or nut pine (*Pinus cembroides* v. *edulis*). The seed is enclosed by a hard, stony protective coat (the remnant of the sporangium wall). **Break** open the seed coat. **Cut** the seed lengthwise. The white, oily material is the RESERVE FOOD (the remnant tissue of the female gametophyte). Embedded in the food material is the EMBRYO. It consists of an axis with one end, the EPICOTYL, surrounded by about eight fingerlike COTYLEDONS. Below the cotyledons is the HYPOCOTYL, at the very end of which is the RADICLE or root. The radicle is not visibly different from the hypocotyl.

Label the diagram of the dissected seed in Figure 27.4C, using the terms capitalized in the preceding description.

Examine the seeds and the cones for any features they might have to aid in seed dispersal. Is the seed propelled from the cone? *Yes No.* Is the cone or the seed attractive to animals by color? *Yes No;* by aroma? *Yes No;* by taste? *Yes No.* Do you think the seed itself can become readily airborne? *Yes No.* Could a seed-eating bird easily break open the seed coat? *Yes No.* What's your conclusion? Do gymnosperms exhibit specialized dispersal mechanisms? *Yes No.* How does this affect the range of the populations of these plants? _____.

What potential is there for gymnosperms exploiting new habitats, limited or unlimited? _____ _____.

EXERCISE 27

Student Name _____

QUESTIONS

1. Gametophytes of the lower vascular plants develop and live as independent plants. Give the specific structure in which the gametophytes of seed plants develop and live. _____

_____

2. The spores of lower vascular plants are dispersed, and then they germinate someplace away from the parent sporophyte plant. Where do the spores of seed plants germinate?

_____

3. The space limitations of the sporangium restrict the size of the gametophyte. How many cells make up the prothallus of the young male gametophyte of pine? _____

4. Are there any antheridia in pine gametophytes? _____

   What represents the antheridia? _____

5. The seed of pine is housed in the remnants of what structure of the cone? _____

6. What tissue in the seed serves as the food supply for the embryo's growth? _____

_____

7. What is the dispersal unit of gymnosperms? _____

8. Which has greater survival potential, a spore or a seed? _____

9. In seed plants, the male gametophyte plant has become extremely reduced and is referred to as the _____

10. What pollen structure has eliminated the risk of survival to the sperm? _____

_____

11. What is the functional advantage for the extreme reduction in size of the male gameto-phyte? _____

12. Are the gymnosperms the first land plants to have a woody growth habit? _____

13. Approximately how long (in months) is the interval between pollination and fertilization?

_____

14. Give two ways that the seed is a more reliable dispersal unit than a spore. (a) _____

_____; (b) _____

15. Since water is not used in fertilization in gymnosperms, by what means are sperm conducted to the egg over distances of a mile or more? _____

_____

16. Name the two cell types in the life cycle that undergo meiosis. (a) _____

_____; (b) _____

17. Are the cells of the food reserve tissue in the seed haploid or diploid? _____

_____

18. Are populations of gymnosperms likely to have their ranges extended as a result of the distribution of their seeds? _____ Explain. _____

_____

19. Give two reasons why you think gymnosperms are of little value as food crops to humans.

    (a) _____

    _____; (b) _____

_____

20. The assimilative survival phase of the gymnosperm is the _____

21. Do gymnosperms have an asexual reproductive phase? _____

22. What is the only means by which they increase their number of individuals in a population? _____

23. Spores of lower vascular plants grow into (a) the gametophyte or (b) the sporophyte? Circle (a) or (b).

24. The seed germinates (a) into the gametophyte or (b) into the sporophyte? Circle (a) or (b).

25. Besides having a better food supply and being better protected all around, give another reason why the seed is a better dispersal unit than a spore when it comes to establishing the next generation. _____
_____

EXERCISE 27   **Student Name** _____

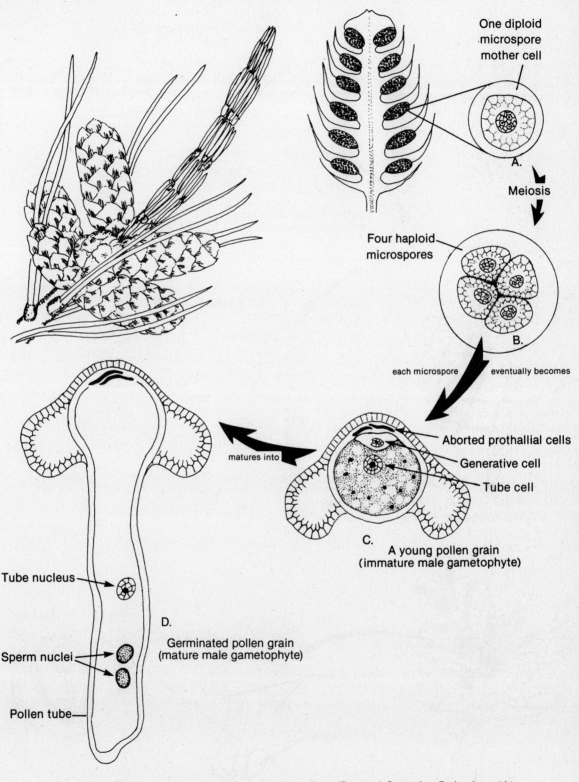

One diploid microspore mother cell

A.

Meiosis

Four haploid microspores

B.

each microspore   eventually becomes

Aborted prothallial cells

Generative cell

Tube cell

C.

A young pollen grain
(immature male gametophyte)

matures into

Tube nucleus

D.

Sperm nuclei

Germinated pollen grain
(mature male gametophyte)

Pollen tube

Figure 27.3   The development of the male gametophyte of pine (*Pinus* sp.). Stages A to C take place within the microsporangium. Pollen is shed from the cone when it is in the stage shown at C. Stage D occurs after the pollen has fallen onto the seed cone and when the eggs in the seed cone are ready to be fertilized. In most species, there is a time span of 1 year between stage C (pollination) and stage D (fertilization).

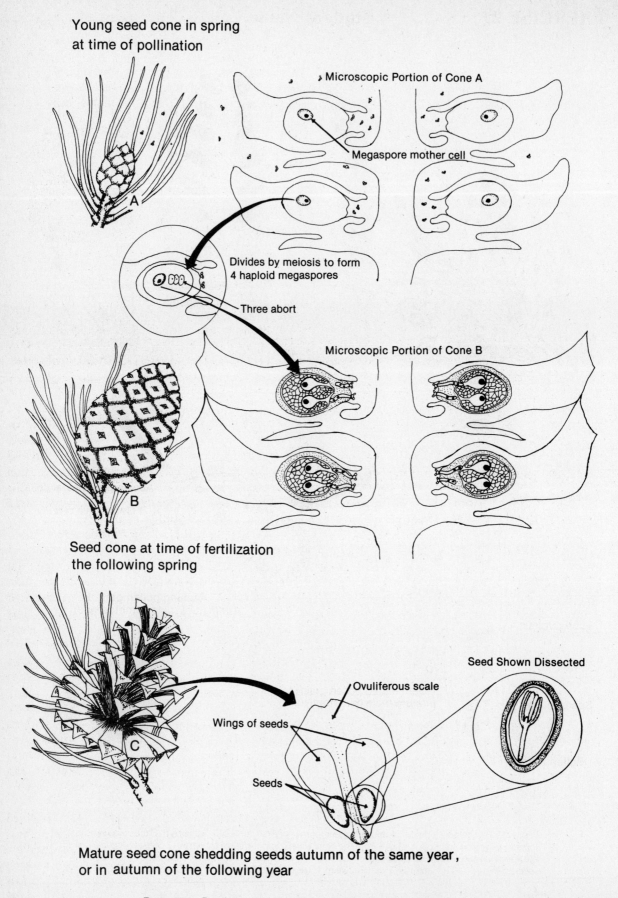

Young seed cone in spring
at time of pollination

Microscopic Portion of Cone A

Megaspore mother cell

Divides by meiosis to form
4 haploid megaspores

Three abort

Microscopic Portion of Cone B

Seed cone at time of fertilization
the following spring

Seed Shown Dissected

Ovuliferous scale

Wings of seeds

Seeds

Mature seed cone shedding seeds autumn of the same year,
or in autumn of the following year

Figure 27.4   Development of the female gametophyte and seed formation.

FLOWERING SEED PLANTS

EXERCISE 28

INTRODUCTION

The flowering plants are the higher vascular plants that have developed the most efficient mode of sexual reproduction, and they have done it with the "invention" of the flower. Details of the origins of the flower are still very much a mystery.

With the advent of the flower there evolved the most modern group of plants in the world, the angiosperms—flowering plants (division Anthophyta).

For most humans, the flower is simply a thing of beauty, but in the plant kingdom it is a highly refined mechanism, the design of which optimizes precise pollen transfer and reception. Pollen transfer agents are insects, birds, bats, wind, and water. Through these pollen transfer agents, there is greater success in cross fertilization. This leads to more hybridizing among plants, and consequently to more genetic diversity in populations.

By comparison, the more primitive seed plants (the gymnosperms, such as pines, spruce, hemlocks, cycads, and so on) have haphazard, inefficient pollination. The 725 gymnosperm species are outnumbered 400 to 1 by the 350,000 or so known species of flowering plants. With few exceptions, they are the dominant flora found anywhere on land. It is the flowering plants that have sustained humanity. Between 1000 and 2000 angiosperms are utilized by people. About 10 percent of these are important in world trade, and 15 species provide the bulk of the world's food crops.

In the previous exercises, you have spent much time learning the details of the life cycles of plants and the role played by the gametophyte and sporophyte generations. In Exercise 27 you have seen how the sporophyte generation of the gymnosperms has become the large woody tree or shrub, while the gametophyte is reduced in size, and, in the case of the male, exists as the tiny pollen grain; in the case of the female it is a tiny microscopic kernel of tissue. See Table 27.2 of Exercise 27. Both male and female gametophytes are borne inside the cones of the plant.

In the angiosperms, as with the gymnosperms, the gametophyte is reduced to a form microscopic in size. And where are these tiny gametophytes? They are inside the flower. The male gametophyte is the pollen grain and the female gametophyte is a small kernel of cells called an embryo sac and located inside the ovule. The flower is borne on the sporophyte plant, which may be a tree, a shrub, or a soft-stemmed, herbaceous type of plant.

Parts of the flower are converted into the fruit, an organ whose design promotes seed protection and dispersal (in most cases). Protection of the seed and its efficient dispersal has given the angiosperms greater capacity for spread of their populations. Genetic diversity of these populations (as achieved by greater hybridizing among angiosperms) allows for adaptation to a wide variety of habitats and hence a dominant place in the world's landscape.

The success of the flowering plants is due mainly to a combination of six features that give them tremendous competitive advantage over all other plant groups. These features are (1) flower, (2) fruit, (3) endosperm, (4) herb growth form, (5) vessels and sieve tubes, and (6) vegetative regenerative abilities (see Table 28.1).

You have already studied all the features of the sporophyte generation of the flowering plant in some detail. This exercise emphasizes the gametophyte generation of angiosperms. It has undergone even further reduction than in gymnosperms and has two new features associated with it—the endosperm and the fruit.

Table 28.1 SUMMARY OF PROGRESSIVE FEATURES OF ANGIOSPERMS, GIVING THEM SELECTIVE ADVANTAGE OVER GYMNOSPERMS

PROGRESSIVE FEATURE	SELECTIVE ADVANTAGE
1. Flower	1. An organ that functions in precise pollen transfer and whose co-evolution and alliance with insects as agents of pollen transfer led to the origin of many new species adapted to wide varieties of habitats.
2. Fruit	2. An organ that protects the seed and aids in its dispersal.
3. Endosperm	3. A special triploid tissue in the seed with nutrients and growth-promoting hormones for the rapid vigorous growth of the embryo.
4. Herb	4. A nonwoody plant form that matures quickly to reproduce itself in a matter of months (compared with the several years required by most trees). The rapid life cycle results in more generations of plants per unit of time. Any genetic mutations that have occurred, coupled with any gene recombinations, are more readily expressed. This generates variation that allows for adaptations to a wide variety of habitats within any given time span.
5. Vessels and sieve tubes	5. Large water and food-conducting tubes capable of rapid transport and mobilization of materials when needed for growth of buds, cambium, leaves or developing fruits and seeds.
6. Vegetative regeneration	6. The ability of vegetative organs (stems, roots, and leaves) to regenerate whole new plants provides a rapid way to maintain and renew a population.

Note the diagrams of the generalized life cycle of the angiosperm shown in Figure 28.1 and the pictorial life cycle in Figure 28.2. **Look** at them as a guide as you work through the exercise.

I. MALE GAMETOPHYTE

The male gametophyte is the pollen grain, as in gymnosperms. Remember from your study of the flower that the egg is in the ovule. But since the ovule is enclosed in the ovary, it is quite some distance from the stigma, where the pollen grain is deposited. Hence the pollen grain must be able to produce a very long tube to reach the ovule and deliver the sperm to the female gametophyte. The evolution of the increase in pollen tube growth was probably closely correlated with the evolution of the enclosure of the female gametophyte.

ACTIVITY 1 **Transfer** some pollen from the anthers of impatiens (*Impatiens* sp.) or some species of spiderwort (*Tradescantia* sp.) to a drop of sugar solution (10 percent cane sugar) on a slide. Include a few brush bristles or some such small bits of material before adding the cover slip to prevent the grains from being crushed. Put the slides in petri dishes containing several layers of filter paper saturated with water. Place the petri dishes in a warm place where the temperature can be kept at about 21–25°C. **Examine** the slides at intervals of 1 hour to see the development of the tube.

ACTIVITY 2 **Examine** a stamen of a lily (*Lilium* sp.) or other flower. **Note** the stalk or FILAMENT. It is a modified microsporophyll. At the tip of the filament is the ANTHER, composed of four microsporangia fused together. Each microsporangium is a POLLEN SAC.

ACTIVITY 3 **Deposit** some pollen into a drop of water on a slide and examine under the microscope. The coat (wall) of the pollen grain will have a particular surface pattern. This pattern varies and is useful in identifying species, genera, or families. It can be smooth or consist of spines, ridges, pores, plates, and so on.

ACTIVITY 4 **Examine** prepared slides of young lily (*Lilium* sp.) flower buds that show a cross section through the anther. The anther is four-parted. Each part is a POLLEN SAC or MICROSPORANGIUM. If the anther is very young, the entire anther mass will be filled in by tissue, some of which is sporogenous tissue. You can identify this tissue because it has numerous cells in stages of nuclear division. Some of these are MICROSPORE MOTHER CELLS. What kind of cell division is going on, (a) mitosis or (b) meiosis? Circle (a) or (b).

If the anther on the slide is older, you will see a more loose assemblage of cells in the microsporangia. Usually you can find single cells and cells clustered in fours (tetrads). The single cells are MICROSPORE MOTHER CELLS. Each of the cells in a cluster of four is a MICROSPORE.

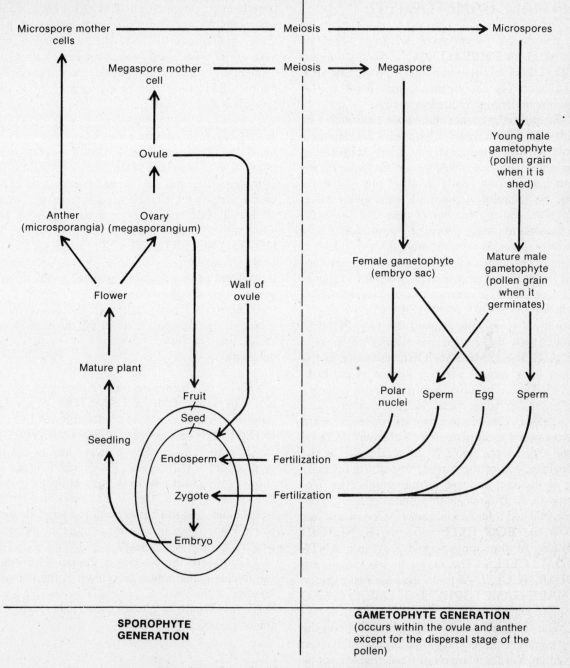

Figure 28.1   Generalized life cycle of the flowering plant.

Label the diagram of the young anther as shown in Figure 28.2 using the terms capitalized in the preceding description.

ACTIVITY 5   **Examine** prepared slides of mature anthers of lily (*Lilium* sp.) or of lily pollen grains. If your slide shows anthers, the chambers of the **TWO MICROSPORANGIA** or **POLLEN SACS** on either side of the central axis will be confluent and also open to the outside. **POLLEN GRAINS** at this stage consist of only two cells, a **GENERATIVE CELL** and a **TUBE CELL**. In this stage they are shed.

Label the mature anther and the young pollen grain just after its release as shown in Figure 28.2, using the terms capitalized in the preceding description.

After pollination, the generative cell forms two **SPERM NUCLEI**, and the tube cell elongates into the pollen tube with a **TUBE NUCLEUS**. In this three-nucleate state, the pollen grain represents the **MATURE MALE GAMETOPHYTE**.

Label the mature male gametophyte as shown in Figure 28.2, using the terms capitalized in the preceding description.

II. FEMALE GAMETOPHYTE

The female gametophyte is a small kernel of cells called the **EMBRYO SAC**. It develops within the **OVULE** (megasporangium) inside the **OVARY** of the flower. As with the male gametophyte, the female has undergone reduction in size.

Study the sequence numbered 1 through 7 of Figure 28.2, which shows the stages of development of the female gametophyte. All these stages take place within the ovule. Note that, in comparison with gymnosperms (see Figure 27.4B of previous exercise), the gametophyte has undergone even further reduction, and the number of cells that make up the final mature stage (No. 6 of Figure 28.2) is only seven (read details in next paragraph).

In the majority of flowering plants, the development of the female gametophyte proceeds as shown in the sequence numbered 1-7 in Figure 28.2. A single **MEGASPORE MOTHER CELL** divides by meiosis to form four haploid **MEGASPORES**. Three of these degenerate (abort) and the remaining **FUNCTIONAL MEGASPORE** gives rise to the female gametophyte, which is known as the **EMBRYO SAC**. The functional megaspore enlarges greatly, its wall becoming the wall of the embryo sac. By a series of successive nuclear divisions, eight haploid nuclei are formed. (No. 6 of Figure 28.2). Of these eight, three come to lie in the sac at the micropylar end of the ovule, three at the opposite end of the sac, and two in the center. (No. 7 of Figure 28.2). The nuclei at the ends of the sac become invested by cell membranes. At the micropylar end is the **EGG CELL** with two **SYNERGID CELLS**. At the opposite end are three **ANTIPODAL CELLS**. The nuclei in the center are **POLAR NUCLEI**. All these together make up the **FEMALE GAMETOPHYTE** or **EMBRYO SAC** (No. 7 of Figure 28.2).

The sequence No. 1-7 of Figure 28.2 shows the most common type of embryo sac and of embryo sac formation. But there are other types that will not be discussed here. (Actually, in lily—*Lilium*—the antipodal nuclei and one of the polar nuclei are triploid instead of haploid.)

Whatever these minor differences, the important new development in the flowering plant is the presence of the polar nuclei, which are fertilized by a sperm nucleus to form an **ENDOSPERM NUCLEUS** that has anywhere from three to five (polyploid) sets of chromosomes (depending on the species). This polyploid endosperm nucleus divides several times. Accompanying cell wall formation leads to the nutrient tissue known as the endosperm. As with most polyploid conditions (multiple chromosome sets), the polyploid endosperm is a vigorous tissue, supplying high concentrations of hormones for the embryo. This has great selective advantage for the flowering plants because it provides guarantees for speedy germination and survival of the seed and seedling. **Label** the sequence of embryo sac development (No. 1-7 of Figure 28.2) using the terms capitalized in the preceding description.

ACTIVITY 6 **Examine** slides or preserved specimens that show transverse sections through a young ovary of lily (*Lilium* sp.). **Note** that the **OVARY** of lily is three-parted and has three chambers of **LOCULES**. Only two **OVULES** are visible in each locule on the side, but many more are actually present. The ovule with its supporting stalk is crook-shaped. If a **MEGASPORE MOTHER CELL** is present in an ovule, it appears as a large, ovoid, vacuolated, translucent cell surrounded by much smaller, more square-shaped cells.

Label the cross section of the lily ovary as shown in the cut-away view of the ovary in Figure 28.2, using the terms capitalized in the preceding description.

ACTIVITY 7 **Examine** slides showing the ovules with MATURE FEMALE GAMETOPHYTES. **Locate** an OVULE and note the INNER and OUTER INTEGUMENTS enclosing it. The EMBRYO SAC (or mature female gametophyte) makes up the central portion of the ovule. It is difficult to find a sac showing all eight nuclei (**look** again at No. 6 of Figure 28.2). This is because the sac is an oval, three-dimensional mass, and the nuclei do not all lie in the same plane. Hence it is impossible to cut a thin section in which all nuclei will appear.

Complete the labeling of the development of the female gametophyte as shown in the enlarged figures of the ovules (No. 1-7) in the life cycle diagram of Figure 28.2, using the terms capitalized in the preceding description.

III. FERTILIZATION

Study illustration No. 7 in Figure 28.2, which shows the pollen tube passing through the **MICROPYLE**, a small opening between the **INTEGUMENTS** at the base of the **OVULE**. This diagram and the two that follow it illustrate **FERTILIZATION**. **SPERM** are shown in solid circles. Nuclei of the embryo sac are shown as open circles. The **EGG** is at the micropylar end of the embryo sac between the two synergids.

One sperm nucleus fuses with the egg nucleus to form the diploid **ZYGOTE**. The second sperm nucleus fuses with the two polar nuclei to form the triploid **ENDOSPERM NUCLEUS**. This union of egg and sperm and of polar nuclei and sperm is called **DOUBLE FERTILIZATION** and occurs only in angiosperms.

Label the zygote and endosperm nucleus in No. 8 of Figure 28.2.

The endosperm nucleus divides many times and eventually the triploid **ENDOSPERM** tissue is formed, as noted earlier. The **ZYGOTE** cell divides many times and develops into the **EMBRYO**. Maturation stages follow that produce the **SEED**. **Label** the seed in Figure 28.2.

IV. SEED

ACTIVITY 8 **Examine** various seeds of flowering plants. **Note** that most of them tend to be small and lightweight. **Observe** the diversity of size, and surface features, such as spines, hairs, hooks, and so on. The seeds of angiosperm trees, more so than herbs, are comparable with seeds of gymnosperms in size, weight, and ease of dispersal.

It is in the herbs that we see the tremendous modifications of both the seed and the fruit for dispersal. Seeds of herbs may be very small. Note, however, that seeds of some herbs used as food sources for humans may have been selected for

millennia for large size, and these are therefore many times larger than their wild relatives.

V. FRUIT

ACTIVITY 9 **Examine** the various edible fruits made available to you and **note** the enclosure of the seeds. Good examples are found in produce that most people call vegetables but that botanically are fruits. These are green pepper, cucumber, fresh peas in pods, soybeans, melons, squash, and similar foods.

Note that the seeds are attached to the wall of the FRUIT and in different ways depending on the species. The tissue of the fruit is the matured wall of the OVARY of the flower (and often includes other parts of the flower). As it matures, the seed produces growth-promoting hormones that stimulate the growth of the ovary and related parts. Remember (from Exercise 14) that the fruit isn't only fleshy but may also be papery or a dry container.

Examine also the fruits of trees, such as ash (*Fraxinus* sp.), oak (*Quercus* sp.), walnut (*Juglans* sp.), maple (*Acer* sp.), sweet gum (*Liquidambar*), elm (*Ulmus* sp.), buckeye (*Aesculus* sp.), and so on.

Examine also the fruits of some of your local weeds such as thistle (*Cirsium* sp.), cocklebur (*Xanthium* sp.), or others that are available.

EXERCISE 28   **Student Name** _____

QUESTIONS

1. How many male gametophytes develop from one microspore mother cell? _____

2. How many female gametophytes develop from one megaspore mother cell? _____

3. Are archegonia present in the angiosperm female gametophyte? _____; in the gymnosperm female gametophyte? _____

4. How many cells make up the mature female gametophyte in angiosperms? _____; how many in the gymnosperms? _____

5. The home gardener or commercial producer of strawberries, white potatoes, or tulips never uses seeds for planting. How then is it possible that they can get a crop or a garden of flowers? _____

   _____

6. What does the angiosperm endosperm provide for the embryo? (a) _____

   _____; (b) _____

   _____

7. Is the fruit unique to the angiosperms? _____ State the functions of the fruit.

   _____

8. List the six features that have given angiosperms superiority over all other plant groups.

   (a) _____; (b) _____

   (c) _____; (d) _____

   (e) _____; (f) _____

9. What are the two functions of the fruit? (a) _____;

   (b) _____

10. Why is the growth of an angiosperm seedling usually so much faster than the growth of the young plant of any other group of plants? _____

    _____

11. The gymnosperm embryo utilizes nutrients supplied by what tissue? _____

    _____ Is this issue diploid or triploid?

    _____

12. The rapid life-cycle turnover of herbs allows more chance for expression of genetic variation. New variations can only be kept and built up in a population if they are concentrated among a few individuals at the outset. This is done by precise pollen transfer among plants of the new mutant types. What pollen transfer agents are responsible for precise transfer between plants? _____

13. Is wind a very precise pollen transfer agent? _____ Does it indiscriminately blow pollen from one plant to any other plant? _____

14. Do you think the new mutant genes of a few individuals could become established in a population if the plants were wind-pollinated? _____

15. Many bees are pretty consistent in their activities. If one starts out the day visiting and picking up pollen from a purple-colored, mint-scented flower, it visits only that same type flower as long as it can find them. All these similar plants make up what is called a small gene pool. New mutants don't tend to become lost or "swamped out" in such small gene pools. Is it possible for a new species to become established with the help of this constancy of insect visitation? _____

16. All gymnosperms and many angiosperm trees are wind-pollinated. Most herbs are insect-pollinated. In which group or groups, therefore, would you expect to find the greatest number of species? _____

17. Where would you expect to find the greater number of angiosperm species, in the tropics or in the cooler temperate zones? _____ Explain. _____

_____

EXERCISE 28 **Student Name** _____

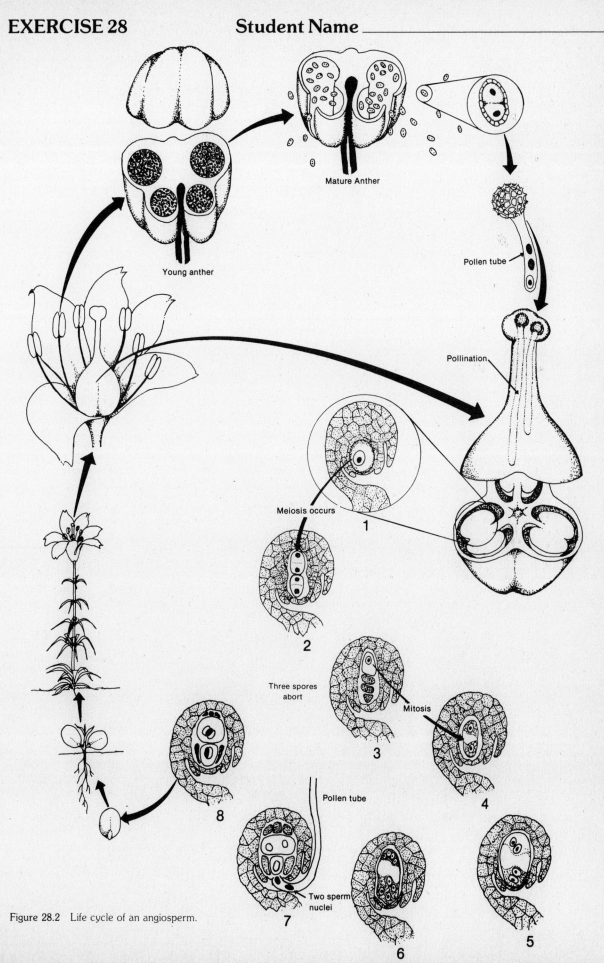

Figure 28.2 Life cycle of an angiosperm.

PART III

Genetics, Ecology, and Taxonomy

MEIOSIS AND MONOHYBRID AND DIHYBRID CROSSES

EXERCISE 29

In preceding exercises you have learned that *meiosis* is associated with sexual reproduction. Meiosis is a special kind of cell division that has three effects:

1. It ensures that eggs and sperm (gametes) have only half the number of chromosomes as the body cells (those of roots, leaves, and so on). This prevents continued multiplication of the chromosomes with each successive generation.

2. It maintains the identity of the species by keeping the chromosome number constant.

3. It scrambles the chromosomes and rearranges genes into different combinations so that new sets of chromosomes are passed into the gametes. Meiosis is analogous to shuffling and dealing cards in a card game. After each meiotic "shuffle" (division), the resulting nuclei are "dealt" (receive) new combinations of the individual chromosomes of the same basic chromosome set characteristic of the species.

The sperm carries one set (the *haploid number*) of chromosomes that match the set of chromosomes carried by the egg. Their fusion creates a new individual with two sets of chromosomes (the *diploid number*).

The haploid sets match because for every chromosome in the sperm there is a matching chromosome in the egg. We call such chromosomes *homologous. Homologous chromosomes* have the same kinds of genes in the same arrangement.

I. TERMS AND DEFINITIONS USED IN GENETICS

ALLELE One of two or more mutant (i.e., chemically different) forms of a gene. Example: R = the normal or wild type allele; r = a mutant allele (form) of R.

CENTROMERE The specialized part of a chromosome where the spindle fibers are attached during cell division.

CHROMOSOME A thread of DNA (deoxyribonucleic acid)—the molecular basis of heredity.

DIHYBRID A cross between parents differing with respect to two specified pairs of allelic genes. Examples: *RRTT* crossed with *rrtt*, or *rrTT* crossed with *RrTt*.

DOMINANT ALLELE An allele that masks the expression of another allele of the same gene. Example: R masks (is dominant to) r.

F_1 The first filial generation; the offspring resulting from the first experimental crossing of plants or animals.

F_2 The offspring produced by the crossing of two F_1 individuals.

GENE The basic hereditary unit that occurs at a certain place (locus) on a chromosome. It can mutate to various allelic forms. Also, a specific DNA coding for one function.

GENOTYPE The genetic composition or formula for one or more genes (allelic pairs). Examples: formula (genotype) is *Rr, RR,* or *RRTt*.

HOMOZYGOUS Having a pair of like alleles for any one gene. Examples: *RR, rr,* or *tt.*

HETEROZYGOUS Having a pair of unlike alleles for any one gene. Examples: *Rr* or *Tt.*

MONOHYBRID A cross between parents differing with respect to one specified pair of alleles of a given gene. Examples: *RR* crossed with *rr*, or *Rr* crossed with *Rr*.

PHENOTYPE The visible or detectable properties of an organism produced by the combined effect of the genotype and the environment.

PUREBRED Having all specified alleles homozygous. Examples: *RR*, *rr*, *RRTT*, *rrtt*, or *rrTT*.

RECESSIVE ALLELE An allele that is masked by another allele of the same gene. Example: *r* is masked by (is recessive to) *R*.

SISTER CHROMATIDS Two genetically identical daughter strands of a replicated chromosome, joined by a single centromere.

II. COMPARISON BETWEEN MEIOSIS AND MITOSIS

Note Figure 29.1, which shows the essential differences between meiosis and mitosis. Mitosis involves *one* nuclear division; meiosis involves *two* nuclear divisions—referred to as **MEIOSIS I** and **MEIOSIS II**. Meiosis occurs during sporogenesis, the formation of haploid spores that (in plants) ultimately produce the gametes.

ACTIVITY 1 **Look** at stage 3 (metaphase) in Figure 29.1. **Follow** the description and **answer** these questions.

1. In meiosis the homologous chromosomes are side by side. Are they side by side in mitosis? *Yes No.*
2. In meiosis each centromere has a spindle fiber attached only to one side. In mitosis each centromere has a spindle attached to _____ side(s).

Look at what has occurred between stages 3 and 4.

3. In mitosis, what has been separated, (a) homologous chromosomes; (b) sister chromatids? Circle (a) or (b).
4. In meiosis I, what has been separated, (a) sister chromatids; (b) homologous chromosomes? Circle (a) or (b).

In meiosis I you see illustrated *Mendel's law of segregation*. This law states that the members of a pair of homologous chromosomes are separated (segregate) during meiosis." This law is not applicable in mitosis.

5. What is separated during mitosis? _____

Look at stage 4. Remember that sister chromatids are exact replicas of each other. Homologous chromosomes match each other gene for gene, but the alleles may vary.

6. In stage 4 of mitosis each cell shows (a) chromosomes; (b) chromatids. Circle (a) or (b).
7. In each cell of stage 4 of mitosis, there are now (a) exact replicas; (b) a matched pair. Circle (a) or (b).
8. In stage 4 of meiosis each cell shows (a) a pair of sister chromatids; (b) a pair of homologous chromosomes. Circle (a) or (b).
9. In each cell of stage 4 of meiosis, there are now (a) exact replicas; (b) matched pairs. Circle (a) or (b).

Look at what has occurred during meiosis II. In stage 5, spindle fibers are attached to both sides of the centromere.

10. During meiosis II there occurs a separation of (a) sister chromatids; (b) homologous chromosomes. Circle (a) or (b).
11. How many *different* kinds of gametes are produced? (a) 4; (b) 2. Circle (a) or (b).
12. **Compare** the cells of stage 6 with the cells in stage 1. Those of stage 1 represent (a) the haploid number; (b) the diploid number. Circle (a) or (b). Those of stage 6 represent (a) the haploid number; (b) the diploid number. Circle (a) or (b).
13. Is there any chromosomal difference between the cell in stage 1 of mitosis and that of stage 4 in mitosis? *Yes No.*
14. Is there any chromosomal difference between the cell in stage 1 of meiosis and that of stage 6 in meiosis? *Yes No.*
15. Does any one gamete in stage 6 have homologous chromosomes present? *Yes No.*

III. A REVIEW EXERCISE IN MEIOSIS

ACTIVITY 2 **Look** at Figure 29.2. This figure shows two cells (No. 1 and No. 2) that are identical in genotype (*AaBb*). Stages 1 to 6 are the same as stages 1 to 6 in Figure 29.1. But in this case, the cells are shown as each having two pairs of chromosomes, *Aa* and *Bb*. The same changes that occurred with the chromosome pair *Aa* in Figure 29.1 will occur with each of the chromosomes pairs *Aa* and *Bb* here.

Work out the sequence for meiosis for cell No. 1; then work out the sequence for meiosis in cell

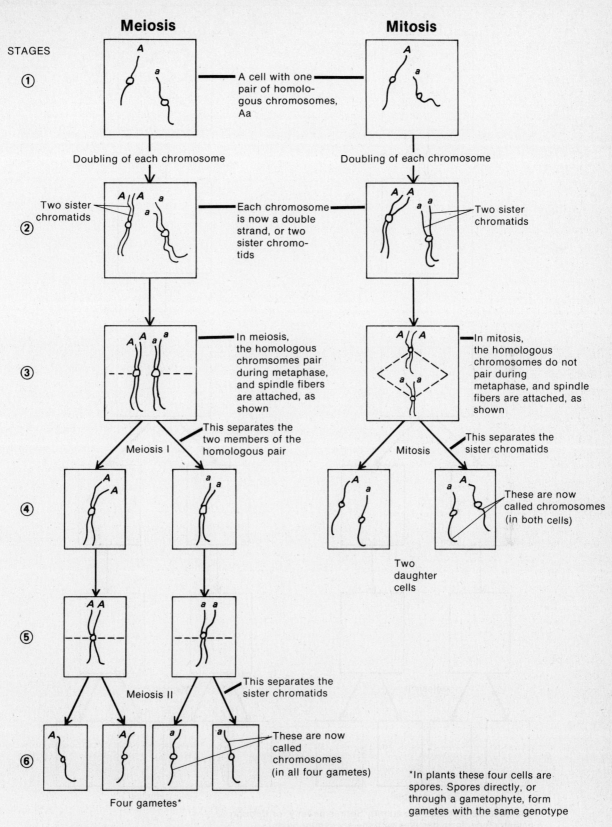

Figure 29.1 Meiosis and mitosis compared. The boxes represent cells. Only those stages showing essential changes are included. Nuclear membrane not shown. The figures in the boxes represent a pair of homologous chromosomes A and a. The letters can also be used to represent genes in the chromosomes.

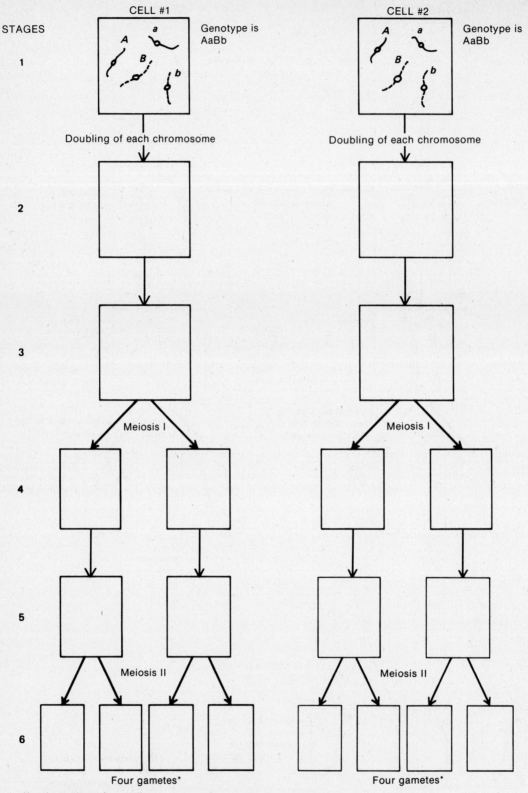

STAGES

CELL #1   Genotype is AaBb

1

Doubling of each chromosome

2

3

Meiosis I

4

5

Meiosis II

6

Four gametes*

CELL #2   Genotype is AaBb

Doubling of each chromosome

Meiosis I

Meiosis II

Four gametes*

*In plants these four cells are spores. Spores directly, or through a gametophyte, form gametes with the same genotypes.

Figure 29.2   Meiosis involving two homologous pairs of chromosomes Aa and Bb.

No. 2. Before you work out the series for cell No. 2, look at stage 3 (metaphase) of cell No. 1. Note that you have drawn one possible assortment of chromosomes. Answer the following questions:

1. Is there any other way you could have arranged the chromosomes?   *Yes   No.*

2. Can *AA* and *aa* be arranged so that they both go to the same cell in stage 4?   *Yes   No.*

3. Can *BB* and *bb* be arranged so that they both go to the same cell in stage 4?   *Yes   No.*

4. Can *AA* and *BB* be arranged so that they both go to the same cell in stage 4?   *Yes   No.*

5. Can *AA* and *bb* be arranged so that they both go to the same cell in stage 4?   *Yes   No.*

6. Can the same be said for *aa* and *BB?*   *Yes   No.*

The *conclusion* drawn from answering the preceding set of questions is that different pairs of chromosomes assort themselves at metaphase independently of the arrangement of the other homologous pairs. This is the basis for *Mendel's law of independent assortment.* This law states that "the members of different pairs of homologous chromosomes are assorted independently of each other into gametes." This independent assortment "scrambles" the chromosomes. It gives new chromosome combinations to gametes and contributes to variation in the species.

Now **complete** the drawings for cell No. 2. In stage 3 of cell No. 2 arrange the chromosomes so that when they segregate, the cells produced in stage 6 will be different from those produced in stage 6 of cell No. 1.

IV. MEIOSIS AND GAMETE FORMATION

Look back at Figure 29.1 and note stage 1 of meiosis. The genotype of the cell of stage 1 is *Aa.* The individual plant's genotype is also *Aa.* Traditionally, this is written as shown:

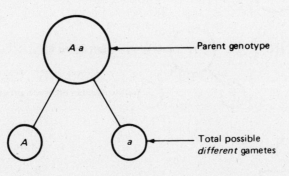

Look at Figure 29.2. Stage 1 represents two cells of one plant that undergo meiosis. The genotype of each cell is *AaBb.* The plant's genotype is also *AaBb.* We illustrate the gametes produced by this one plant as shown:

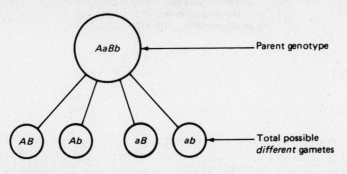

ACTIVITY 3 **Practice** figuring the *possible* genotypes for the gametes produced by the following parent genotypes. If you think more gamete circles are needed, put them in.

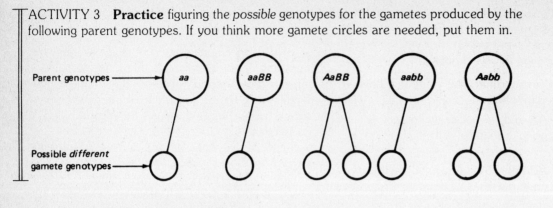

Parent genotypes ⟶ *aa*   *aaBB*   *AaBB*   *aabb*   *Aabb*

Possible *different* gamete genotypes ⟶

V. MONOHYBRID CROSSES

A. A MONOHYBRID CROSS WITH SIMPLE COMPLETE DOMINANCE

A plant that is homozygous or purebred for tall is crossed with one that is homozygous for dwarf. Let T represent the allele for tall and t the allele for dwarf.

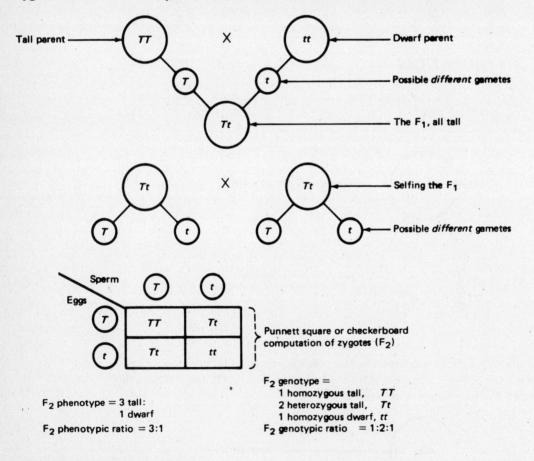

Tall parent ⟶ *TT*   X   *tt* ⟵ Dwarf parent

T   *t* ⟵ Possible *different* gametes

Tt ⟵ The F_1, all tall

Tt   X   *Tt* ⟵ Selfing the F_1

T   *t*   *T*   *t* ⟵ Possible *different* gametes

Sperm
Eggs

	T	*t*
T	*TT*	*Tt*
t	*Tt*	*tt*

Punnett square or checkerboard computation of zygotes (F_2)

F_2 phenotype = 3 tall:
1 dwarf
F_2 phenotypic ratio = 3:1

F_2 genotype =
1 homozygous tall,   *TT*
2 heterozygous tall,   *Tt*
1 homozygous dwarf, *tt*
F_2 genotypic ratio   = 1:2:1

B. THE MONOHYBRID BACKCROSS (OR TEST CROSS)

In crosses involving simple dominance, such as T and t of the preceding section, both *TT* and Tt appear tall. Often it is necessary to test which of the tall individuals is Tt and which is *TT*. This is determined by means of a **BACKCROSS**. The backcross (also called a test cross) involves crossing a homozygous recessive plant (tester stock) with plants showing the dominant trait and whose genotype is to be determined.

1. If:

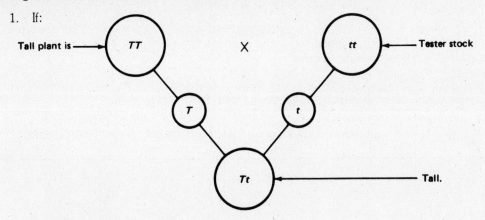

If only tall plants are produced, then the tall parent can be assumed to be homozygous dominant for tall.

2. If:

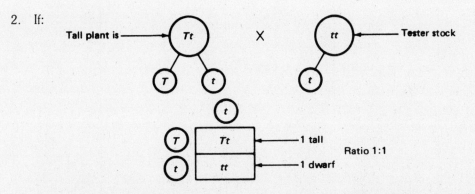

When half the offspring are tall and half are short, then the tall parent can be assumed to be heterozygous. This is a 1:1 ratio. Whenever data can be described in this ratio, you can assume that you are dealing with a backcross and that the individual showing the dominant trait is heterozygous and carries a recessive allele as well as the dominant one.

VI. MONOHYBRID OBSERVATIONS AND ANALYSES

ACTIVITY 4 **Examine** the F₂ generation seedlings that show color differences representing simple monohybrid inheritance. Suggested plants are corn (*Zea mays*) showing green and albino seedlings, tobacco (*Nicotiana tabacum*) showing green and albino seedlings, and sorghum (*Sorghum vulgare*) showing red and green seedlings.

1. How many of the seedlings are green? _____ How many are nongreen (albino)? _____

2. What is the ratio of green to albino seedlings? _____

3. Are all the green seedlings homozygous? Yes   No.

4. Are all the green seedlings heterozygous? Yes   No.

5. Among these green seedlings you would expect the homozygous dominants to be (a) twice as numerous as the heterozygous dominants; (b) half as numerous as the heterozygous dominants? Circle (a) or (b).

6. What proportion of these seedlings will breed true if self pollinated? _____

ACTIVITY 5 **Obtain** an ear of corn (*Zea mays*) from the instructor and note the presence of both colored and white grains on the ear. These grains represent the F₂ generation of a certain cross. Which color is dominant? _____

1. **Work** in teams and count the number of colored grains on the ear. **Count** only five rows. **Use** a toothpick or straight pin to mark your starting row as a reference point.

2. **Count** and **record** the number of white grains. If counting only five rows, they must be the *same* rows as before.
3. **Compile** the data for all teams on the blackboard and in Table 29.1. **Figure** the ratio and the parent genotypes and **record** in Table 29.1.

ACTIVITY 6 **Obtain** another ear of corn from the instructor in which the grains are the result of a test cross (backcross). **Count** the number of colored grains and white grains in five rows. **Compile** and **record** the data in Table 29.1.

Table 29.1 INHERITANCE OF GRAIN COLOR IN CORN (*ZEA MAYS*)

STUDENT TEAM	CORN EAR WITH F₂ GENERATION GRAINS		CORN EAR WITH GRAINS FROM A TEST CROSS	
	Colored	White	Colored	White
Class totals				
Phenotypic ratio				
Parent genotypes				

VII. DIHYBRID CROSSES

An example is shown below. In the garden pea, the allele *T* for tall is dominant to the allele *t* for dwarf. The allele *R* for red flowers is dominant to the allele *r* for white flowers. A purebred tall red plant is crossed with a purebred dwarf white.

ACTIVITY 7 **Fill in** the genotypes of the gametes below:

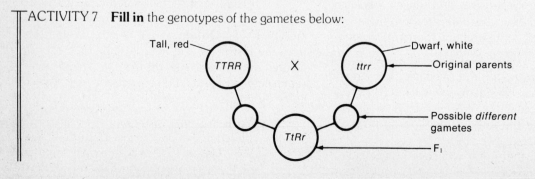

Tall, red — *TTRR*   X   *ttrr* — Dwarf, white — Original parents

TtRr

Possible *different* gametes

F₁

Give the F₁ phenotype: _____

Fill in the genotypes of the gametes below.

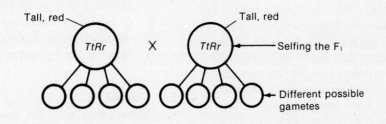

Fill in the following 16 squares with the zygote genotypes, using the gametes you have entered in the illustration directly preceding.

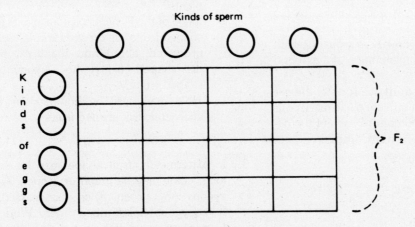

The phenotypic ratio of this F₂ is _____

List the phenotypes from the most abundant to the least abundant and give the proportion of each:

PHENOTYPE	PROPORTION

The cross shown above gives the classical F₂ ratio of 9:3:3:1. Whenever data of a given problem can be described in this ratio, you can assume that you are dealing with a dihybrid cross.

ACTIVITY 8 **Complete** Table 29.2, which shows other dihybrid crosses. **Fill in** the genotypes as indicated.

VIII. DIHYBRID OBSERVATIONS AND ANALYSES

ACTIVITY 9 **Obtain** from the instructor an ear of corn (*Zea mays*) in which the grains represent the F_2 generation of a certain cross. You will see four types of grains: purple full, purple wrinkled, white full, and white wrinkled. The full endosperm is starchy and the wrinkled endosperm is sugary.

Work in teams and **count** five rows of the ear of corn and tabulate the number of each of the four types of grains. **Use** a toothpick or straight pin to mark your starting row. **Count** the number of each of the four grain types found in these five rows. **Compile** and **record** all the figures from all individuals or teams in the class in Table 29.3. Use the letters *A* and *a* for the color and *S* and *s* for the endosperm.

Obtain another ear of corn (*Zea mays*) in which the grains represent the F_2 generation of a dihybrid test cross.

Work in teams and **count** five rows of the ear of corn. **Use** a toothpick or straight pin to mark your starting row. **Count** the number of each of the different grain types in these rows. **Compile** and **record** all the data in Table 29.3.

IX. SOME GENETICS PROBLEMS

For additional practice, you might want to work out the following problems or others provided by the instructor.

1. In summer squash, white fruit color dominates over yellow. A white-fruited squash plant, when crossed with a yellow-fruited one, produced 54 white- and 49 yellow-fruited plants.
   a. What are the genotypes of the parents?
   b. If the white-fruited parent is self-fertilized, what will be the fruit color of the offspring?
   c. What is the genotypic ratio?

2. In squash, disk-shaped fruit type dominates over the recessive allele for sphere-shaped fruits. A squash plant with disk-shaped fruit is crossed with another squash plant. The F_1 seeds when planted give rise to 86 plants with disk-shaped fruits and 79 plants with sphere-shaped fruits.
   a. What are the genotypes of the parent plants?
   b. If two of the F_1 heterozygous plants for disk-shaped fruits are crossed and 464 seeds are produced, about how many will produce disk-shaped fruits; how many sphere-shaped fruits?
   c. How many will be homozygous for the character disk-shaped fruits?

3. In watermelon, fruit shape may be short or long; color may be green or striped. A homozygous long, green melon was crossed with a homozygous short, striped variety. Work out the F_2 generation, and show the phenotypic ratio and the genotypic ratio.
   a. Which factors are dominant?
   b. Which are recessive?

EXERCISE 29   **Student Name** _____

QUESTIONS

1. Suppose the heterozygous condition produced a different trait from either the homozygous dominant or the homozygous recessive. Would this increase or decrease the number of phenotypes in the F_2 of a monohybrid cross? _____

2. What would be the F_1 phenotypic ratio if you crossed two such heterozygous individuals? _____

3. For each of the following species, the diploid number is given. In the space provided, give the number of homologous pairs this represents.

   Rye *(Secale cereale)* = 14; _____ pairs

   Tobacco *(Nicotiana tabacum)* = 48; _____ pairs

   Bean *(Phaseolus vulgaris)* = 22; _____ pairs

   Corn *(Zea mays)* = 20; _____ pairs

4. Suppose you examine the cells of a species of plant and found eight chromosomes. The homologous pairs looked as follows: one long straight pair, one short straight pair, one long bent pair, and one short bent pair. You then breed several generations of plants of this species. At the end of this time, would you expect to find
   a. Some plants with all the straight chromosomes and none of the bent ones? _____

   b. Some plants with all the long chromosomes and none of the short ones? _____

   _____ Explain. _____

   _____

5. In plants like those described in Question 4, what proportion of the gametes would have
   a. Two straight chromosomes (one long and one short) and two bent chromosomes (one long and one short) _____

   _____

   b. Three straight chromosomes and one bent one? _____

   c. Four straight chromosomes? _____

6. An individual plant has two sets of chromosomes, a maternal set and a paternal set. When this plant produces gametes, do the chromosomes of the paternal set segregate from the chromosomes of the maternal set? _____

7. If the F_2 of a cross shows a 3:1 phenotypic ratio, how many *pairs* of alleles are involved? _____

8. What is the *purpose* of a test cross? _____

   _____

9. Plant traits such as vigor, yield, hardiness, and so on are the result of the combined effect of numerous genes on different chromosomes. How does use of a tester stock help plant breeders develop lines with desired combinations of these traits? _____

   _____

   _____

   _____

10. Does meiosis in plants give rise directly to eggs and sperm? _____

Table 29.2 DIHYBRID CROSSES*

PARENTS	$TTrr$ $\times$ $ttRR$		$TtRr$ $\times$ $ttrr$		$ttRr$ $\times$ $TTrr$	
Genotypes of all possible *different* gametes						
Genotypes of offspring						

*The number of answer blanks does not necessarily reflect the number of possible genotypes.

EXERCISE 29

Student Name _____

Table 29.3 INHERITANCE OF GRAIN COLOR AND ENDOSPERM IN CORN (*ZEA MAYS*) AS SEEN IN THE PHENOTYPES OF AN F₂ GENERATION AND A TEST CROSS

STUDENT TEAM	F₂ GENERATION				TEST CROSS			
	Purple, Full	Purple, Wrinkled	White, Full	White, Wrinkled	Purple, Full	Purple, Wrinkled	White, Full	White, Wrinkled
Class totals								
Phenotypic ratio								
Parent genotypes								

PLANT ECOLOGY

INTRODUCTION

Plant ecology deals with the interrelationships of plants and their environment. These relationships may be obvious, as in the case of species that will not survive freezing temperatures, or very complex, as when one forest type gradually gives way to another over the centuries. Since many of these principles must take the form of examples that will be explained to you, there will be only one laboratory investigation comparable to those of the previous exercises.

I. COMPETITION

Observe a demonstration showing a series of five or six sets of plants that have been growing for several weeks. The sets differ only in the number of plants that are in each pot. The soils, watering regimen, and light level were the same for each pot. The number of seeds initially planted has been calculated (using expected germination percentages as you learned in Exercise 3) to result in eventual plant populations of 1, 2, 4, 8, and 16 (or 16 and 32) plants per pot. Suitable species for such a demonstration are corn (*Zea mays*), grain sorghum (*Sorghum* sp.), jimson weed (*Datura* sp.), or other fast-growing herbaceous plants.

ACTIVITY 1 **Calculate** the average height of the plants in each pot and record this in Table 30.1. Weigh the total mass of plant material produced above the soil level in each pot. Carefully wash the roots out of the soil in each pot. Or, if using peat cubes instead of pots of soil, weigh the entire mass of plants, including stems, leaves, roots, and peat cube. The dry weight may be obtained by drying the plants

Table 30.1

POT CODE LETTER	NUMBER OF PLANTS IN POT	AVERAGE HEIGHT (CM)	TOTAL WEIGHT PER POT (G)	WEIGHT PER PLANT (G)
a	1			
b	2			
c	4			
d	8			
e	16			
f	32			

in an oven at about 70°C for 48 hours. The demonstration may have the weights and heights already calculated for you. In that case, just record them in Table 30.1.

The demonstration is intended to show the principle of competition, which is very important in ecology. In which pot(s) were the individual plants the largest?_____. In which were the plants the smallest? _____. Is this the result you expected? *Yes No.*

ACTIVITY 2 Using the data in Table 30.1, **construct** a graph based on the weights per plant (Figure 30.1). Enter the values for weights of single plants and then the values for all the plants. Connect the weights of single plants with a dashed line, and the total weight of all plants per pot with a solid line. (For this example it doesn't make any difference whether you weighed the tops or the whole plants or whether they represent fresh weights or dried weights, so long as you are using weights of the same type, that is, all dried tops or all fresh weights of the whole peat cube with plants.) Also, if for some reason, you ended up with 14 rather than 16 plants or even 6 plants instead of 8, just enter on the figure the number you actually had. The important point is to remember to divide the total weight by 14 or 6 (rather than by 16 or 8), so that the value of weight per plant is correct. **Enter** all values obtained in Figure 30.1.

OBSERVATIONS

Examine the shape of the graph you have just made. How would you describe the trend in the weight of the average plant?_____

_____ What about the trend in the total weights? Which pots had the greatest total weight of plants? _____. Was the pot with two plants twice the weight of the one with one? *Yes No.* How great was the difference?_____. Was the pot with 16 plants twice that of the one with 8? *Yes No.* Was it 16 times as heavy as the single plant? *Yes No.* What was the actual difference? _____. The weights per plant generally decrease. Does this decrease show signs of leveling off? Is there a point where all plants are the same size, regardless of the number per pot? *Yes No.*

What about the trend in the total weight per pot? Generally, the maximum weight is attained with about eight plants. Was this true in your example? *Yes No.* Does the graph show a leveling off at about this point? *Yes No.* Does the graph level off sooner? *Yes No.* Where? _____. Does the total weight continue to increase throughout all the examples? *Yes No.* If a leveling off, or plateau, is visible in the

graph of total weights per pot, what does this represent in terms of the ecological principle of competition? _____

_____.

Your laboratory instructor will discuss this with you at this time. This discussion will also cover the environmental factors that are important in controlling competition. In other words, for what environmental factors are these plants competing?

II. SUCCESSION

The seeds planted for the competition example were selected from a uniform lot of genetically identical (or very similar) characteristics. This demonstration is typical of the way that agricultural crops compete with each other. Plant breeders of vegetable, grain, and forest crops determine that a certain maximum density of plants will result in the greatest yield per plot (we used total weight to show this). Below this density, single plants are larger, and above this density, individual plants are small and stunted.

In nature the same basic principle applies. There is only a certain amount of plant growth that may be supported by a given amount of soil, moisture, and light. But in nature there are plants of thousands of different genetic backgrounds, belonging to dozens of species, in any one area. Some are bound to outcompete the others. They grow and flourish and reproduce themselves while the less successful species and individuals are lost from that spot.

Over the years, though, this competition causes changes in the successful species in an area. The vigorously germinating weedy species that may have been dominant initially on bare soil may not be able to sustain themselves indefinitely in the same place. There are many reasons for this phenomenon, but the most common one is that the originally successful species have, simply by growing successfully, so changed the environment for germination and growth that the new conditions of shade, moisture, and nutrients became better suited for the germination and growth of seeds of other species.

ACTIVITY 3 **Observe** photographs or a slide presentation or film showing the change in plant species over the years in a single area that is typical for your geographical location.

If the organization of your laboratory permits, the best way to examine these changes is through a field trip to a variety of nearby plant communities in various stages of succession. If you must rely on photographs or a film, be sure to ask the laboratory instructor for assistance whenever you have a question about any of the principles of succession that are demonstrated to you.

OBSERVATIONS

What is the nature of the first colonizers on a patch of bare soil in your area? Are they (a) large or (b) small plants. Circle (a) or (b). Are they (a) annuals or (b) perennials? Circle (a) or (b). Are they (a) woody or (b) herbaceous? Circle (a) or (b). Do they typically grow in (a) full sun, (b) partial sun, (c) full shade? Circle (a), (b) or (c). Do they usually occur (a) as mixed stands of dozens of species, or (b) in masses or clusters of two or three species only? Circle (a) or (b). Do they (a) start to grow first around the edges of the bare area, or (b) seem to start more or less all over the new habitat? Circle (a) or (b). Do they produce (a) many or (b) few seeds? Circle (a) or (b). Are these seeds (a) small or (b) large? Circle (a) or (b).

ACTIVITY 4 **Fill** in Table 30.2. **Follow** the same line of inquiry, either from personal observation if you are able to go on a field trip, or from quizzing the laboratory instructor and the special assistants.

OBSERVATIONS

What sorts of changes can you note in this succession of stages? What change is there in the size of the plants? _____.
Are the later stages characterized by woody plants? *Yes No.* Do the young seedlings of the dominant species germinate and grow well in the shade created by their parents? *Yes No.* What do you suppose would happen if you had a chance to observe a plant community where the seedlings of the dominant species grew and developed normally in the environment created by their mature parents? Would there be a further change in the dominant species? *Yes No.* Explain why or why not. _____

_____.

The name given to an unchanging community is *climax.* Such a climax community perpetuates itself. Are there examples of climax communities near your school? *Yes No.* Why or why not? Discuss the reasons with your instructor and special assistants.

III. PLANT GEOGRAPHY

As has been brought out in the laboratory discussion, areas of climax vegetation are pretty scarce in most of North America. The settlers cut the forests, plowed the prairies, and drained the swamps. Nevertheless, examples still exist of most of the important natural plant formations of the continent. The term biome is used to refer to the combination of plant formation and the associated important animal species. Even before you studied botany as a subject, you surely had some general idea of which of the major plant formations was potentially dominant in your part of the country, such as deciduous forest, coniferous forest, or grassland. Why *are* there different dominant plant formations? As you studied the structure of the stem and leaf in previous exercises, several adaptations to adverse conditions were pointed out. These include mainly modifications that allow the plant to live in areas in which temperature and moisture are not ideal, especially the combination of high temperature and low moisture. Other adaptations are not so readily correlated to the appearance of the plant. Cold hardiness is one of these.

ACTIVITY 5 **Observe** photographs or a slide presentation or a film that show examples of the major plant formations of North America, as mapped and named in Figure 30.2. This may be a continuation of the films or slides you observed before to see succession, or it may be a special class meeting or lecture just for this purpose.

Locate your area on the map. Does the name of the potentially dominant vegetation region agree with your own observations? *Yes No.* If not, why not? _____

_____. Discuss this with your laboratory instructor or the special lecturer. Many terms will be used in these discussions, and they are defined in Table 30.3.

ACTIVITY 6 Taking the given information about existing North American plant formations (Table 30.4) combined with the definitions presented in Table 30.3, **fill in** the formation types represented in simplified fashion in Table 30.5. Not every possible set of conditions is represented here.

OBSERVATIONS

Are there any North American plant formations whose characteristics are not given in the table? *Yes No.* Which one or ones? _____
_____. **Fill in** their combination(s) of conditions in line(s) seven or eight of Table 30.5. Can you dream up another theoretical set of conditions not given here? There are many possible sets not shown. **Enter** one or two more sets of hypothetical conditions in lines nine or ten. **Discuss** with your lab partners the overall effect of these conditions as they lead to stresses on the plants that may attempt to live in such an environment. Have you invented an environment in which nothing can survive? Most of the combinations not already given amount to just that—an area of little or no plant growth. North America has representatives of almost all typical plant formations, even the tropical forests.

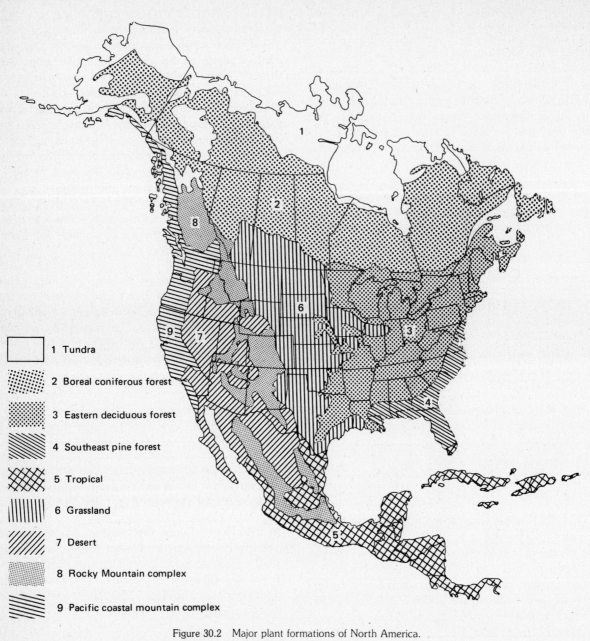

Figure 30.2 Major plant formations of North America.

Legend:

1 Tundra
2 Boreal coniferous forest
3 Eastern deciduous forest
4 Southeast pine forest
5 Tropical
6 Grassland
7 Desert
8 Rocky Mountain complex
9 Pacific coastal mountain complex

Table 30.3 GENERAL CLASSIFICATION OF PRECIPITATION, TEMPERATURE, AND GROWING SEASONS

ANNUAL PERCIPITATION		TEMPERATURE DURING GROWING SEASON		GROWING SEASON	
Classified as	(mm)	Classified as	(°C)	Classified as	(Months)
Low	75 to 500	Cold	2 to 10	Short	1 to 3
Medium	501 to 1750	Cool	11 to 20	Medium	4 to 8
High	1751 to 4000	Warm	15 to 25	Long	9 to 12
	or more	Hot	25 to 40		

Table 30.4   CHARACTERISTICS OF NORTH AMERICAN PLANT FORMATIONS

NO. TAKEN FROM FIG. 30.2	FORMATION	ANNUAL PRECIPITATION (MM)	EVAPORATION	TEMPERATURE (°C)	GROWING SEASON (MONTHS)	CHARACTERISTIC PLANT GROUPS	OTHER FEATURES
1	Tundra: arctic— high latitude alpine— high altitude	125 to 375	Low	Summer: 2 to 10 Winter: –40 to –20	1 to 3, limited by cold	Grasses, dwarf shrubs, lichens, mosses	Treeless; soil thaws 20–50 cm in summer; permanently frozen below
2	Boreal coniferous forest	375 to 900	Low	Summer: 10 to 20 Winter: –55 to –20	3 to 4	Spruce, fir, birch, aspen, larch	Many bogs and lakes, poorly drained, few trees species, many mosses and lichens
3	Eastern deciduous forest	Eastern: 1200 to 1800 Western: 650 to 900	Medium	Summer: 15 to 25 Winter: –5 to +5	4 to 7	Oaks, hickories, maples, beech, basswoods, elms, buckeye	Many tree species, some shrubs and herbaceous plants
4	Southeast pine forest	1500 to 1800	Medium to high	Summer: 22 to 30 Winter: 5 to 10	5 to 8	Pines, cypress, oaks, hickories, magnolias	Pines successional and of high economic value
5	Tropical forest	Wetter areas: 1800 to 4000 or more Dry areas: 750 to 1200	Medium to high	Summer: 25 to 33 Winter: 20 to 25	Wet: 12 Dry: 3 to 5	Mahogany, palms, figs, live oaks, grasses in some areas	Large number of species
6	Grassland	Eastern: 650 to 900 Western: 300 to 500	High	Summer: 15 to 25 Winter: –15 to 10	North: 4 South: 8	Grasses and many broadleaved herbs	Most of precipitation in spring and early summer, fall droughts; tall grasses eastern part, mid and short grasses western part
7	Desert	75 to 250	High	Summer: 25 to 40 Winter: 2 to 10	1 to 3, limited by moisture	Cacti, many shrubs	Many annuals when rains come

Table 30.5 THE TYPES OF FORMATIONS ASSOCIATED WITH PARTICULAR COMBINATIONS OF ENVIRONMENTAL CONDITIONS

AREA	PRECIPITATION	EVAPORATION	TEMPERATURE	GROWING SEASON	FORMATION TYPE
1	Low	High	Hot	Short	
2	Low	Low	Cold	Short	
3	Medium	Medium	Warm	Medium	
4	High	Medium to high	Hot	Long	
5	Low to medium	Low	Cool	Short	
6	Low to medium	High	Warm	Short to medium	
7					
8					
9					
10					

IV. HUMAN USE OF THE PLANT FORMATIONS

It is easy to think of some of the direct uses that we make of plant communities. The coniferous forests are cut for lumber and pulpwood and left to seed another generation of the same species. Are there any other formations that are used directly? *Yes* *No*. Which one(s) can you think of? (a) _____; (b) _____ _____. It is normal for us to use the natural vegetation in one of four ways: (1) direct use, as with the lumber from trees; (2) agricultural adaptation of former plant communities; (3) removal of vegetation for building sites for homes, industry,

and transportation corridors; and (4) intrusion into natural plant formations for extraction of underground resources, such as minerals or oil. Each of these uses disturbs the natural successional processes, and many of them totally destroy the formation over a wide area.

ACTIVITY 7 **Fill in** Table 30.6 showing the type of human use typically made in North America of each of the original plant formations. When wide variation is known to you, indicate the range of use. Use text readings, outside reading assignments, class discussions, and evening bull sessions to assist in the formulation of these ideas. It is not intended that you be able to complete this table in the remainder of a single laboratory period.

SECTION 30

Student Name _____

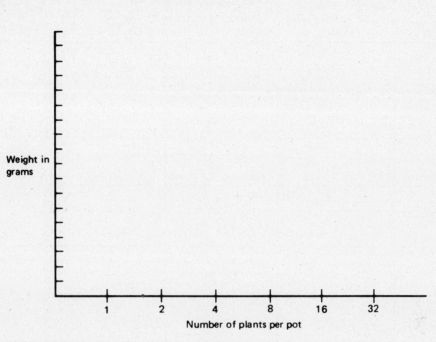

Figure 30.1

Table 30.2

CHARACTERISTIC	PIONEER STAGE	SECOND STAGE	THIRD STAGE	OLDEST STAGE
Size of plants				
Annual or perennial				
Woody or herbaceous				
Sunlight adaptation				
Mixed species or masses of one kind				

Table 30.6

FORMATION	DIRECT USE OF PLANT FORMATION IF ANY	TYPICAL AGRICULTURAL USES IF ANY	OTHER HUMAN INTRUSIONS COMMONLY SEEN	RELATIVE ABUNDANCE OF UNDISTURBED AREAS
Eastern deciduous forest				
Tundra and high latitude				
Boreal forest				
Southeastern coniferous forests				
Grasslands				
Desert				
Pacific coastal forests				
Tropical forests				

KEY FOR IDENTIFICATION OF SOME COMMON TREE GENERA

A dichotomous key is given here which will serve to identify several dozen of the most commonly planted ornamental and native trees. Obviously, not all genera can possibly be included. A very wide selection of trees has been made, and it includes most of the trees on most campuses in temperate North America. A few campuses in very mild climates may have a wide variety of other genera—so wide that it is considered beyond the scope of this manual to try to include them all. In a few places in the key, the phrase "frost-free" is used. The term is used rather loosely to describe species grown only in areas where hard killing frosts are infrequent. Examples might be the palms and *Araucaria*. **Compare** descriptions with illustrations in Figure 31.1.

Directions for Using Key: This key consists of pairs or couplets of choices, 1A–1B, 2A–2B, and so on. Compare your plant specimen with the descriptions given in couplet 1A–1B. Choose whichever statement (1A or 1B) is more applicable to your specimen and go on to the next statement as indicated by the number at the extreme right. Continue through the key, each time choosing between a pair of choices until you terminate at the name of a plant.

1A. Leaves needle- or scalelike; plants without flowers and fruits; commonly bearing naked seeds in woody cones (except *Taxus*) .... **Division Coniferophyta**
1B. Leaves flat, broad, not needlelike; seeds never in woody cones 2
2A. Leaves fan-shaped, with parallel venation (Figure 31.1A); seed with bad-smelling fleshy outer layer; not enclosed in a fruit .... **Division Ginkgophyta**
2B. Leaves with netted or parallel venation; seeds always enclosed in a fruit that develops from the ovary of a flower .... **Division Anthophyta**

I. DIVISION GINKGOPHYTA

1. Flat, fan-shaped leaves (Figure 31.1A); contains the single species ...... **Ginkgo biloba**
(maidenhair tree)

II. DIVISION CONIFEROPHYTA

1a. Leaves borne in clusters 2
1b. Leaves borne singly along the stem...4
2a. Leaves more than five in a cluster...3
2b. Leaves two to five in a cluster, persistent for several seasons .... **Pinus**
(the pines—at least forty species native and cultivated in the continental United States)
3a. Leaves deciduous, soft, linear, flattened .... **Larix**
(larch or tamarack—native to Northern states and mountains)

3b.   Leaves persistent, stiff, quadrangular .... **Cedrus**
                           (true cedar—cultivated in warmer areas without severe winters)

4a.   Leaves (at least some of them) scalelike to triangular, opposite or whorled 5

4b.   Leaves needlelike or somewhat flattened, alternate or scattered along stem...........7

5a.   Branchlets flattened; cones longer than wide, brown and woody at maturity; all leaves
      scalelike, opposite .... **Thuja**
                                  (arbor vitae)

5b.   Branchlets not flattened...6

6a.   Leaves mostly in whorls of three, many of them needlelike; cones round, bluish, and berry-
      like at maturity .... **Juniperus**
                                  (juniper)

6b.   Leaves needlelike to triangular, growing in tight spirals and forming a rounded branch 2–8
      cm in diameter; cones large and woody (frost-free areas).................... **Araucaria**
                              (monkey puzzle tree)

7a.   Leaves soft, on green branchlets, the whole superficially similar to a compound leaf (Figure
      31.1Z) 8

7b.   Leaves rigid, on brownish branches, solitary 10

8a.   Leaves dark green, persistent, the branchlets on a green stem, larger stem sometimes bear-
      ing solitary leaves .... **Sequoia sempervirens**
                                  (redwood)

8b.   Leaves light, yellow-green, deciduous together with the branchlets, larger stems usually
      brownish...9

9a.   Leaves up to 2.5 cm long, subopposite; leaf-bearing branchlets often over 10 cm long;
      youngest woody branches reddish brown with subopposite axillary buds 2–3 mm long
      .... **Metasequoia glyptostroboides**
                                (dawn redwood)

9b.   Leaves not over 1.5 cm long, alternate; leaf-bearing branchlets 4–8 cm long; youngest
      woody branches yellowish tan with spirally arranged inconspicuous buds .... **Taxodium**
                                (bald cypress)

10a.   Leaves four-sided in cross section; woody petiole persistent on stem .... **Picea**
                                  (spruce)

10b.   Leaves flattened in cross section 11

11a.   Shrubs; leaves without lines of stomata on lower surface; seed at maturity enveloped by
      fleshy red tissue.. **Taxus**
                                  (yew)

11b.   Trees; leaves with two lines of stomata on lower surface 13

12a.   Leaves 8–20 mm long, round-tipped, two-ranked (forming two rows along stem); cones not
      more than 2.5 cm long.. **Tsuga**
                                  (hemlock)

12b.   Leaves 2–6 cm long with pointed tips, usually not two-ranked; cones larger 13

13a.   Leaves more or less straight, dark green; cones pendulous, falling off intact; bracts con-
      spicuous, three-lobed ....**Pseudotsuga menziesii**
                                (Douglas fir)

13b.   Leaves mostly strongly curved, light bluish green; cones upright, disintegrating on the tree;
      bracts hidden .... **Abies**
                                  (fir)

III. DIVISION ANTHOPHYTA

1a.   Trees without branches, a single stem bears very large leaves directly (the palms)
      .... **Section D**

1b.   Trees with normal branching, several orders of branches and twigs bearing smaller leaves
     2

2a.   Branches (at least the younger ones) with spines .... **Section A**

2b.   Branches without spines 3

3a.   Leaves simple (Figure 31.1J)...4

3b.   Leaves compound (Figure 31.1K through M) 7

4a.   Leaves alternate 5

4b.   Leaves in whorls or opposite .... **Section E**

5a.   Leaf-margin entire (Figure 31.1B)................................... **Section B**

5b.   Leaf-margin toothed or lobed, not entire (Figure 31.1C through I) 6

6a.   Margin toothed, often doubly toothed, or appearing shallowly lobed (Figure 31.3C through
      G) .... **Section C**

6b.   Margin deeply lobed (Figure 31.1H or I) .... **Section D**

7a.   Leaves alternate .... **Section F**
7b.   Leaves opposite... **Section G**

SECTION A. Branches with Spines

1a.   Leaves simple 2
1b.   Leaves pinnately compound (Figure 31.1K or L) 3
2a.   Leaf margins entire .... **Maclura pomifera**
(Osage orange or hedge apple)
2b.   Leaf margins toothed and in some species shallowly lobed.................**Crataegus**
(hawthorn)
3a.   Spines modified stipules (Figure 31.1O), up to 2 cm long; leaves singly compound
Figure 31.1K with 7–17 leaflets .... **Robinia pseudoacacia**
(black locust)
3b.   Spines modified branches (Figure 31.1P), up to 7 cm long, often forked (trunks often with
clusters of very long, branched, adventitious spines); leaves singly or doubly compound, mostly with more than 18 leaflets .... **Gleditsia triacanthos**
(honey locust)

SECTION B. Branches without Spines; Leaves Simple, Alternate, with Entire Margins

1a.   Leaves heart-shaped.. **Cercis canadensis**
(red bud)
1b.   Leaves not heart-shaped 2
2a.   Leaves 20 cm or longer when fully grown 3
2b.   Leaves not more than 18 cm long 4
3a.   Stipular scars encircling twigs at nodes; terminal bud large, with a single budscale covered
with gray hairs (Figure 31.1N) .... **Magnolia soulangeana**
(saucer magnolia)
3b.   No stipular scars or stipules present; terminal bud small, thin, naked, covered with short
brown hairs .... **Asimina triloba**
(pawpaw)
4a.   Leaves bristle-tipped, leathery; fruit an acorn .... **Quercus**
(oak—several oaks key out here, of which the more common are the shingle oak, *Q.
imbricaria* and the evergreen live oak, *Q. virginiana*)
4b.   Leaves not bristle-tipped, not leathery 5
5a.   Veins markedly incurving toward the tip of leaf (Figure 31.1X); flowers perfect; usually
shrubs .... **Cornus alternifolia**
(pagoda dogwood)
5b.   Veins not incurving; flowers unisexual; usually trees 6
6a.   Leaves ovate with rounded or emarginate apex (Figure 31.1W) .... **Cotinus**
(smoke-tree)
6b.   Leaves ovate or oval; leaf apex acute or acuminate (pointed) 7
7a.   Young twigs smooth; pith of twig chambered (Figure 31.1Q); three bundle scars ....
....**Nyssa sylvatica**
(sour gum)
7b.   Young twigs hairy; pith of twig not chambered; one bundle scar **Diospyros virginiana**
(persimmon)

SECTION C. Branches without Spines; Leaves Simple, Alternate, with Toothed Margins

1a.   Leaf margins singly toothed (Figure 31.1C or F).................................2
1b.   Leaf margins doubly toothed (Figure 31.1D or E)................................10
2a.   Sap milky .... **Morus**
(mulberry)
2b.   Sap not milky..3
3a.   Leaves triangular or heart-shaped, about as broad as long.......................4
3b.   Leaves longer than broad, widest near the middle 5
4a.   Leaves asymmetrical with oblique base (Figure 31.1V), several (usually three) large veins
arising from petiole .... **Tilia**
(linden, or basswood)

4b.   Leaves symmetrical, pinnately veined (Figure 31.1S), with prominent midrib from petiole to apex of leaf...**Populus**
    (the many species of poplars, aspens, and cottonwoods belong to this genus)

5a.   Teeth of leaf margin coarse, two or fewer per centimeter (Figure 31.1F), or margin undulate (Figure 31.1G); buds mostly 15–25 mm long, pointed ....**Fagus**
    (beech)

5b.   Teeth of leaf margin fine, three or more per centimeter; buds shorter 6

6a.   Petioles with one to four small glands near leaf blade ....**Prunus**
    (plum, peach, cherry)

6b.   Petioles lack glands...7

7a.   Leaves at least three times longer than broad.................................**Salix**
    (willow)

7b.   Leaves not more than two times longer than broad.............................8

8a.   Leaves asymmetrical (Figure 31.1V), with three to five large veins arising from petiole; pith of twig chambered (Figure 31.1Q) ....**Celtis**
    (hackberry)

8b.   Leaves symmetrical, with one main midrib; pith of twig not chambered 9

9a.   Buds elongate with many scales (Figure 31.1R); flowers greenish; bark of trunk scaling off in slender vertical strips ....**Ostrya virginiana**
    (hop-hornbeam)

9b.   Buds ovoid with few scales; flowers white to purple, showy; bark of trunk not scaling off in strips ....**Malus**
    (apple)

10a.   Leaf blade asymmetrical at base with one side smaller (Figure 31.1U) ....**Ulmus**
    (elm)

10b.   Leaf blade symmetrical at base 11

11a.   Petioles with one to four small glands near leaf blade ....**Prunus**
    (plum, cherry)

11b.   Petioles lacking glands 12

12a.   Bark of trunk and larger branches chalky or silvery white, peeling horizontally in thin papery strips ....**Betula**
    (birch)

12b.   Bark brown, not peeling in horizontal strips 13

13a.   Pistillate catkins conelike, woody, persistent until the second year; buds short-stalked; two bud scales...**Alnus**
    (alder)

13b.   Pistillate catkins with thin papery scales, not woody, soon deciduous, not persistent; buds sessile with several scales ....**Ostrya virginiana**
    (hop-hornbeam)

SECTION D. Branches without Spines; Leaves Simple, Alternate, Margin Deeply Lobed

1a.   Trees lacking branches, the single trunk bearing relatively few very large leaves; frost-free areas; the palms 2

1b.   Trees with definite branches bearing numerous leaves; not palmlike 7

2a.   Leaves palmately divided..3

2b.   Leaves pinnately divided 5

3a.   Petioles commonly lacking thorns or spines; usually not treelike................**Sabal**
    (palmetto)

3b.   Petioles usually with thorns or spines; plants becoming treelike 4

4a.   Leaves glossy green; trunks marked with rings from old leaf bases ....**Livistonia**
    (Australian fan palm)

4b.   Leaves light green or gray-green; trunks with persistent leaf bases and often dead leaves ....**Washingtonia**
    (Washington fan palm)

5a.   Leaf bases and petioles forming a green sheath the top 1–2 m of the trunk   **Roystonea**
    (royal palm)

5b.   Leaf bases not forming a conspicuous green sheath 6

6a.   Leaf bases deciduous, forming a smooth trunk decidedly swollen at the lower end and often leaning at a considerable angle from the vertical....................**Cocos nucifera**
    (coconut)

6b.   A portion of the leaf bases persistent on the trunk; stem of constant diameter and rarely leaning..**Phoenix**
    (date palm)

7a.   Leaves palmately veined, with more than one main vein arising from petiole (Figure 31.1T)
....8

7b.   Leaves pinnately veined (Figure 31.1S) 10

8a.   Sap milky; many of the leaves only toothed .... **Morus**
(mulberry)

8b.   Sap colorless 9

9a.   Leaves star-shaped, cleft into five to seven wedge-shaped lobes with evenly toothed margins; terminal bud present; bark of trunk deeply furrowed, persistent; twig lacks encircling scars at nodes .... **Liquidambar styraciflua**
(sweet gum)

9b.   Leaves not star-shaped, three to five lobed, the margins not evenly toothed; terminal bud absent; bark thin, more or less smooth, eventually peeling off in large pieces; twigs with encircling scars at nodes (Figure 31.1N)............................... **Platanus**
(sycamore, plane tree)

10a.   Branchlets yellowish green; leaves and bark aromatic .... **Sassafras albidum**
(sassafras)

10b.   Branchlets gray or grayish brown; leaves and bark not aromatic 11

11a.   Stipular scars encircling twigs at nodes (Figure 31.1N); leaf broadly notched at apex, four lobed, about as broad as long (Figure 31.1Y) .... **Liriodendron tulipifera**
(tulip tree)

11b.   No encircling stipular scars present; leaf longer than broad with five or more marginal lobes; *Quercus*—oaks 12

12a.   Lobes pointed (Figure 31.1I), bristle-tipped .... **Quercus**
(black oak group)

12b.   Lobes rounded (Figure 31.1H or G), not bristle-tipped .... **Quercus**
(white oak group)

SECTION E. Branches without Spines; Leaves Simple, in Whorls or Opposite

1a.   Leaves three (sometimes two) at a node, ovate to ovate-oblong, with a long, pointed tip, densely hairy beneath .... **Catalpa**
(catalpa)

1b.   Leaves opposite, sparingly pubescent or glabrous when fully grown.................2

2a.   Leaves with entire margins and pinnate veins that are incurved toward apex (Figure 31.1X)
.... **Cornus**
(the dogwoods)

2b.   Leaves lobed, palmately veined (Figure 31.1T) .... **Acer**
(maple)

SECTION F. Branches without Spines; Leaves Compound, Alternate

1a.   Leaves doubly compound (Figure 31.1L) 2

1b.   Leaves singly compound (Figure 31.1K) 4

2a.   Base of leaflets very asymmetric, midrib of leaflets decidedly off center........ **Albizzia**
(silk tree, "mimosa")

2b.   Base of leaflets symmetric, midrib in center of leaflet 3

3a.   Leaflets ovate, entire, with pointed apex .... **Gymnocladus dioica**
(Kentucky coffee tree)

3b.   Leaflets oval or lanceolate, somewhat toothed, with rounded apex .... **Gleditsia**
(honey locust)

4a.   Pith chambered (Figure 31.1Q); fruit a large, nearly spherical drupe with indehiscent exocarp; leaflets 11 to 33 .... **Juglans**
(walnut)

4b.   Pith not chambered; fruit not as above...5

5a.   Leaves 30–90 cm long with 13–41 leaflets, of unpleasant odor when crushed
.... **Ailanthus altissima**
(tree of heaven)

5b.   Leaves usually less than 30 cm long, no unpleasant odor 6

6a.   Rachis of leaves zigzag, the leaflets alternate, entire .... **Cladrastis lutea**
(yellowwood)

6b.   Rachis of leaves straight, the leaflets opposite...................................7

7a.   Leaflets less than 5 cm long with rounded or emarginate apex (Figure 36.1W), entire or inconspicuously toothed; fruit a legume 8

7b.   Leaflets more than 5 cm long with pointed apex, conspicuously toothed (Figure 31.1C) through E); fruit drupelike 9

8a.   Leaves with 7 to 17 leaflets, the terminal one usually present; stipules spiny (Figure 31.1O) . . . . **Robinia pseudoacacia**
(black locust)

8b.   Leaves mostly with more than 18 leaflets, the terminal one usually missing; stipules absent . . . . **Gleditsia**
(honey locust)

9a.   Small trees or shrubs; sap milky; buds naked; drupes red, indehiscent, less than 1 cm in diameter, in pyramidal clusters . . . . **Rhus**
(sumac)

9b.   Large trees; sap watery; buds with bud scales; drupes with dehiscent exocarp, more than 12 mm in diameter, singly or two to four in a cluster . . . . **Carya**
(hickories)

SECTION G. Branches without Spines; Leaves Compound, Opposite

1a.   Leaves palmately compound (Figure 31.1M) 2
1b.   Leaves pinnately compound (Figure 31.1K) 4

2a.   Leaflets usually seven, coarsely doubly toothed (Figure 31.1E); buds sticky . . . .
. . . . **Aesculus hippocastanum**
(horse chestnut)

2b.   Leaflets usually five, finely singly toothed (Figure 31.1C); buds not sticky; buckeyes . . .3

3a.   Fruit smooth; petals dissimilar, the upper ones longer than the stamens; to 25 m tall . . .
. . . . **Aesculus octandra**
(sweet buckeye)

3b.   Fruit echinate (prickly); petals nearly equal, shorter than stamens; small tree, to 15 m tall . . . . **Aesculus glabra**
(Ohio buckeye)

4a.   Leaflets three to five, coarsely and irregularly toothed and lobed . . . . **Acer negundo**
(box elder)

4b.   Leaflets 5 to 11, entire or finely toothed; *Fraxinus*—ashes 5

5a.   Young twigs four-angled . . . . **Fraxinus quadrangulata**
(blue ash)

5b.   Young twigs round in cross section 6

6a.   Leaf scars concave at top; petioles glabrous or nearly so . . . . **Fraxinus americana**
(white ash)

6b.   Leaf scars straight at top or nearly so; petioles velvety pubescent . . . .
. . . . **Fraxinus pennsylvanica**
(red ash)

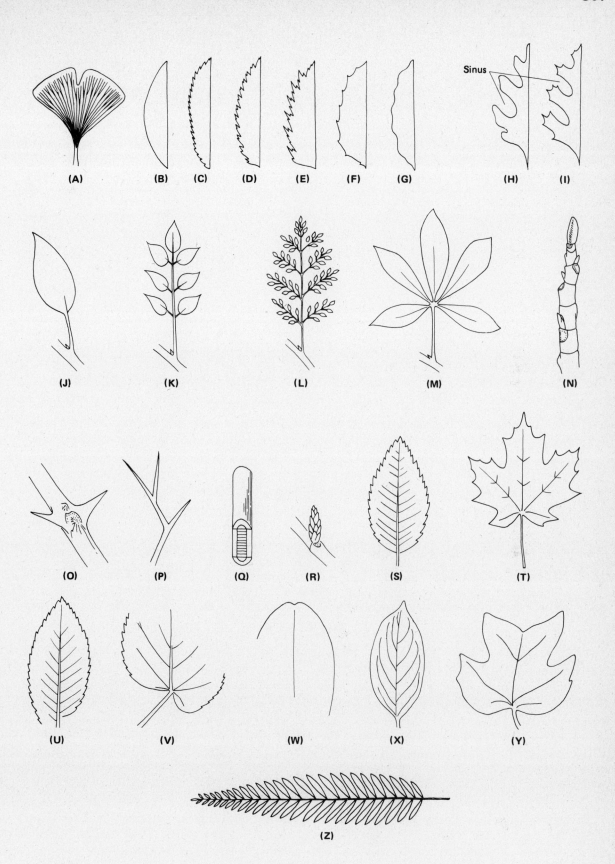

Figure 31.1 Illustration of characters used in the identification of trees.

INSTRUCTOR'S GUIDE:

MATERIALS NEEDED, SOURCES OF MATERIALS, DIRECTIONS, AND GENERAL SCHEDULING

INTRODUCTION

This section is designed to assist the instructor in planning the materials required for each activity in the exercises. Successful completion of an exercise often requires a classroom sequence different from that logically used in the text presentation, and guidance is given when this is recommended. Students may also use the Instructor's Guide to aid in scheduling themselves during "free" laboratory periods used in regular or remedial instruction. True self-study is also enhanced by the additional information in this guide, and its use is recommended when such self-instruction is undertaken.

STUDENT SUPPLIES

Depending on the individual school policy, the school may supply slides, forceps, and so on to the class, or each student may have to purchase a personal set. If the latter is the case, the following items are suggested as adequate for ordinary needs during one semester:

- 6 1″ × 3″ glass slides
- 12 cover slips
- 2 dissecting needles
- 2 single-edge razor blades
- 1 fine-point forceps
- 1 small 15 cm metric ruler
- 1 pipette (eye-dropper)
- 1 hand lens (6—10×)

No drawing paper is needed.

EXERCISE 1—THE WHOLE PLANT

Activity 1

Potted plants for demonstration of vegetative parts. Plants not to be used up. One plant per four students. Two-month-old bean (*Phaseolus vulgaris*); two-month-old cucumber (*Cucumis sativus*); any dwarf variety of tomato (*Lycopersicon* sp.); coleus (*Coleus* sp.); ornamental pepper (*Capsicum annuum*); Christmas cherry (*Solanum pseudocapsicum*); rubber plant (*Ficus elastica*); or any suitable species of house plant or potted woody dicot; or any leafy shrub such as lilac (*Syringa vulgaris*), privet (*Ligustrum* sp.), or forsythia (*Forsythia* sp.)

Activity 2

Plants bearing flowers for demonstration. Not to be used up. One plant per four students. Three-month-old bean plant (*Phaseolus vulgaris*); house plants such as *Oxalis* sp., *Fuchsia* sp., flowering maple (*Abutilon* sp.); or any available garden plants with simple flowers such as *Petunia* sp. or snapdragon (*Antirrhinum* sp.).

Activity 3

Plants bearing fruit or just samples of fruit for demonstration. Not to be used up. About one specimen per four students. Three-month-old bean plant (*Phaseolus vulgaris*); dwarf varieties of tomato (*Lycopersicon* sp.); ornamental pepper (*Capsicum annuum*); Christmas cherry (*Solanum pseudocapsicum*); or any field, garden, or landscape plant seasonably available or dried, such as tulip (*Tulipa* sp.), milkweed (*Asclepias* sp.), cherry (*Prunus* sp.), shepherd's purse (*Capsella bursa-pastoris*), oak (*Quercus* sp.), dogwood (*Cornus* sp.), mountain ash (*Sorbus* sp.), maple (*Acer* sp.), catalpa (*Catalpa*), elm (*Ulmus* sp.), tree of heaven (*Ailanthus altissima*), peanut (*Arachis hypogaea*), ragweed (*Ambrosia* sp.), velvet leaf (*Abutilon theophrasti*).

Activity 4

One- or two-week-old seedlings grown in glass or plastic jars, tumblers, and so forth. They should be old enough to show at least the first pair of true leaves. For demonstration. Not to be used up. Start them from seeds placed between the side of the jar and paper toweling. First line the jar with paper toweling. Stuff the center with crumpled toweling. Moisten the papers, then insert two or three seeds between the side of the jar and the paper lining. Keep moist and in a sunny place. Suggested plants: bean (*Phaseolus vulgaris*); cucumber (*Cucumis sativus*); tomato (*Lycopersicon* sp.). One plant per four students.

To reduce the possibility of fungus growth on the germinating seeds, use a solution of 9 parts water and 1 part household bleach for moistening the paper toweling. Use plain tap water thereafter for keeping the seeds and seedlings moist. See also Activity 2 of Exercise 2 in Instructor's Guide for recommendations for sterilizing seeds.

Activity 5

Monocot plants for demonstration of vegetative parts. Not to be used up. Some should be in flower for use in flower examination in Activity 6. About one plant per four students. Wandering Jew (*Zebrina pendula*); spiderwort (*Tradescantia virginiana*); purple heart (*Setcreasea pallida*); day-flower (*Commelina* sp.); *Philodendron* sp.; dumbcane (*Dieffenbachia* sp.); water hyacinth (*Eichhornia* sp.); any lawn or pasture grass; any seedling or mature plant of corn (*Zea mays*); or oats (*Avena sativa*).

Activity 6

Monocot plants with flowers. For demonstration. Not to be used up. One plant per four students. See list of suggested plants in the exercise under this activity.

Activity 7

Seedlings of monocots that are one to two weeks old (old enough to show at least one or two leaves). Grow these in glass or plastic tumblers or jars. Follow same directions as for dicot seedlings, in Activity 4. One plant per four students.

EXERCISE 2—SEED STRUCTURE AND GERMINATION

Activity 1

Purchase seeds of the pinto, kidney, or lima bean (*Phaseolus* sp.) by the pound at any local grocery store. Soak them in water overnight or 6–8 hours before class. Enough seeds for at least two per student for dissection.

Activity 2

Three or four beakers or plates containing dry seeds of bean, pea, or other dicots—just for display.

Another display of seeds that have been kept moist for several days and have imbibed water. These seeds can be prepared in several ways. Here are three suggestions. (a) Place some seeds on moist filter paper (or paper toweling) in petri dishes. Do not flood the dish. (b) Put seeds in a beaker or container with moistened vermiculite. (c) Make up a "rag doll"— this is a wet paper towel rolled into a cylinder and with the seeds wrapped in the layers of the roll. Tie each end of the rag doll with string or twist tape. Store in plastic bag or carton to keep moist.

Mold is always a problem. Treat seeds with a fungicide such as captan, thiram, botran, ferbam, zineb, benomyl. Put a little of the liquid fungicide solution in an 8 or 16 oz. margarine tub and swish seeds around in it. Paper toweling for rag doll can be wet with water that has 1 part commercial bleach for every 9 parts water. Fungicides are available from garden supply centers, seed stores, or any of the seed, nursery, or horticultural supply catalog firms listed in this Instructor's Guide.

Activity 3

This is a demonstration of the pressure exerted by the swelling of seeds that have imbibed water. Mix some plaster of paris with enough water to make a mixture the consistency of stiff pancake batter. Pour mixture into paper cups. Plant dry bean seeds in the mixture while it's still liquid. Allow plaster to set up hard. Tear off the paper cup. Let dry for 2–3 days. When needed, set the hardened plaster cup in a saucer of water. Seeds inside the plaster will imbibe the water and swell up. The pressure exerted by the swelling seed will crack open the plaster of paris cup. All this should be prepared several days before class. The cracked plaster of paris is put on display for the class.

Activity 4

Plant bean seeds in sand (preferably) or soil. Stagger your plantings so that three age categories of seedlings (3-day-old; 6- to 8-day-old; and 10-day-old) are available to the class. These "ages" are based on about 22–23°C germinating temperatures. Have enough seedlings for one of each age category per two students. Students will dig (lift) the seedlings from the growing medium. All seedlings will therefore be used up in each class. An easy way to "dig" the seedling is to spread apart the fingers, plunge them in the medium under the seedling, then lift the seedling out. Gently shake off excess soil or sand. Prepare enough seedlings for all classes.

Activity 5

Prepare seeds of pea (*Pisum sativum*) the same as described for bean in Activity 4. Prepare enough seedlings for all classes.

Activity 6

The seed coat of honey locust (*Gleditsia triacanthos*) is impermeable to water unless it has been fractured or partially eroded away after several months' exposure to the elements of weather. If you have seeds with intact coats, you can scarify (scratch or cut) the coat with a triangular metal file. Scarify until you see a small white hole in the coat. This job is time-consuming. It's best to get several files and let the students do it several days prior to the class. If you have a grinding wheel and use pliers to hold the seed against the wheel, the job can be speeded up. If seeds are not available locally, they can be purchased from Herbst Brothers Seedsmen Inc., 1000 N. Main St., Brewster, New York 10509. Soak the seeds on moist paper in petri dishes or in wet rag dolls for 2–3 days before class. Many other legumes exhibit this property, and locally available species may be substituted.

Activity 7

Soak grains of corn (*Zea mays*) overnight. In urban areas where whole, dry corn is not readily available, order untreated sweet corn seed by the pound from any supplier of vegetable seeds.

Activity 8

Prepare grains of corn (*Zea mays*). Follow the same directions as for bean in Activity 4.

EXERCISE 3—FACTORS THAT INFLUENCE SEED GERMINATION

Activity 1

Students will work in groups of four. Each group will need the following: (a) 100 seeds of one kind of plant; (b) one beaker to soak these seeds; (c) enough bleach type disinfectant for soaking the seeds. Make up 1 part commercial bleach (5 percent sodium hypochlorite) with 9 parts water; (d) two paper towels per student; (e) string or twist tape.

Suggested seeds: lettuce (*Lactuca sativa*); radish (*Raphanus sativa*); spinach (*Spinacia oleracea*); corn (*Zea mays*); bean (*Phaseolus vulgaris*); pot marigold (*Calendula officinalis*); celosia (*Celosia argentea* v. *cristata*); dahlia (*Dahlia pinnata*); nasturtium (*Tropaeolum majus*); or morning glory (*Ipomoea purpurea*).

Activity 2

Staining with tetrazolium dye is facilitated when seeds are softened in water beforehand. Place seeds (corn, bean, and so forth) on top of or between moist paper towels overnight. Alternatively, corn can be put in a beaker of water for 3–4 hours at about 30°C. Don't soak beans this way, however.

Seeds of small-seeded legumes and some other kinds do not need softening, namely, bean, pea, soybean, and so on.

The pericarp of grasses (corn, sorghum, and so on) is not permeable to tetrazolium, so the seed should be bisected with a razor blade before testing. Bean and soybean do not need to be sectioned. Use a 1.0 percent tetrazolium solution for legumes (beans, soybeans, and so on) and a 0.1 percent to 0.25 percent solution for grasses and cereals (corn) that are bisected.

Corn ideally requires ½ to 1 hour of staining at 35°C, but often results satisfactory for classroom accuracy are faster, even at room temperature (20°C). Soybean and bean ideally require 3–4 hours at 35°C, but results (for classroom purposes) are obtainable at less than 1 hour.

To prepare a 1.0 percent solution, dissolve 1 gram of tetrazolium powder in 100 ml distilled or tap water. To prepare a 0.1 percent solution, mix 1 part of 1.0 percent solution with 9 parts water or dissolve 1 gram tetrazolium powder in 1000 ml water.

Store the solutions in the dark or in an amber-colored bottle. Solutions may be kept for several months at room temperature. The solution used in a test should be discarded after each test.

The tetrazolium powder (2,3,5-triphenyl tetrazolium chloride, or TTC) is available from several chemical supply houses. Two such are Nutritional Biochemical Corporation, 26201 Miles Road, Cleveland, Ohio 44128, and Fisher Scientific Company, 1458 North Lamon Avenue, Chicago, Illinois 60651.

Activity 3

No materials required.

Activity 4

Seeds of honey locust (*Gleditsia triacanthos*) are needed—about 10–20 per student. Paper towels for each student to make a rag doll. Commercial bleach and fungicide as described in Activity 2 of Exercise 2 for wetting paper towels and sterilizing seeds. Twist tape or string. Many locally available legumes exhibit the same response to scarification and may be used.

Activity 5

Have students work in pairs. Supply about 10 seeds of corn (*Zea mays*) per student. Paper towels—one per student. Two small jars (at least one with a screw cap). Use baby or junior-food jars. Sterilize or have students sterilize the seeds with fungicide as described in Activity 2 of Exercise 2. Also, use dilute bleach water as described in Activity 2 of Exercise 2 for the water required in this germination test.

Activity 6

Enough seeds of any one type for each student to have 20–25 seeds. One paper towel per student to make rag doll tester. Twist tape. Seeds of peas, turnips, cabbage, cucumber, green pepper, eggplant, and sweet corn. Plastic bags for storing about 10 rag doll testers. Or use freezer cartons. Refrigerator and germination chamber (if possible).

EXERCISE 4—USE OF THE MICROSCOPE FOR STUDY OF PLANT CELLS

Activity 1

For each student, one compound microscope, with at least two objectives, 10X and 40X, and a 10X ocular.

Activity 2

Prepared slides of commercial fibers—cotton, silk, and so on. One slide per student, or per two students. Available from most biological supply houses. See list of these. May be called "Three Threads" or "Colored Threads."

Activity 3

Living elodea (*Anacharis canadensis*) plants—available from biological supply houses or locally from pet supply stores. One clean microscope slide, cover slip, and eye dropper per student.*

Activity 5

Onions—one or two of 5–8 cm diameter are sufficient for entire class and two to three bottles of IKI (iodine-potassium-iodide) solution. For each student*: 1 slide, 1 cover slip, 1 forceps.

Activity 6

Four or more microscopes with slides set up, as described in the exercise.

---

*Part of recommended student lab kit.

EXERCISE 5—THE ROOT AND THE APICAL MERISTEM

Activity 1

Collect plants for demonstration of different types of root systems. Not to be used up. Dandelion (*Taraxacum officinale*); any lawn grass or grassy weed; a mature (or at least 2-month-old) corn plant (*Zea mays*) and/or barley plant (*Hordeum vulgare*); turnip (*Brassica rapa*); beet (*Beta vulgaris*); carrot (*Daucus carota* v. *sativa*); bean (*Phaseolus vulgaris*); *Coleus blumei*; English ivy (*Hedera helix*); nasturtium (*Tropaeolum majus*); African marigold (*Tagetes erecta*); *Rhoeo* sp; wandering Jew (*Zebrina pendula*); *Kalanchoe* sp.

About 4–6 weeks before class, start cuttings in water, sand, or peat cubes, pellets, or blocks. The peat products are produced under various trade names, for example, Jiffy-7, Kys-Kube, Solo-Gro, Fertl-Cubes, and so forth. Most are available locally in garden supply centers, but can be purchased more cheaply in large quantities from horticultural supply houses and some biological supply houses. See list at end of Instructor's Guide.

Suggested plants for cuttings: geranium (*Pelargonium* sp.); *Coleus blumei*; *Kalanchoe* sp.; bloodleaf (*Iresine* sp); *Impatiens* sp.; *Peperomia* sp.; African violet (*Saintpaulia ionantha*); snake plant (*Sansevieria* sp.). Plants are for demonstration and not to be used up.

Activity 2

Radish seedlings about 4 days old germinating on moist filter paper in petri dishes. One dish per four students.

Additional material suggested but not absolutely necessary: other plants or stem cuttings rooted in peat cubes. In many ways these are to be preferred over the radish seedlings because of ease of handling by the students without disturbing the root hairs. Cuttings should be started a month before the class; seeds started 2–3 weeks before needed.

Activity 3

Water hyacinth (*Eichhornia crassipes*), water lettuce (*Pistia* sp.) to show root caps. Plants are for demonstration and not to be used up.

If green alder (*Alnus crispa*) is to be used, you will probably have to grow young seedlings from seed. One source of seed is Herbst Brothers Seedsmen, Inc., 1000 N. Main St., Brewster, New York 10509. *Alnus* seed must be stratified for 60 days in a moist medium at 1°–5°C to overcome dormancy (unless this has already been done by the seed supplier). Germination requires 30–40 days, preferably with 8-hour days of 30°C and 16-hour nights of 20°C.

Once you have these young plants available, they are well worth the work. The red root caps are striking.

Activity 4

Prepared slides of longitudinal section of root tip of onion (*Allium cepa*). One slide per student. Slides available from biological supply houses. See list. One compound microscope per student.

Activity 5

Uses same materials as in Activity 4.

Activity 6

Germinated seeds of bean (*Phaseolus vulgaris*) or pea (*Pisum sativum*), grown in sand or vermiculite, and whose primary roots are 3–4 cm long. Enough seeds for one to two per two students. Each pair of students to have available: paper toweling, 12 inches or so of thread, India ink, a metric ruler, a straight (common) pin, a small bottle, and a cork to fit the bottle. A small hole in the cork is recommended, although not absolutely necessary (see text).

Activity 7

Prepared slides of orchid roots showing mycorrhiza. Available from biological supply houses. One slide per one or two students. One compound microscope per student.

EXERCISE 6—INTERNAL STRUCTURE AND FUNCTION OF ROOTS

Activity 1

Prepared slide of cross section of young dicot root such as buttercup (*Ranunculus* sp.). One slide per student. Available from biological supply houses. Compound microscopes— one per student.

Activity 2

Prepared slide of young willow (*Salix* sp.) showing origin of branch root. One slide per student. Available from biological supply houses. Compound microscopes—one per student. Other species may be used, such as the water hyacinth (*Eichhornia*).

Activity 3

Prepared slide of a cross section of a monocot root. Slides available from biological supply houses are wheat (*Triticum* sp.); *Iris* sp.; corn (*Zea mays*); or greenbrier (*Smilax* sp.). One slide per student. Compound microscopes—one per student.

Activity 4

Prepared slide of cross section of a woody root, such as basswood (*Tilia* sp.) or tulip tree (*Liriodendron tulipifera*).

EXERCISE 7—HERBACEOUS STEMS

Activity 1

Representative dicot herbaceous plants for students to examine. Not to be used up. One plant per two to four students. In addition to *Coleus blumei* and *Plectranthus*, other suggested plants are geranium (*Pelargonium* sp.); any 1-month-old plant of bean (*Phaseolus vulgaris*), cucumber (*Cucumis sativus*), or squash (*Cucurbita* sp.), all of which you can start from seed.

Activity 2

Prepared slides of longitudinal section of stem tip of *Coleus blumei*. One slide per student. Available from biological supply houses. See list. Compound microscopes—one per student.

Activity 3

Monocot plants for examination (with some plants to be stripped of some leaves by students). Students are to see how the leaf and stem arrangement differs from that of dicots.

Suggested plants: young (preferably 1-month-old) corn plants (*Zea mays*), about one per student for dissection; also for dissection or demonstration any of the common monocot house plants, such as wandering Jew (*Zebrina pendula*); spiderwort (*Tradescantia* sp.); purple heart (*Setcreasea pallida*); *Rhoeo* sp.; dumbcane (*Dieffenbachia* sp.); or *Philodendron* sp.

Activity 4

No materials required.

Activity 5

For display purposes only, have a few dicot plants (preferably the same kinds as used in Activity 1) showing the effect of 2,4-D injury. Two to four days beforehand spray the plants with 2,4-D or related **phenoxy** herbicide such as Mecoprop or Banvel. These herbicides are widely available in garden supply centers or from horticultural supply houses. See list. Do not spray in the greenhouse or classroom. For this purpose, it is safer and just as effective to apply the herbicide with a paintbrush.

Activity 6

Prepared slides of cross section of stem of a dicot herb such as sunflower (*Helianthus* sp.) or geranium (*Pelargonium* sp.). One slide per student. Compound microscopes—one per student.

Activity 7

Samples of twine, rope, bags, or cloth made from jute, hemp, or linen fibers. Available from arts and crafts shops, hardware stores, or your grandmother's linen closet. Most macramé cord is jute fiber. For display only, not to be used up.

Activity 8

Prepared slides of cross section of stem of a monocot such as corn (*Zea mays*) or lily (*Lilium* sp.). One slide per student. Compound microscopes—one per student.

Activity 9

Samples of products made from sisal—for example, twine, rope, shopping bags, hammocks—or from Manila hemp (abaca)—for example, rope—or any textile products, like table mats, sandals, or cloth. For display only.

Activity 10

No materials required.

EXERCISE 8—WOODY STEMS I. INTRODUCTION TO WOODY STEM ANATOMY—THE GYMNOSPERM STEM

Activity 1

Prepared slides of macerated wood of pine (*Pinus* sp.) or some other conifer. One slide per student. Available from biological supply houses. Compound microscopes—one per student.

Activity 2

Prepared slides of cross section of 3-year-old (or older) stem of pine (*Pinus* sp.). One slide per student. Available from biological supply houses. Compound microscopes—one per student.

EXERCISE 9—WOODY STEMS II. ANATOMY OF THE ADVANCED WOODY STEM—THE ANGIOSPERM STEM

Activity 1

Prepared slides of macerated angiosperm wood. One per student. Available from biological supply houses. Oak (*Quercus* sp.) is most commonly used for this purpose, but buckeye (*Aesculus* sp.) and chestnut (*Castanea*) are also widely available. Compound microscopes—one per student.

Activity 2

Prepared slides of oak (*Quercus* sp.), basswood (*Tilia* sp.), or other hardwood showing three sections of cut: cross, radial, and tangential. One per student. Slides with three square sections of mature wood are preferable to those which use a 6–8 mm twig cross section to represent the cross section. Compound microscopes—one per student.

Activity 3

Prepared slides showing cross section of woody angiosperm stem. Many genera available from biological supply houses, for example, *Tilia, Liriodendron, Platanus,* and *Magnolia.* In this manual, Figure 9.4 and text description are based on *Tilia.*

Activity 4

No materials required.

EXERCISE 10—WOODY STEMS III—EXTERNAL AND INTERNAL FEATURES OF GROWTH

Activity 1

Collect leafless twigs (with at least 3–4 years' growth) of woody plants in mid winter when the buds are fully developed. Store these in some dry place and use for this exercise. One twig per student is adequate. If the class can have available more than one species, it would be preferable. Recommended plants: horse chestnut or Ohio buckeye (*Aesculus* sp.); walnut (*Juglans* sp.); tree of heaven (*Ailanthus altissima*); ash (*Fraxinus* sp.); and tulip tree (*Liriodendron tulipifera*).

Activity 2

In mid winter collect twigs of plants with thorns, spines, or prickles. Suggested plants: honey locust (*Gleditsia triacanthos*); hawthorn (*Crataegus* sp.); black locust (*Robinia pseudoacacia*); crabapple (*Malus* sp.); and rose (*Rosa* sp.).

Activity 3

Bring in fairly large branches for the students to practice pruning. Supply one pair of pruning shears per four students. These are widely available from hardware stores, department stores, garden supply shops, and horticultural supply houses.

Collect branches that show the pollard effect from poor pruning.

Have slides or photos to show poorly pruned trees.

Activity 4

Blocks of wood measuring about 5 × 5 to 10 × 10 cm and cut so they show radial, tangential, and cross section planes. One block per one to two students. Prepare these yourself or obtain from biological supply houses. Either gymnosperm or angiosperm wood is adequate. Suggested material: red cedar (*Juniperus* sp.); almost any pine (*Pinus* sp.); Douglas fir (*Pseudotsuga menziesii*); white oak (*Quercus alba*); hickory (*Carya* sp.).

If you can, obtain blocks or rounds of wood cut from larger limbs; these are helpful for display purposes.

Activity 5

The same material may be used as in Activity 4.

Activity 6

"Hough" slides are no longer available from biological supply houses. They are thin cut sections of wood mounted between glass. They are suggested here because some schools may still own a set. Carolina Biological Supply has available a modern substitute for the original Hough slides, which they term "Wood Sections." These are smaller, 5 × 5 cm cardboard mounts. Several species are available. In lieu of mounted slides, you could have available for the students (and your own reference) the following publications: (1) *What Wood Is That?* by Herbert L. Edlin, Viking Press, New York, 1969. This contains thin sections of wood glued to the pages. (2) *Wood Handbook* (1974), U.S. Department of Agriculture Handbook No. 72, Forest Products Laboratory. Available from the Superintendent of Documents, Washington, D.C. 20402. (3) *Textbook of Wood Technology,* Vol. I, by A. V. Panshin, Carl DeZeeuw, and H. P. Brown (3rd ed.), McGraw-Hill Book Co., New York, 1970. This has a key for wood.

Activity 7

Pieces of Douglas fir or pine plywood about 10 cm × 10 cm should be cut from larger sheets so as to have one piece per two students. Grades A-C, A-D, and CDX are all adequate. Veneered panels are interesting to the students, but be certain that genuine hardwoods are used, and not printed vinyl.

Activity 8

Requires wood cores, one pair per two students, who should work together. They are not available commercially at this time. Two techniques are recommended for preparation for class use.

Technique No. 1. This requires access to woodworking equipment able to saw 1.25 cm (½ in) thick boards out of 15–20 cm (7–10 in) diameter tree trunks, branches, or fence posts. Following Part A of Appendix Figure 1, cut two to six 1 to 1.25 cm thick boards out of each tree section, being sure that the centerline of each board points toward (or touches) the exact center of the stem. Long tree sections aren't necessary; a piece 30 cm (12 in) long is adequate and is more easily handled. Seasoned wood is also preferable to "green" (fresh) wood. The larger, flat faces of these boards are then sanded smooth. Finally, cut the boards into sections 1 to 1.5 cm (⅜ to ⅝ in) wide, as shown in Part B of Appendix Figure 1. These will become the wood "cores" or sticks used in Activity 8. For the comparisons required, call the "cores" made from one tree section "A" and make a second set "B" from a

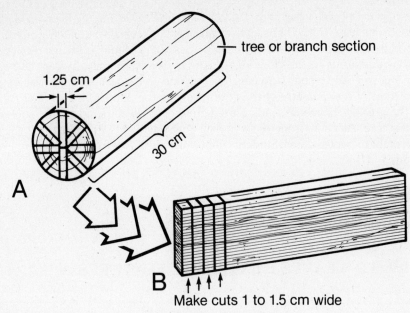

Appendix Figure 1. Technique No. 1.

different tree from the same general growing area. Native "wild" growth from uncultivated forests is more satisfactory than lawn trees, and higher-elevation locations show more seasonality than those from subtropical locations.

Technique No. 2.   True increment cores may also be used as follows. Choose two (or more) softwood trees growing under natural conditions in an environment where distinct seasonality may be expected. Extract a series of 20–25 cm (8–10 in) long cores with an increment borer—available from some biological supply houses or from Forestry Suppliers, PO Box 8397, Jackson, Mississippi 39204, or from Ben Meadows, 3589 Broad St., Atlanta (Chamblee), Georgia 30366. As each core is extracted, slip it into a large-diameter soda straw. If you can find paper straws, the ends can be simply folded shut. A small cotton plug may also be used, and will be required for plastic straws. Keep cores from different trees separate by marking the straws or by using different color straws. Allow the cores to dry without removing them from the straws for several weeks.

Prepare enough strips of 6 mm (¼ in) plywood to make one for each member of the class, following the pattern in Part A of Appendix Figure 2. The groove is cut by passing the strip over a circular saw with the blade tilted to 45° and set to cut only 3 mm (⅛ in) deep. Slide the dry core out into the groove. If some rings have separated, slide them together

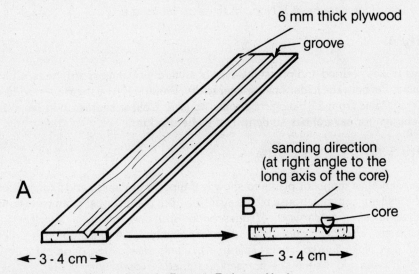

Appendix Figure 2. Technique No. 2.

gently with your fingers. Glue with Duco Household Cement (TM) or an equivalent non-water-soluble glue. Do NOT use white, casein glue, because it contains water that will be absorbed by the core. Using a straight pin or dissecting needle, push any displaced sections close together before the glue dries. Allow to dry thoroughly, at least 48 hours.

Using a power sanding belt with very fine grit, and on a slow speed, if possible, sand off about one third of the diameter of the core following the pattern in Part B of Appendix Figure 2. Rings will be blurred if sanding is lengthwise, so always sand at right angles to the length of the core. All cores from one tree may be labeled so that the instructor is able to keep track of the expected dates. Students then work with cores from different trees in their comparisons.

Activity 9

Uses the same materials as Activity 8.

EXERCISE 11—LEAVES I: EXTERNAL FEATURES

Activity 1

Twigs or branches of almost any conifer such as juniper (*Juniperus* sp.); pine (*Pinus* sp.); spruce (*Picea* sp.); fir (*Abies* sp.); hemlock (*Tsuga* sp.); or any small-leaved plant native to dry habitats such as sagebrush (*Artemisia* sp.) or creosote bush (*Larrea divaricata*). Enough specimens for each student or pair of students. Not to be used up. Try to have a variety of specimens.

Activity 2

Specimens of cacti; or spurges such as crown of thorns (*Euphorbia splendens*) or redbird cactus (*Pedilanthus tithymaloides*), both common house plants. Also any succulents (the stonecrop family) commonly available in plant shops, viz., species of *Sedum: Escheveria; Crassula; Kalanchoe;* and *Haworthia.* Not to be used up.

Activity 3

These specimens are also all common house plants and can be bought or rented (for nominal amount) from plant shops. Obtain rubber plant (*Ficus elastica*); fiddle-leaf fig (*Ficus lyrata*); dumbcane (*Dieffenbachia* sp.); caladium (*Caladium* sp.); rex begonia (*Begonia rex*); banana (*Musa* sp.) (a leaf of this would suffice rather than the entire big plant); staghorn fern (*Platycerium* sp.). None of these to be used up.

Activity 4

Have boxes of dried and pressed leaves of a variety of angiosperm trees. Collect the leaves, place flat between folds of newspaper to dry. Preferably, dry them in a standard plant press as available from all biological supply houses. Collect simple and compound leaf types—enough for several per student. Not to be used up.

Activity 5

Representative monocot plants to show leaf types. For display and examination only. Many of these are common house plants available from plant shops. Examples: dumbcane (*Dieffenbachia* sp.); spiderwort (*Tradescantia* sp.); wandering Jew (*Zebrina* sp.); *Sansevieria* sp.; *Dracaena* sp.; *Rhoeo discolor*; pineapple (*Ananas comosus*); gladiolus (*Gladiolus* sp.); lily (*Lilium* sp.); tulip (*Tulipa* sp.); or any grass, sedge, or orchid (*Orchis* sp.). Leaves of grasses and iris may be pressed and dried as for Activity 4.

EXERCISE 12—LEAVES II: INTERNAL STRUCTURE

Activity 1

Prepared slide of cross section of dicot leaf such as lilac (*Syringa* sp.) or privet (*Ligustrum* sp.). One per student. Compound microscopes—one per student.

Activity 2

Leaves of plants for students to peel off the epidermis. Suggested plants: wandering Jew (*Zebrina* sp.); any *Sedum* species; German ivy (*Senecio mikanioides*); or corn (*Zea mays*). Razor blades,* microscope slides* and cover slips,* forceps,* pipette.* Enough living plants are needed to supply one living leaf for every one or two students. These will be used up. Compound microscopes—one per student.

Activity 3

Any thin, clear plastic ruler with millimeter markings* can be used to measure the field diameter. More accurate measurement can be made if a stage micrometer slide is used. Ocular micrometer discs are not suitable for this purpose. If the microscopes are all of one type, it will save time if the instructor has measured the field diameter accurately and posted it in the laboratory.

Activity 4

Zebrina plants for display. One plant (or set of plants) that's been grown in a well lighted place. Another plant (or set of plants) that's been grown for a month or more in a shadier place. The new leaves (if formed after the plant was placed in the shady place) will be the ones to examine. One set of plants per class; not to be used up. Prepared slides; cross sections of sun and shade leaves of any woody plant. One slide per one or two students. Compound microscopes—one per student.

Activity 5

Prepared slides of cross section of corn leaf (*Zea mays*). One slide per one to two students. Available from biological supply houses. Compound microscopes—one per student.

Activity 6

Prepared slides of cross section of leaf (needle) of pine (*Pinus* sp.). One slide per one to two students. Available from biological supply houses. Compound microscopes—one per student.

Activity 7

For display only. Living plants of rubber plant (*Ficus elastica*); fiddle-leaf fig (*Ficus lyrata*); and *Philodendron*—any of the large-leaved varieties. These are all common house plants. Optional: prepared slides of cross section of leaf of rubber plant or fig. One per two to four students, or for demonstration only.

---

*Part of recommended student lab kit.

EXERCISE 13—FLOWERS

Activity 1

This activity is much more interesting with living flowers, but 35-mm slides may partially substitute. Good examples to use are *Petunia* sp.; *Rhododendron* sp.; *Hibiscus* sp.; snapdragon (*Antirrhinum* sp.); daylily (*Hemerocallis* sp.); *Clivia* sp.; most orchids; larkspur (*Delphinium* sp.); *Fuchsia* sp.; *Nicotiana* sp.; and many honeysuckles (*Lonicera* sp.), especially *Lonicera tartarica*. Some examples of each of these general types are available from florists year round. If sufficient numbers are available that the students may dissect a flower and examine the location of anther, stigma, and nectary, they will see the relationship of flower structure to pollinator specificity much more clearly.

Activity 2

Grasses, including grains such as wheat and barley, may be pressed in the manner of a herbarium specimen—though there is no need to mount them—and used for this activity. Fresh specimens may also be used, either collected or grown in the greenhouse. Allow at least 15 weeks for flowering for most grain crops.

Activity 3

Pollen of abundant producers, such as *Zea mays*, ragweed (*Ambrosia* sp.), *Plantago* sp., and others may be collected by collecting flowers, in season, and drying in a closed paper bag. The bag will then contain the dried pollen for class use. A few types of pollen are available from biological suppliers as prepared slides. In general, any plant may be used, but check before assigning it to a class, as a number of commonly available florist's crops have misshapen, abortive pollen.

Activity 4

Simple, single, regular flowers—to be used up. One per one to two students. Suggested: petunia (*Petunia hybrida*); tulip (*Tulipa* sp.); gladiolus (*Gladiolus* sp.); lily (*Lilium* sp.); primrose (*Primula* sp.); poppy (*Papaver* sp. or *Eschscholtzia*); *Browallia* sp.; *Amaryllis* sp.; *Hoya* sp.; *Clivia* sp.; *Azalea* sp.; or *Magnolia* sp. Many of the flowers used in Activity 1 may be reused here.

Activity 5

Flowers of composites—to be used up. Dandelion (*Taraxacum officinale*); aster (*Aster* sp. or *Cineraria* sp.); daisy (*Chrysanthemum leucanthemum*); sunflower (*Helianthus* sp.). Available from fields if in season and some from florists. Specimens to be used up. One head per four students.

Activity 6

Inflorescence of oats (*Avena sativa*). One floret per student. May or may not be used up, depending on whether or not the students dissect it. Allow about 15–20 weeks to flower from seed if you are growing your own. Bluegrass (*Poa* sp.) is also suitable, but is much smaller. Many species of *Bromus*, especially *Bromus inermis*, are almost as usable as oats.

EXERCISE 14—FRUITS

Activity 1

This demonstration requires a series of four to six preserved specimens showing the stages of development of a fruit from pollination (the ovary) to maturation. A few are occasionally available from biological suppliers, often called the "life history" of the plant. Some of these series are embedded in plastic for permanent display.

Many examples, such as *Malus, Fragaria, Pisum, Phaseolus,* and *Capsicum,* are easy to collect in quantity at the proper season of the year and preserve for class use. The usual preservative is FAA (48 percent ethanol; 2 percent glacial acetic acid; 10 percent formalin; 40 percent water), though materials may be transferred to 50 percent alcohol after a few days in the FAA.

Activity 2

Many examples of each of these fruit types may be purchased locally at the grocery store. Some, such as the grains, should be collected from the wild or from croplands, and preserved for class use. As for Activity 1, a few examples are available from biological supply houses either preserved in formalin or embedded in clear plastic blocks. These latter are very expensive, but may be the only way to obtain certain examples out of season. If fresh materials are to be dissected, allow one per six to eight students.

Activity 3

All species suggested are common weeds or native trees, and local collection is normally used. Some biological supply houses do offer displays or sets showing these and other species under the heading of "seed dispersal" or a similar title.

EXERCISE 15—WATER RELATIONS: OSMOSIS AND DIFFUSION

Activity 1

Either prepare before class or have the students prepare cellulose acetate dialysis tubing cut so as to make up 10–15 cm (4–6 in) long tubes ("sausages") tied at both ends with a string, and filled with 10 percent sugar solution dyed with a tiny amount of eosin or methylene blue (0.005 percent). The size called "1 inch flat width" is best for this work. Available from biological supply houses.

Prepare enough of a 20 percent sugar solution to partly fill beakers into which the "sausages" will be put.

Need two beakers per each group of students. Students work in groups of one to four. One beaker has tap water. One has the 20 percent sugar solution. One triple-beam balance (scales) per 10 students. Paper toweling to dry "sausages."

Activity 2

Potatoes, turnips, or rutabagas—one of any plant 5–8 cm (2–3 in) in diameter per four students; will be used up.

"French-fry" cutters, one per 20 students, or cork borers, 6–10 mm diameter, about one per four students.

One triple-beam balance (scales) per 10 students. Two beakers (100–250 ml or so) per pair of students. Tap water in one, strong (10 percent) salt (NaCl) solution in the other— enough to cover "plug" with liquid.

Activity 3

Set up the apparatus as shown in Figure 15.1. More elaborate "Osmometer Demonstration" apparatus may be purchased especially for this purpose. Fill one "sausage" with a glucose solution and the other with a starch solution. Weigh and record each sausage, then insert the glass tubing into each and insert in tap water in the beakers as shown in Figure 15.1

Two test tubes
Bottle of IKI (iodine-potassium-iodide)
Bottle of Benedict's solution—allow about 10 ml per class for demonstrations. Biological supply houses have dry concentrate or solution for sale.
Roll of Tes-Tape, a glucose test paper available from biological supply house or local pharmacy
Bunsen burner
Tripod
Asbestos mats or asbestos wire gauze mats
500 ml beaker
Tap water

Activity 4

Leaves of plants such as elodea (*Anacharis canadensis*); *Rhoeo* sp.; or *Zebrina* sp; one per one or two students. Will be used up.

One microscope slide* and cover glass* per student. Small dropping bottle with 5 percent salt (NaCl) solution—one per four to eight students. Paper toweling. Compound microscopes—one per student.

Activity 5

Take stem cuttings of the following plants and place them in water of different salt concentrations, for example, no salt, 0.5, and 3.0 percent. Or use some garden or lawn fertilizer and make up different concentrations in water. Since most fertilizers, as sold, are partly insoluble, use about 3.0 percent and 6.0 percent fertilizer by weight. Use test tubes for setting the cuttings in the solutions. Set in test tube racks or blocks.

Suggested plants: almost any non-woody plant, such as geranium (*Pelargonium* sp.); tomato (*Lycopersicon* sp.); coleus (*Coleus blumei*); and so on.

Activity 6

Bottle of crystals of potassium permanganate ($KMnO_4$)
Petri dishes—one per two to four students
Non-nutrient agar or water in petri dishes
Dishes with tap water (if petri dishes are unavailable)
Forceps*—one per two to four students

Activity 7

Two to three hours before class, put stalks of celery (*Apium graveolens*), or stems of white flowered carnation (*Dianthus* sp.) or *Chrysanthemum* sp. in beakers containing solutions of colored food dyes or methylene blue. Make sharp, clean cuts just before placing the stem into the beaker, so the dye will easily enter the plant.

Activity 8

Two glass tubes each about 60–70 cm (24–28 in) long and 5 cm (2 in) in diameter
Four rubber or cork stoppers to fit these tubes
Cotton
Dropper bottle with 1 N ammonium hydroxide (NH_4OH)
Dropper bottle with 1 N hydrochloric acid (HCl)
Watch with a second hand

---

*Part of recommended student lab kit.

Activity 9

Set up demonstration overnight or at least 4–5 hours before class.

Three bell jars per classroom
Three glass plates to set jars on
Plant in pot with soil covered with plastic wrap or aluminum foil to seal escape of moisture from it. Or use plant growing in a peat cube and wrap the cube in plastic wrap or aluminum foil to prevent escape of moisture
Wet sponge

Activity 10

For each group of (2–4) students: assemble materials needed for apparatus shown in Figure 15.3.

Shallow pan
A micropipette (capacity 0.1 ml or 0.5 ml, graduated in 0.01 ml is ideal) or a section of glass capillary tube about 15 cm long
Glass or plastic elbow
Rubber tubing to fit elbow to pipette or tube
Rubber tubing to fit elbow to plant stem
Razor blade*
Metric ruler 15 cm,* reading in mm, if capillary tubes are used
Plants for taking leafy stem sections, geranium (*Pelargonium hortorum*); tomato (*Lycopersicon* sp.); or potato (*Solanum tuberosum*)

Activity 11

Set up plants under bell jars and water the soil heavily so as to induce guttation. Suggested plants: two-week-old seedlings of barley (*Hordeum vulgare*); wheat (*Triticum* sp.); tomato (*Lycopersicon* sp.); or almost any soft-tissued young plant 10–15 cm (4–6 in) tall.

Activity 12

Dry woody material and seeds to be put in water by the students to observe imbibition of water. Use large seeds, such as beans, mature cones of any pine, spruce, hemlock, and so on, or pieces of wood veneer.

EXERCISE 16—PHOTOSYNTHESIS

Activities 1 and 2

2 test tubes per student
2 strips of filter paper per student to fit lengthwise in test tubes
1–2 bottles of water-soluble ink for class. Any brand name non-permanent black writing ink is satisfactory, but office supply businesses are probably the only common source today. Ink may be applied with toothpicks; pens are not necessary
Test tube racks or blocks
Mortar and pestle or electric blender
Fresh leaves (use spinach or geranium leaves if outdoor plant material is not available)
Enough ether-acetone solvent to grind up the leaves and have a sufficient volume for class use (few ml per student). Make up solvent with 95 parts ether and 5 parts acetone. Distribute leaf extract in dropper bottles—one per 6–12 students

---

*Part of recommended student lab kit.

Activity 3

Enclosed-element electric hot plates—one per four to eight students
600 ml glass beakers—one per four to eight students
90–95 percent alcohol (either ethanol or methanol)—enough for about 100–200 ml for four to eight students
Small beaker (100 or 250 ml)—one per two students
Bottle of IKI (iodine-potassium-iodide)—one per four to eight students
Petri dish—one per two students
Plant material to supply each one to two students with one leaf. Leaves must be variegated. Suggested: *Coleus blumei*; Algerian ivy (*Hedera canariensis* v. *variegata*); geranium (*Pelargonium hortorum* v. *marginatum*).

Note: Variegated leaves *must* first be boiled in plain water to break down cell walls.

Activity 4

Demonstration spectroscope—one per class, or per group of four to eight students. Available from biological and physical science supply houses.
Bottle or test tube of leaf pigments in solution—one per class, or per group, according to availability of the spectroscopes. Use the same solution prepared for Activity 2, diluted to allow passage of more light.
Optional—If light source is very strong, e.g., full sunlight, use a thin, green leaf from *Senecio mikanioides*; one per group.

Activity 5

Cut 4 cm^2 (2 cm × 2 cm) squares of heavy black paper and attach (by using "sprung" or weakened paper clips) to leaves of plants of geranium (*Pelargonium* sp.) or bean (*Phaseolus vulgaris*). Often, a 2 × 4 cm strip folded gently around the leaf margin so as to cover both surfaces works best to demonstrate this point. Store plants in darkness for 72 hours. Then, expose plants to full sunlight for 8–12 hours just before class, without removing the black paper light shields. Enough plants to supply one leaf per one to two students.

Enclosed-element electric plate—one per two to four students (just as used in Activity 3)
Petri dish—one per two students
Bottle of IKI—one per eight students
90–95 percent alcohol—enough for about 50 ml per two students

Activity 6

Boil some tap water (enough for three test tubes) an hour or more before class. Let it cool in a covered pan and pour into the tubes. Label tubes 1, 2, and 3, and have available for class.
One dropping bottle with phenol red indicator
A soda straw or comparable glass or rigid plastic tubing
A sprig or two of elodea (*Anacharis canadensis*) or parrot's-feather (*Myriophyllum brasiliense*). These are usually available through aquarium suppliers or biological supply houses.
Some lighting to supply bright light to illuminate the test tubes

Activity 7

Set up two beakers with plants, funnels, and tap water as shown in Figure 16.4; one for demonstration. Add several drops of 1 percent sodium bicarbonate (NaHCO$_3$) to the water in the beaker or blow your breath into it with a straw or glass tube. Use same kind of plant material as in Activity 6. Wood splinters (available from supply houses).

Activity 8

White potatoes—one piece about $2 \times 4 \times 4$ cm per four to eight students—will be used up.
Scalpel—one per four students; a paring knife will be adequate.

Microscope slide* and cover slip*—one per student. Compound microscopes—one per student.
Dropping bottle of IKI—one per four to eight students
Sample batches of flour made from rice (*Oryza sativa*); oats (*Avena sativa*); sweet potato (*Ipomoea batatas*); soybean (*Glycine max*); and so on. One sample (about 250 g) of any one of the above is sufficient for entire class. "Health food" stores often carry these unusual flours.

EXERCISE 17—DIGESTION AND RESPIRATION

Activity 1

Corn grains that have been kept wet for 48 hours at about 20-22°C with good aeration.

Cut in half lengthwise on the flat side so as to produce right and left halves. These germinating grains are placed, cut face down, on the surface of a fresh sterile plate of starch agar, poured about 2-3 mm deep in a 15 cm petri dish. Let the plates sit, covered, in the dark for about 36-48 hours before class. Needed as a demonstration, one plate per class.
Starch agar may be purchased as a dry concentrate from biological supply houses, or may be made up by adding about 1 g soluble starch (or ordinary cornstarch) to each 100 ml of water in a standard 1.5 percent *non*-nutrient agar recipe.

Bottle of dilute IKI for each class—about 5 ml will be used in each demonstration

Activity 2

About 250 g of wheat should be germinated for each class. This is best done by placing about 1 kg of grain in a 4 or 5 liter (or 1 gallon) jar and covering with water for 6-8 hours. Then pour off the water while shaking the jar vigorously. Half fill with more lukewarm tap water and shake again. Pour off all liquid water, and let jar stand either upright or on its side with the cap very loosely in place. Repeat the washing and shaking every day for about 4-5 days before using as described. "Health food" stores may sell whole grains as "wheat berries."
An electric blender is needed for each class. The types that can use household canning jars are preferable, as a supply of clean jars can be made ready each day.
A 500 ml flask or bottle is needed for each class to hold the extract after filtering
A large (100-200 mm) funnel with a cheesecloth pad for filtering the extract
Bottle of Benedict's solution—about 10 ml will be used in the demonstrations
Hot water bath on hot plate or bunsen burner for Benedict's tests
Bottle of IKI solution—about 15 ml will be used in the demonstrations

Activity 3

One 125 ml Erlenmeyer flask for each working group of two to four students
About 50 ml of 1 percent starch solution for each two to four student group
At least 2 test (or culture) tubes for each student group. Sizes 10×75 mm to 15×125 mm are satisfactory
Dropper bottles of IKI—one 30 ml bottle for each student group
Dropper bottles of Benedict's solution—one 30 ml bottle for each student group

---

*Part of recommended student lab kit.

Hot water bath on hot plate or bunsen burner for each four-student group
Test tube rack for each two-student group

Activity 4

This may be either a demonstration or may be performed by a group of two to eight students if time permits.

5 test (or culture) tubes per group—any size from 10 × 75 through 15 × 125 will work
Dropper bottles of 1 percent starch solution—one per group
Dropper bottles of amylase (available from ICN Pharmaceutical, Inc., or from biological suppliers. The amylase derived from bacteria is satisfactory and by far the least expensive)
Hot water bath and test tube rack as for Activity 3

Activity 5

This demonstration is started by soaking about 50–100 g of dry seeds such as corn or beans for about 12 hours. The imbibed seeds are then tied loosely in cheesecloth and suspended in a tightly sealed jar. A 250–400 ml wide-mouth specimen bottle is useful, and the cheesecloth bag is usually suspended over (but not touching) about 10 mm of water simply by catching the string tying the bag under the bottle cap or stopper. One of the jars should contain a smaller jar or several open vials filled to the 20 mm level with a 5 percent solution of potassium pyrogallate or pyrogallol. Do not open the jars after setup. Allow 4–6 days to show suppression of germination.

Activity 6

Use any easy-to-root species, such as coleus (*Coleus blumei*); tomato (*Lycopersicon* sp.); wax begonia (*Begonia semperflorens*); geranium (*Pelargonium hortorum*); wandering Jew (*Zebrina pendula*); or purple heart (*Setcreasea purpurea*). If the season of the year is suitable, that is, late winter through late spring, many species of willow (*Salix*) or poplar (*Populus*) show this effect markedly. Use any tall, narrow container such as a graduated cylinder (500 or 1000 ml) or a 32 × 300 mm culture tube. Any small vibrator-type aquarium aerator pump will work, so long as the tube is kept at the bottom of the rooting container.

Activity 7

Set up the demonstration apparatus as shown in Figure 17.4. The bottles may be any type of specimen bottles from 200 to 500 ml (or 6 to 10 oz) which may be tightly stoppered. The sizes of the stoppers and tubing illustrated should be altered to match the bottles utilized. Several extra bottles numbered 3 and 4 should be available for the lots of seeds held at the three different temperatures. It may be desirable to have different sets of seeds at each temperature if different class sections are close together and time for evolution of CO_2 is short.

Common wheat, barley, beans, or corn (maize) are preferable for this demonstration, as their temperature sensitivity is clear. Their general treatment should be as outlined for Activity 2.

The illustration shows a small rubber squeeze-bulb positive displacement hand pump for air transfer. With this pump, squeezes may be counted and gas flow standardized easily. Use 10, 15, or 20 squeezes each time, for example. Almost any hand-operated tire or air-mattress pump could be adapted to this system. If none of these is available, an aquarium aerator pump might be used, with its time of operation measured exactly. In this last case, a time of about 15–30 seconds is about right, depending on the volume of the pump.

The NaOH or KOH in Bottle 1 is about 7 normal, and should be adequate for an entire day's classes. Add more fresh solution if any precipitate begins to collect in Bottle 2. The $Ba(OH)_2$ in Bottles 2 and 4 is about 0.5 normal, and several liters should be prepared at once, in advance. About 300–500 ml will be used for each class, depending on the size of the bottles used.

A funnel (about 75 mm) is needed for each of the three sample temperatures used. Circles of medium-coarse filter paper are needed to remove the $BaCO_3$ precipitate from each Bottle 4. Number the paper in pencil *before* filtering with the temperature (5°, 20°, 30°) of the germinating seeds tested. Weigh another circle of filter paper to determine the weight for subtracting from the weight of dried precipitate plus paper.

A drying oven set at 80°C and a good balance are needed. One oven for all classes is adequate. If an oven is not available, the wet filter papers and precipitate can be dried directly on a covered-element (smooth surface) hot plate set to about 50°C. If a sensitive balance for each class is not available, a student could be sent to a central prep room where an assistant is stationed to weigh the precipitate on a torsion or electronic balance.

Activity 8

Prepare about 2 hours before each class period enough yeast suspension to allow about 20 ml per student. To make the suspension, mix one cake of compressed yeast or two packages of dry yeast with one liter of 25°C water and about 50 g of sucrose (or glucose). After 1–2 hours, bubbles of CO_2 should be evident. After 4–6 hours, the sugar will be used up. More may be added, or a new culture started.

> Fermentation tubes, one for every one or two students. The type graduated in ml is preferable, but the ungraduated type may be used if a 15 cm metric ruler* is available
> Dropper bottles of phenol red, one per every four to eight students
> Foam plastic ice bucket with crushed ice or ice water, capacity approximately 2 liters (½ gal). One per 8–12 students
> Access to a refrigerator

EXERCISE 18—PLANT MOVEMENT AND GROWTH RESPONSES TO STIMULI

Activity 1

> Pots of seedlings of oats (*Avena sativa*) or barley (*Hordeum vulgare*) containing 4–8 seedlings each. Use 1½–2 inch plastic or clay pots or compressed organic media such as Jiffy-7, Kys-Kube, Fertl-Cube, or Solo-Gro, all of which are available through horticultural supply houses. The seeds should be started in the dark 5–10 days before they are needed. Some experimentation may be needed to determine the exact time span desirable under your temperature regimen. A coleoptile length of 2–4 cm is good, but seedlings are too old if the first leaf breaks through the coleoptile. One pot will be used by every group of two to four students
> Boxes of ordinary toothpicks. Each student will use three or four
> Small jars of 1:10,000 indole acetic acid (IAA) mixed with lanolin. IAA is available from biological supply houses. One 50–100 g jar or cup per four to eight students
> A DARK box or cabinet for in-class coleoptile development. It is a help if the temperature is at least 25°C but not more than about 32°C inside the box

Activity 2

A demonstration should be prepared about 36–96 hours before class of several susceptible dicot plants such as tomato (*Lycopersicon* sp.), garden bean (*Phaseolus* sp.),

---

*Part of recommended student lab kit.

dandelion (*Taraxacum officinale*), or jimson weed (*Datura* sp.). Any lawn or field weed could also be used by transplanting them to pots either before or after treatment with the herbicide. Use any phenoxy herbicide such as 2, 4-D, Mecoprop, or Banvel mixed according to label directions. Wet the foliage of the test plants, but do not soak them as if watering the plant. Do not spray in the greenhouse, or even return the sprayed pots to the greenhouse at any time, as many of the other greenhouse plants are very susceptible to residual fumes of the herbicide. Applications by paintbrush are acceptable and safer than sprays.

Activity 3

The radish seedlings should be planted about 4–8 per 1½–2 inch pot or compressed organic cube as were the seedlings for Activity 1, above. The pots should be started about 4–7 days before needed, and should be kept the entire time in their assigned box, with monochromatic light if specified. Only two monochromatic filters are needed, corresponding to wavelengths in the blue (450 nm) and red (650 nm) portions of the spectrum. They are available from at least one biological supply house. Any small desk lamp will provide adequate light. If an incandescent bulb is used, check 3–4 hours after setup to see that temperature does not exceed 30–31°C in the growth area.

Activity 4

This demonstration calls for two 8–10 cm (3–4 in) pots planted with any fast-growing herbaceous plants. Seedlings of oats (*Avena sativa*), barley (*Hordeum vulgare*), radish (*Raphanus sativus*), and cucumber (*Cucumis sativus*) about 3 to 6 cm tall are the most successfully used. Potted geraniums (*Pelargonium hortorum*) or *Coleus blumei* may also be used, but will require 5–10 days to show effects. The seedlings will demonstrate phototropism in 36–48 hours if placed near any unidirectional source of light.

The clinostat, itself, is available as an item of scientific apparatus designed especially for this experiment. It revolves at a rate of 2 revolutions per hour (rph). Any clockwork or motor-driven turntable may be used, and any speed of rotation from 1 revolution per minute (rpm) to 1 rph is satisfactory. Speeds much faster than 2–4 rpm lead to centrifugal effects in auxin distribution; hence, unmodified record turntables (33 rpm) are not satisfactory.

Activity 5

This demonstration requires a mature (10–15 day old) *asexual* culture of *Phycomyces blakesleeanus* growing on potato-dextrose agar in the bottom of a 500–1000 ml Erlenmeyer flask or any bottle of similar size. The culture has been allowed to develop with the flask or bottle covered with aluminum foil or black construction paper to exclude light. About 72 hours before class, or when asexual sporangiophores have begun to develop, open a small (2–3 cm) "window" in the covering about one half or two thirds of the way up the side of the flask. Place the culture so that moderate light may enter through the opening. No especially strong spot lights are needed, and a north window is ideal as to intensity. Living cultures are available from biological suppliers, but be sure to use only "+" or "−" to start this demonstration, and not *both* strains.

Activity 6

This demonstration requires placing imbibed corn grains on the bottom of a petri dish, embryo side down to the glass. Four to six grains are adequate per dish. Cover the grains with a circle of heavy filter paper; then add more filter paper or a crumpled paper towel to fill the dish a little beyond the top rim. Now place the cover on the petri plate and secure with one or two heavy rubber bands. The seeds are then allowed to germinate as the plate is held vertically, either in a special rack or by embedding one edge in a lump of modeling clay. The paper towel is kept moist, but not dripping wet. A solution of 10 percent household bleach may be used for the first wetting of the paper towels. Allow 4–5 days of development before required for class.

Activity 7

Any healthy, fast-growing potted plant may be used for this demonstration. As suggested in the Activity, specimens growing in compressed organic media such as Jiffy-7, Kys-Kubes, and so on show exposed root tips more easily. If growing a plant on its side for several days, provide an aluminum foil cover for the root mass to retain moisture and exclude light except when being examined by the class.

Activity 8

This activity requires that several healthy potted *Mimosa pudica* plants be available in each classroom for observation of the sensitive reaction. They require about 3–6 months to attain a suitable size from seed, but may be propagated by cuttings after this. Seed is available from major seed houses. If the plants are healthy and well-watered, they will "recover" from the sensitive reaction in 15–30 minutes.

Insectivorous plants may be purchased from major biological supply houses and, occasionally, through sellers of house plants. All types are best kept in a bog terrarium of at least 35 liter (10 gallon) size and maintained according to the instructions supplied with the plants.

The demonstration slides of grass leaf cross sections are available from all major biological supply houses. Some supply corn (*Zea mays*) for this purpose; some stock only *Poa*; and some supply an unnamed xerophytic grass. Any of these is satisfactory. One demonstration set-up, consisting of two paired microscopes, should be available for every 8–10 students.

EXERCISE 19—BACTERIA AND BLUE-GREEN ALGAE

Activity 1

Prepared slides of bacteria illustrating the three cell shapes. One slide per student. Available from biological supply houses; called "Bacteria Types" or "Bacteria-3 Types."
Compound microscopes—one per student

Activity 2

Plants of soybeans (*Gycine max*); beans (*Phaseolus vulgaris*); or other legume, at least 2–3 months old. You can grow these in sand (preferably) or soil and pull them up to show the roots to the class. Class will examine the nodules on the roots. Fertilizing pots with nitrogen fertilizer may suppress nodule development
Microscope slide* and cover glass.* One per student
Compound microscopes—one per student

Activity 3

All major biological suppliers have cultures of *Gloeocapsa* and prepared slides, as well. It is also a common moist soil alga, and you may be able to collect your own from rocks, tree bark, flower pots, bricks, and greenhouse walks.

One microscope slide* and cover slip* per student (or one prepared slide per student)
One compound microscope per student
One dropper bottle of methylene blue stain per four to six students
Small pieces (1 × 2 cm) of paper towel—one per student; cut them before class from any institutional brown paper toweling.

---

*Part of recommended student lab kit.

Activity 4

Oscillatoria is available from biological suppliers as living cultures, preserved cultures, or prepared slides. It is also one of the most common naturally occurring algae of stagnant water.

One microscope slide* and cover slip* per student (or one prepared slide per student)
One compound microscope per student

Activity 5

Anabaena is commercially available from supply houses as living material, preserved material, or prepared slides. It is also a common aquatic alga in more or less permanent ponds. If present in a pond at all, it is likely to be very abundant.

One microscope slide* and cover slip* per student (or one prepared slide per student)
One compound microscope per student.

Activity 6

Nostoc is available from biological suppliers as living cultures, preserved cultures, or prepared slides. It may also be found in colonies in all types of fresh water as well as on moist soil and in moist moss mats.

One microscope slide* and cover slip* per student (or one prepared slide per student)
One compound microscope per student

Note: If prepared slides of *Oscillatoria, Anabaena,* and *Nostoc* are purchased for class use, it is necessary to have only enough of each genus for one third of the expected students. The class may then work simultaneously on all three types, with the students sharing slides.

EXERCISE 20—FUNGI I: PRIMITIVE FUNGI AND SLIME MOLDS

Activity 1

Water cultures of living *Saprolegnia* or *Achlya* available from biological supply houses; best cultured on cracked hemp seeds. Nonviable hemp seeds are also available from supply houses, either separately or as part of a "water mold culture kit"
Compound microscopes—one per student
Microscope slide* and cover glass.* One per student

Activity 2

Living culture of bread mold (*Rhizopus stolonifera*), available from biological supply houses
For best results, culture the mold as follows: Make up potato-dextrose agar, autoclave it, and then pour into sterile petri dishes. Cut squares or rounds of cellulose acetate paper to fit the petri dish, and lay on top of the solidified agar. Inoculate the agar with the *Rhizopus* organism. Store covered in a warm 20–25°C, (preferably) dark place. In about 4–7 days the mycelium and sporangia will be abundant. For class distribution, let each student cut out 1 cm^2 of the acetate, using a razor blade. If you use commercially made bread to culture the mold, you will have to use brands that do not contain preservative (usually it's calcium propionate).

---

*Part of recommended student lab kit.

Activity 3

Make up potato-dextrose agar petri dishes as for Activity 2. Inoculate half of each dish with positive strain *Phycomyces blakesleeanus,* and the other half with the negative strain of *P. blakesleeanus.* Set in 20–25°C dark (or relatively so) place. Zygospores should be well formed in about 10–14 days. *P. blakesleeanus* and potato-dextrose agar are available from biological supply houses.

Set up 2–3 plates on dissecting microscopes for student examination.

Activity 4

Culture the slime mold *Physarum polycephalum* on non-nutrient agar or on moist filter paper in petri dishes. The organism is available from biological supply houses. Grind up some rolled oats in a mortar and sprinkle over the culture periodically. Avoid overfeeding. Every second day rinse the agar surface with tap water. This is essential to reduce bacterial growth. Add grains of oatmeal when old ones have been consumed.

Keep culture in low light.

One plate per four or more students.

EXERCISE 21—FUNGI II: THE HIGHER FUNGI

Activity 1

Dried or preserved specimens of *Peziza* (cup fungus). Available preserved from biological supply houses. Or collect and dry them yourself, if you have a local source. One specimen for examination per four students

Prepared slides of cross section for the ascocarp of *Peziza* sp. One per one to two students. Available from biological houses

One compound microscope per student

Activity 2

Dried or preserved specimens of morels (*Morchella esculenta*). One per four students. Preserved material available from biological supply houses. Collect and dry them yourself, if you have a local source.

Activity 3

Dried leaves of lilac (*Syringa* sp.) infected with powdery mildew. You can collect these leaves during late summer from lilac bushes. Dry them flat in newspaper (in plant press if available) and store in cardboard boxes. Have enough leaves (for example, one per two students) for students to scrape off bits of the powder onto a drop of water on a slide. Infected leaves may also be purchased from biological suppliers

One microscope slide,* cover glass,* dissecting needle* (or something comparable for scraping)

Activity 4

Prepare cultures of *Sordaria fimicola* (an ascomycete) on non-nutrient agar in petri dishes and have available for class—one dish per four to eight students. Culture by autoclaving 6–10 flakes of oatmeal in the petri plate, then pouring sterile non-nutrient agar *over* the sterile oats. Inoculate with the living material when agar cools *or* follow special instructions packed with the culture.

---

*Part of recommended student lab kit.

Sordaria is available from biological supply houses. Depending on conditions, if the culture is grown at 20–25°C, it should be mature and have perithecia ready for class use, if you start the culture 2 weeks before class and inoculate the new plates with a generous (5 × 5 mm) chunk of the old culture.

Prepared slides are also available from supply houses.

Activity 5

No materials necessary.

Activity 6

Dissolve one yeast cake or two 7-g packages of dry yeast and 50–100 g of sucrose in 1000 ml of water. Let stand for 2–3 hours at 25–28°C to assure budding formation by the time students examine it

One compound microscope, one slide,* one cover glass* per student

Activity 7

Prepare cultures of *Penicillium* and/or *Aspergillus* on potato-dextrose agar. Follow same directions as for culturing *Rhizopus* for Activity 2 of Exercise 20. Start the cultures 4–5 days before class use. One culture plate per four to eight students.

30 ml dropper bottle of 5 percent potassium hydroxide (KOH). One per four students
30 ml dropper bottle of 70 percent ethanol. One bottle per four students
One compound microscope, 2 slides,* 2 cover glasses* per student

Prepared slides of either *Penicillium* or *Aspergillus* are available from biological supply houses. Allow one slide of each between two students.

Activity 8

Mushrooms—*Agaricus campestris* or *A. bisporus*—available fresh from grocery, or canned, or available preserved from biological supply houses. About one per two students.

Activity 9

Fresh specimens of mushrooms in the spore shedding stage
White paper or index cards
Finger bowls, glass beakers, or similar containers, to cover the mushrooms. Set up materials 3–5 days before class, depending on the maturity of the mushroom. An overly mature mushroom will deliquesce rather than shed spores
One or two demonstrations per class will be needed

Activity 10

Prepared slides of cross section of the cap of any gill-type mushroom, such as *Coprinus* sp. One per student
One compound microscope per student

Activity 11

Specimens may be collected locally during the summer, but demonstration materials of wheat rust (*Puccinia graminis*) and grain smut (*Ustilago* sp.) are available from biological supply houses. Enough for 3–4 demonstrations per class; not to be used up.

---

*Part of recommended student lab kit.

EXERCISE 22—GREEN ALGAE, EUGLENOIDS, AND LICHENS

Activity 1

Cultures of *Chlamydomonas* (or *Carteria*) are needed—sufficient for each student. All major biological supply houses now supply living cultures of *Chlamydomonas*. Each student will use about 2 ml of the culture.

Standard glass microscope slide* and cover slip*—one of each per student.

Dropper bottles of methyl cellulose—one per four to eight students. This is available from all biological supply houses either in solution or as a dry powder. It is usually mixed with water to 10 percent (by weight) and added to the drop on the slide.

Dropper bottles of iodine

Compound microscopes—one per student

Activity 2

Biological supply houses may have these strains of *Chlamydomonas* listed as + and − or as "mating type" kits. They are not always available.

Standard glass microscope slide* and cover slip*—one of each per student
Compound microscopes—one per student

Activity 3

Prepared slides of *Pandorina* are available from all major biological supply houses. Living cultures are sometimes available as well.

Standard glass microscope slide* and cover slip*—one of each per student
Compound microscopes—one per student

Activity 4

Cultures of *Volvox* are available from all major biological supply houses, as are prepared slides.

Standard glass microscope slide* and cover slip*—one of each per student.
Compound microscopes—one per student

Activity 5

Prepared slides of *Ulothrix* are available from all biological supply houses. Living cultures are frequently available as well. One slide per student.

Compound microscopes—one per student

Activity 6

Prepared slides of *Oedogonium* are available from all biological supply houses. The activity describes, and Figure 22.5 is based upon, the macrandrous species of this genus. Most unspecified examples of *Oedogonium* are macrandrous, whether slides, living cultures, or preserved material are offered. Do not use specimens described as, or found to be, nanandrous. One slide per student. Compound microscopes—one per student.

Activity 7

The same slides or specimens as used in Activity 6 are to be used here as well. Prepared slides may be expected to show the sexual phase more reliably.

---

*Part of recommended student lab kit.

Activity 8

Preserved material of *Ulva* is available from all major biological supply houses. Allow one thallus for every four to eight students for gross observation only. Not to be used up.

Activity 9

Living cultures of *Euglena* are available from all major biological suppliers, as are prepared slides.

One microscope slide* and cover slip* per student (or one prepared slide per student)
Dropper bottle of methyl cellulose, as used for Activity 1, above
Compound microscopes—one per student

Activity 10

Dried or fresh specimens of lichens representing the three growth forms: crustose, foliose, and fruticose. Available from biological supply houses, or collect them yourself. For general demonstration for class; not to be used up.

Activity 11

Prepared slides of *Physcia* showing median section of thallus and apothecia. One per student. Available from biological supply houses either as "*Physcia*" or as unnamed "lichen ascocarp"

One compound microscope per student

EXERCISE 23—IDENTIFICATION OF SOME COMMON FRESHWATER ALGAE

Make up cultures containing a variety of genera of freshwater algae. Genera should correspond to those in the Algae Key of this exercise. All genera are available from biological supply houses as either single cultures or mixtures of genera. Distribute in beakers or test tubes about one per four students. Pipettes or droppers with each beaker or test tube.

Or, if possible, collect water cultures from lakes, ponds, reservoirs, or rivers

One compound microscope per student
Several slides* and cover glasses* per student

EXERCISE 24—BROWN ALGAE, RED ALGAE, AND GOLDEN-BROWN ALGAE

Activity 1

Living or preserved specimens of *Fucus*. Not to be used up. One per two students. Biological supply houses offer both living and preserved material and plastic mounts. Collect your own, if you are near the ocean.

Activity 2

Living or preserved specimens of *Laminaria* and/or *Nereocystis*. Available from biological supply houses as living or preserved material, or in plastic mounts. One or two specimens for entire class. Not to be used up.

---

*Part of recommended student lab kit.

Activity 3

Living or preserved specimens of *Sargassum*. One sprig or so per each student. Not to be used up. Available from biological supply houses as living or preserved material, or in plastic mounts. If you live in the southern Gulf States, you may know where to collect your own material.

Activity 4

Various specimens of red algae available from biological supply houses as living or preserved material. Suggested specimens: *Gelidium* sp.; *Rhodymenia* sp.; and *Stenogramma* sp.

Activity 5

Soil diatoms may be collected from any moist, undisturbed soil. Marine and freshwater species are also sold by biological suppliers as living cultures, preserved specimens, or prepared slides.

One microscope slide* and cover slip* per student (or one prepared slide per student)
One compound microscope per student

Diatomaceous earth and preserved cultures of marine diatoms are available from suppliers, as are prepared slides. Especially selected prepared mounts are often available for demonstration purposes showing large, unbroken centric diatoms, but they are too costly for general class use.

One microscope slide* and cover slip* per student (or one prepared slide per student)
One compound microscope per student

EXERCISE 25—MOSSES AND LIVERWORTS, AND AN INTRODUCTION TO LAND PLANTS

Activity 1

Living and preserved material of *Mnium, Polytrichum*, and similar genera are sold by biological suppliers. The Activity is based upon the availability of a recently collected living moss mat. Allow one plant per student; will be used up.

One microscope slide* and one cover slip* per student
One compound microscope per student

Activity 2

All biological suppliers offer prepared slides showing a longitudinal section through the antheridial head of a moss. The Activity and Figure 25.4 assume a dioecious species of *Mnium* or a similar genus.

One prepared slide per student
One compound microscope per student

Activity 3

This Activity requires living male gametophytes of *Mnium, Polytrichum,* or a similar type of moss. Allow one male gametophyte per two students. Local collection is about the only way to acquire living material at the right stage, and success is, therefore, dependent on the season of the year and your locality.

---

*Part of recommended student lab kit.

One microscope slide,* one cover slip,* and one dissecting needle* per two students
One compound microscope per two students

Activity 4

All biological suppliers offer prepared slides of a longitudinal section through the archeogonial head of a moss. As with Activity 2, above, a species of *Mnium* or a similar type is assumed. Note that slides may be described either as "with archegonia" or as "median section" (or similar wording) of the archegonium. The latter are far better in showing the desired features, but are *several dollars per slide* more expensive. A good compromise is to order one or two of the higher quality mounts per class for demonstration.

One prepared slide per student
One compound microscope per student

Activity 5

Both *Polytrichum* and *Mnium* may be purchased with mature sporophytes attached in the living or preserved state. Local collection of these is so simple in most areas at all times of the year that purchase of specimens should be only rarely required. Allow one sporophyte per student.

One microscope slide,* one cover slip,* and one fine forceps* per student
One petri dish containing 50-100 g of petroleum jelly per 6-10 students
Toothpicks for transferring petroleum jelly—one per student
Pieces of paper towel (about 2 × 4 or 3 × 6 cm)—one or two per student
Dropper bottle of tap water—one per four students
One compound microscope per student

Activity 6

This activity also requires mature capsules of *Polytrichum* or *Mnium,* and these may be purchased from biological supply houses or collected locally. Requires one capsule per student.

One microscope slide* and one cover slip* per student
One compound microscope per student

Activity 7

Moss protonemae are available as whole mounts on prepared slides from all major biological suppliers. They are occasionally offered with buds, which are older plants and are preferable for this Activity. Many species of mosses may be cultured on moist soil, moist clay pots, or agar fortified with mineral nutrients such as Knop's solution. Spores may be taken from locally collected species, or viable spores may be purchased from some supply houses.

One prepared slide per student, or one living culture per four students. The living culture may be a demonstration. Allow about 45-60 days for buds to form on artificial cultures of protonemae.
One compound microscope per student

Activity 8

Living and preserved *Sphagnum* plants are available from all biological supply houses. Dried horticultural *Sphagnum* is sold by the cubic foot in bales or bags. One cubic foot is adequate for 50-100 students.

---

*Part of recommended student lab kit.

Disposable plastic drinking glasses—about 6 oz to 8 oz capacity—one per two to four students—may be re-used in other classes
Triple-beam balance—one per 8–12 students

This activity may be performed as a demonstration, as well.

Activity 9

Living and preserved *Sphagnum* plants are available from all biological supply houses. Individual preserved "leaves" are also sold for microscopic examination. Prepared slides of these "leaves" are also available from some suppliers. Allow one "leaf" per one or two students.

One microscope slide* and one cover slip* per student or one prepared slide of *Sphagnum* "leaf" per two to four students
One compound microscope per student

Activity 10

Gametophytes of *Marchantia* and *Conocephalum* are sold as living or preserved material by all major biological supply houses. Living material may also be fairly easily cultured on noncalcareous stones or coarse gravel in an aquarium or moist terrarium under moderate light levels.

Allow one piece of thallus about 1–2 × 4–8 cm per pair of students. This should be examined grossly only and will not be used up.

Gametophytes of *Marchantia* bearing antheridial and archegonial structures are available primarily as preserved material. Some suppliers are able to provide living material in certain seasons (usually spring only) as well.

Allow one piece of thallus bearing each type of structure per two to four students. They then work in pairs or groups of four sharing samples of both sexes. It should not be used up.
One dissecting microscope per two to four students, or one 6× to 10× hand lens*

EXERCISE 26—LOWER VASCULAR PLANTS: FERNS, CLUB MOSSES, HORSETAILS, AND PSILOTUM

Activity 1

Living or preserved specimens or color transparencies of *Psilotum nudum*. Living specimens available from Ward's Natural Science Establishment, Inc. Also available preserved, on color transparencies, or mounted in blocks of plastic from most biological supply houses.

Many greenhouses, commercial and/or university affiliated, often have living *Psilotum,* or you can grow it as an ordinary pot plant indoors. It's fairly easy to maintain and will grow best in the same pot with some other plant. One or two plants may be about all that you need for the entire class.

Activity 2

Living, preserved, or pressed specimens of *Lycopodium.* Available from most biological supply houses. Carolina Biological Supply Co. offers living material as well as preserved. Try to have species with cones and species without cones. *L. lucidulum* is a common coneless species; *L. clavatum* and *L. complanatum* have cones. You may be able

---

*Part of recommended student lab kit.

to find *L. lucidulum* sold in plant shops as a terrarium plant. Its common names are ground pine and running pine. One plant per four students if possible. Not to be used up. Hand lens, 6–10×*—one per student

Activity 3

Living or preserved specimens of *Selaginella*. Sometimes available from plant shops, sold as a terrarium plant. Living material available from Carolina Biological Supply Co. All supply houses carry preserved and/or plastic mounted material. You can easily grow and maintain it yourself as a pot plant. One plant per four students if possible. Not to be used up. Hand lens, 6–10×*—one per student

Activity 4

Living or preserved specimens of *Equisetum hyemale* and *E. arvense*. *Equisetum* is much more common than *Lycopodium* or *Selaginella* so you may be able to collect your own specimens from old stream banks and roadsides, especially in areas of sandy or infertile soil. Otherwise, plants are available preserved or in plastic mounts from biological supply houses. Plants not to be used up. Hand lens, 6–10×*—one per student.

Activity 5

Have two or three living fern plants for general class demonstration. You can often rent ferns (and other foliage plants) from local plant shops if you don't want to purchase these.

If possible, depending on the season, make available for class observation ferns with new young fiddle-head shoots.

If possible, obtain fern material that can be dug to show its rhizome with roots. Several classes can examine the same material. Suggested ferns: Boston fern (*Nephrolepsis exaltata* var. *Bostoniensis*); Christmas fern (*Polystichum acrostichoides*); maidenhair fern (*Adiantum* sp.); and *Polypodium* sp.

Collect fern leaves sometime before or during the school session and press and dry flat in newspapers in plant press (if available). Collect only leaves with mature sori. Sori will be dark or brown if sporangia are mature. Store leaves in boxes. Usually a student will use up one pinna of a leaf when examining it, so store enough leaves to suit the size and number of your classes.

Hand lens, 6–10×*—one per student

Activity 6

One compound microscope, one slide,* one cover glass,* one dissecting needle* per student
Enough fern leaf material for each 1–2 students to work with one pinna

Activity 7

Uses the same materials as Activity 6. Use of a small (10 cm) petri plate with a dry filter paper hastens the drying and minimizes loss of sporangia to air currents in the room.

Activity 8

Prepared slides of fern prothallus available from biological supply houses. One per student
Compound microscopes—one per student

---

*Part of recommended student lab kit.

Activity 9

Living fern prothalli—available from Carolina Biological Supply Co. Or you can start your own cultures, but you need to do this about a month or more before class.

Soak a Jiffy-7 peat pellet in water. After it has expanded, dust it with fern spores, which you can obtain from your own plant material or purchase from biological supply houses. Place the peat cube in a saucer or petri dish and cover with a plastic drinking glass, beaker, or similar cover. Set it in a cool (14–20°C) place with medium light and don't let it dry out. In a month, several prothalli will have formed. After about three months' growth, young sporophytes will be formed.

Other methods for culturing fern spores are described in Turtox Service Leaflet No. 44 and in *A Sourcebook for the Biological Sciences* by E. Morholt, P. F. Brandwein, and A. Joseph.

Activity 10

Prepared slides of fern prothalli bearing young sporophytes. One slide per student
Or living young sporophytes, which you can prepare yourself as described for Activity 9, or they can be purchased
Compound microscopes—one per student

Activity 11

Microscope slides showing anatomical detail of stems and leaves of fossil plants are offered by Carolina Biological Supply Co. and Ward's Natural Science Establishment, Inc. They may be called "peels," because of the method used to prepare them. Use three to four per class as demonstrations.

Rock compressions or impressions of fossil leaves and/or stems of *Lycopodophyta* and *Arthrophyta* are also available from Ward's. A "fossil plant" set is available from Carolina.

EXERCISE 27—CONE-BEARING SEED PLANTS

Activity 1

Cuttings of different gymnosperms made available for general demonstration to the class. Make available one sample of each genus per three to four students. *Or* take students on a tour to see different gymnosperm trees.

Activity 2

Prepared slide showing longitudinal section of a staminate pine cone. One per student. Available from biological supply houses
One compound microscope
Alternative: Preserved staminate cones. Collect and preserve these yourself or order from biological supply house
Hand lens* or dissecting microscope, one slide,* one cover slip,* and dissecting needle* per student

Activity 3

Young (first-spring) ovulate cones preserved at pollination time. Either collect and preserve yourself or order from biological supply houses. Some suppliers may call

---

*Part of recommended student lab kit.

them "first winter" stage. One per student. Not to be used up, but ultimately they are damaged

Prepared slides of ovulate pine cone showing megaspore mother cell. Available from biological supply houses. One slide per student

Compound microscopes—one per student

Activity 4

One-year-old ovulate cones (collected the second summer). Available from biological supply houses, or collect and preserve yourself. One cone per one or two students.

Activity 5

Mature ovulate cones with woody scales and mature seeds. Available from biological supply houses, or collect yourself. One cone per one or two students. These are not used up at once, but are eventually damaged.

Activity 6

Seeds of pine (*Pinus* sp.). The best kind to use are Colorado Pinyon pine (*Pinus cembroides edulis*). They are large, easy to dissect and the embryo and other seed parts are readily identifiable. However, they are more expensive than other, smaller, seeds. You can get the most Pinyon pine seeds for your money from Herbst Brothers Seedsmen, Inc. Biological supply houses also carry them, but they are more costly. "Health food" stores often carry, or can order, pine "nuts." These are usually the seeds of the Pinyon pine.

Other less expensive pine seeds (not identified as to species by the suppliers) are available from biological supply houses. Mellingers (a horticultural supply house) has several species of pine seed but doesn't always have Pinyon pine.

EXERCISE 28—FLOWERING SEED PLANTS

Activity 1

This activity and Activity 3 can be done together.

Living specimens of impatiens (*Impatiens* sp.) or spiderwort (*Tradescantia* sp.) with flowers in the pollen-shedding stage. If these plants are not available almost any other kind will suffice as long as the flowers have anthers with mature pollen. Examine for viability beforehand. Some varieties of *Impatiens* have abortive, inviable pollen. Students will transfer some pollen to a slide, so three or four plants per class should be adequate. Some of the same pollen material is to be used in Activity 3

30 ml dropper bottles with 10 percent cane sugar solution. One per four students

One slide,* one cover slip,* one dissecting needle* per student

One petri dish lined with 3–4 layers of filter paper. One dish per two students

A few brush bristles or comparable material to hold cover slip up off the slide

Activity 2

Stamens of lily (*Lilium* sp.) flower or other plant. Either preserved or fresh. One or more per student. For examination only. Not to be used up.

Activity 3

This activity can be done simultaneously with Activity 1. The same plant material is used. One slide,* one cover slip,* and one compound microscope per student.

---

*Part of recommended student lab kit.

Activity 4

Prepared slides of young flower buds of lily (*Lilium* sp.) showing a cross section through the young anthers. Available from biological supply houses. One slide per student
Compound microscopes—one per student

Activity 5

Prepared slides of cross sections of mature anthers of lily (*Lilium* sp.) or of lily pollen grains. Available from biological supply houses.
Compound microscopes—one per student
One slide per student

Activity 6

Prepared slides of cross sections of young ovary of lily (*Lilium* sp.) showing megasporocytes (megaspore mother cells). Available from biological supply houses.
One slide per student
Compound microscopes—one per student

Activity 7

Prepared slides of cross sections of ovary of lily (*Lilium* sp.) showing ovule (ovulary) with the mature female gametophyte. Available from biological supply houses. One slide per student
Compound microscopes—one per student

Activity 8

A collection of a wide variety of angiosperm seeds. Have the seeds of herbaceous plants separately grouped from those of trees and shrubs. Include seeds with spines, hairs, hooks, and so on, to show that in many plants the seed itself is the dispersal agent. A kit called "Seed Dispersal Set" is also available from Carolina Biological Supply.

For your own reference, see *Seed Identification Manual* by Alexander C. Martin and William D. Barkley (1961), University of California Press, 1414 So. Tenth St., Richmond, California 94804.

Activity 9

A collection of a variety of edible fruits. Include many of the common vegetables that students know, such as green peppers (*Capsicum* sp.); tomato (*Lycopersicon* sp.); eggplant (*Solanum melongena*); melons, squashes, and pumpkins (*Cucurbita* sp.); okra (*Hibiscus esculentus*); grape (*Vitis* sp.); apple (*Malus* sp.); orange (*Citrus* sp.); and so on. Also fruits of trees such as ash (*Fraxinus* sp.); oak (*Quercus* sp.); walnut (*Juglans* sp.); maple (*Acer* sp.); sweet gum (*Liquidambar styraciflua*); elm (*Ulmus* sp.); and buckeye (*Aesculus* sp.).

Also fruits of weeds locally available such as thistle (*Cirsium* sp.); cocklebur (*Xanthium* sp.); dandelion (*Taraxacum officinale*); goatsbeard (*Tragopogon* sp.); jimson weed (*Datura* sp.); or any other species used in Exercise 14.

---

*Part of recommended student lab kit.

EXERCISE 29—MEIOSIS AND MONOHYBRID AND DIHYBRID CROSSES

Activities 1–3

No materials required.

Activity 4

Flats or pots of F_2 generation seedlings which show color differences representing simple monohybrid inheritance. Will not be used up
Suggested plants: corn (*Zea mays*); tobacco (*Nicotiana tabacum*); and sorghum (*Sorghum vulgare*). Seeds for this are available from biological supply houses. Seeds should be started 6–10 days before class and grown at 20–22°C. Allow more time if temperatures are lower

Activity 5

Ear of corn (*Zea mays*) showing colored and white grains in a 3:1 ratio. Available from biological supply houses
One ear per two students (ideally). Will not be used up
Supply of toothpicks or straight pins

Activity 6

Ear of corn (*Zea mays*) showing the results of a (monohybrid) testcross. Available from biological supply houses
One ear per two students (ideally)
Supply of toothpicks or straight pins

Activities 7 and 8

No materials required.

Activity 9

Ear of corn (*Zea mays*) in which the grains represent the F_2 generation of a cross involving two colors—white and purple—and two endosperm types—full (or starchy) and wrinkled (or sugary). Yellow and purple may be used if white and purple aren't available. Available from biological supply houses
One ear per two students (ideally)
Supply of toothpicks or straight pins

EXERCISE 30—PLANT ECOLOGY

Activity 1

This Activity may be done either by groups of two to four students or by the instructor as a demonstration. In either case, sets of five or six identical pots of soil are planted with varying numbers of seeds. Three to four inch plastic pots are good, and use of a largely synthetic soil medium containing mostly perlite or vermiculite assists calculation of root mass. Use corn (*Zea mays*); grain sorghum (*Sorghum* sp.); jimson weed (*Datura* sp.); or any available fast-growing herbaceous plant with upright growth habit. They should be started 4–12 weeks before use, depending on the species and size of pot. During this period, water each pot of the sets with roughly equal quantities of water, and do not feed any fertilizer unless you are using a totally artificial potting soil. Be sure to plant extra seeds to make up for those which do not germinate, and also to pull up any extra seedlings at about 2–3 cm in height.

One set of five or six pots is needed per group, or per demonstration.

Either fresh or dry weights may be obtained. Dry weights are considered more exact but mean that the plants must be harvested 48–72 hours before class and dried for 40–48 hours at 70°C before weighing. The class will then see a display showing a set of living plants to be measured and a set of dried plants from similar pots with weights shown.

One of the most satisfactory overall methods involves the use of prepared compressed media such as Jiffy-7 or Solo-Gro. To these small "pots," small seeds such as sorghum are better suited than are large corn grains. Some fertilizer is already present, so none need be added. The uniformity of the medium will allow entire masses of plants, roots, and media to be oven-dried easily. The weight of a dry, unused block or pellet may easily be subtracted from the eventual total weight to arrive at the plant biomass. With these smaller rooting masses, too, it may take only 20–30 days to complete maximum growth.

Access to a forced-draft drying oven is needed if dry weights are to be obtained. Paper bags of the type sold for carrying lunches are good to hold the different sets of plants while they are being dried.

Access to a good electronic or torsion balance is necessary. A triple-beam balance may be adequate if large plants are used and if they have been grown for two months or more.

Activity 2

No materials required.

Activity 3

These successional stages are best observed on a field trip, if one can be fitted into the schedule. Exotic communities need not be visited, and abandoned fields, vacant lots, and construction sites may all provide adequate examples of succession. Films, filmstrips, and narrated slide-tape programs on this subject are available from a variety of suppliers, including some biological supply houses. Local examples may also be photographed by the instructor, or a videotape may be prepared by the AV services of your institution.

Activity 4

No materials required.

Activity 5

Requires a film, filmstrip, or slide-tape presentation covering the major environments of North America. Generally suitable programs are available from many sources, including biological supply houses. Rental films from standard educational film bureaus may be the best source for many schools.

Activity 6

No materials required.

Activity 7

No materials required.

BIOLOGICAL SUPPLY HOUSES*

1. Carolina Biological Supply Company
   Main Office and Laboratories
   Burlington, North Carolina 27215
   Telephone: 919-584-0381
   800-334-5551

   Powell Laboratories Division
   Carolina Biological Supply Company
   Gladstone, Oregon 97027
   Telephone: 503-656-1641
   800-547-1733

2. Turtox
   8200 South Hoyne Avenue
   Chicago, Illinois 60620
   Telephone: 312-488-4100
   800-621-8980

3. Ward's Natural Science Establishment
   P.O. Box 1712
   Rochester, New York 14603
   or
   P.O. Box 1749
   Monterey, California 93940

4. ICN Pharmaceuticals Incorporated
   26201 Miles Road
   Cleveland, Ohio 44128

HORTICULTURAL SUPPLIERS*

1. Mellinger's
   2310 West South Range
   North Lima, Ohio 44452
   Telephone: 216-549-9861

2. Herbst Brothers Seedsmen, Inc.
   1000 North Main Street
   Brewster, New York 10509

3. Geo. W. Park Seed Co., Inc.
   S. C. Hwy. 254 N.
   Greenwood, South Carolina 29647

4. Stokes Seeds Inc.
   737 Main Street
   Buffalo, New York 14240

5. J. W. Jung Seed Co.
   Seeds and Nursery
   Randolph, Wisconsin 53956

6. Seedway, Inc.
   Hall, New York 14463

7. Gurney Seed and Nursery Co.
   Yankton, South Dakota 57079

8. Burpee Seed Co.
   Warminster, Pennsylvania 18974
   or
   Clinton, Iowa 52732
   or
   Riverside, California 92502

FORESTRY SUPPLIERS*

1. The Ben Meadows Company
   3589 Broad Street
   Atlanta (Chamblee), Georgia 30366
   Telephone: 404-455-0907
   800-241-6401

2. Forestry Suppliers Inc.
   P.O. Box 8397
   Jackson, Mississippi 39204
   Telephone: 601-354-3565
   800-647-5368

*Suppliers suggested or mentioned in this manual. Several other sources exist.

INDEX

Note: Page numbers in boldface type indicate illustrations.